THREE
WORLD
CUISINES

AltaMira Studies in Food and Gastronomy

General Editor: Ken Albala, Professor of History, University of the Pacific (kalbala@pacific.edu)

AltaMira Executive Editor: Wendi Schnaufer (wschnaufer@rowman.com)

Food Studies is a vibrant and thriving field encompassing not only cooking and eating habits but issues such as health, sustainability, food safety, and animal rights. Scholars in disciplines as diverse as history, anthropology, sociology, literature, and the arts focus on food. The mission of **AltaMira Studies in Food and Gastronomy** is to publish the best in food scholarship, harnessing the energy, ideas, and creativity of a wide array of food writers today. This broad line of food-related titles will range from food history, interdisciplinary food studies monographs, general interest series, and popular trade titles to textbooks for students and budding chefs, scholarly cookbooks, and reference works.

Appetites and Aspirations in Vietnam: Food and Drink in the Long Nineteenth Century, by Erica J. Peters

Three World Cuisines: Italian, Mexican, and Chinese, by Ken Albala

THREE WORLD CUISINES

Italian, Mexican, Chinese

KEN ALBALA

ALTAMIRA
PRESS

A division of
ROWMAN & LITTLEFIELD PUBLISHERS, INC.
Lanham • New York • Toronto • Plymouth, UK

The publisher has done its best to ensure that the instructions and/or recipes in the book are correct. However, users, especially parents and teachers working with young people, should apply judgment and experience when preparing recipes. The publisher accepts no responsibility for the outcome of any recipe included in this volume.

Published by AltaMira Press
A division of Rowman & Littlefield Publishers, Inc.
A wholly owned subsidiary of The Rowman & Littlefield Publishing Group, Inc.
4501 Forbes Boulevard, Suite 200, Lanham, Maryland 20706
www.rowman.com

10 Thornbury, Plymouth PL6 7PP, United Kingdom

Copyright © 2012 by AltaMira Press

All rights reserved. No part of this book may be reproduced in any form or by any electronic or mechanical means, including information storage and retrieval systems, without written permission from the publisher, except by a reviewer who may quote passages in a review.

British Library Cataloguing in Publication Information Available

Library of Congress Cataloging-in-Publication Data
Albala, Ken, 1964-
  Three world cuisines : Italian, Mexican, Chinese / Ken Albala.
      p. cm.
  Includes bibliographical references and index.
    ISBN 978-0-7591-2125-6 (cloth : alk. paper) — ISBN 978-0-7591-2126-3 (pbk. : alk. paper) — ISBN 978-0-7591-2127-0 (ebook)
  1. International cooking. 2. Food habits—Italy—History. 3. Food habits—Mexico—History. 4. Food habits—China—History. I. Title.
TX725.A1A315 2012
394.1'20945—dc2   2012000686

∞^{TM} The paper used in this publication meets the minimum requirements of American National Standard for Information Sciences—Permanence of Paper for Printed Library Materials, ANSI/NISO Z39.48-1992.

Printed in the United States of America

Contents

LIST OF FIGURES		vii
LIST OF RECIPES		xi
A NOTE ON THE RECIPES		xix
INTRODUCTION	A THEORY OF GASTRONOMY	1
CHAPTER 1	HISTORICAL BACKGROUND	29
CHAPTER 2	TECHNOLOGY, TECHNIQUES, AND UTENSILS	65
CHAPTER 3	GRAINS AND STARCHES	101
CHAPTER 4	VEGETABLES	149
CHAPTER 5	FRUITS AND NUTS	177
CHAPTER 6	MEAT, POULTRY, AND DAIRY PRODUCTS	205
CHAPTER 7	FISH AND SHELLFISH	261
CHAPTER 8	FATS AND FLAVORINGS	283
CHAPTER 9	BEVERAGES	315
GLOSSARY		339
BIBLIOGRAPHY		347
INDEX		351
ABOUT THE AUTHOR		365

List of Figures

	Map of Italy.	xxiii
	Map of Mexico.	xxiv
	Map of China.	xxv
Figure 1.1.	*Satyricon* scene.	40
Figure 1.2.	Galen.	42
Figure 1.3.	Preparing chocolate.	54
Figure 2.1.	Draining Peking ducks.	70
Figure 2.2.	*Cittara*.	75
Figure 2.3.	Mortar.	76
Figure 2.4.	Hand quern.	77
Figure 2.5.	Ancient olive oil press.	77
Figure 2.6.	*Batticarne*.	79
Figure 2.7.	Food mill.	80
Figure 2.8.	Duck roasting.	81
Figure 2.9.	*Comal*.	86
Figure 2.10.	Newly seasoned wok.	87
Figure 2.11.	Chinese tools.	90
Figure 2.12.	*Molinillo*.	91
Figure 2.13.	Chinese rolling pin.	92
Figure 2.14.	Tortilla press.	93
Figure 2.15.	Meat grinder.	94
Figure 3.1.	*Panettone*.	105

Figure 3.2.	Sifter.	106
Figure 3.3.	Ravioli.	107
Figure 3.4.	Steamed buns.	110
Figure 3.5.	Chinese noodles.	111
Figure 3.6.	*Jiaozi*.	112
Figure 3.7.	A cut Mooncake.	113
Figure 3.8.	Whole mooncake.	114
Figure 3.9.	Blue corn.	119
Figure 3.10.	Cooking tortillas on a *comal*.	120
Figure 3.11.	Grilled corn.	121
Figure 3.12.	Millet.	125
Figure 3.13.	Lasagne.	134
Figure 3.14.	Pizza.	138
Figure 3.15.	*Cornetti*.	141
Figure 3.16.	Coarse tortilla.	142
Figure 3.17.	*Zeppole*.	144
Figure 4.1.	Broccoli rabe.	151
Figure 4.2.	Fresh water chestnuts.	153
Figure 4.3.	Swiss chard.	157
Figure 4.4.	Variety of bean species.	158
Figure 4.5.	Wood ear mushrooms.	163
Figure 4.6.	*Nopalitos*.	165
Figure 4.7.	Tomatillos.	168
Figure 4.8.	Baby artichokes.	174
Figure 4.9.	Minestrone.	176
Figure 5.1.	Produce market, China.	178
Figure 5.2.	Goji berry.	180
Figure 5.3.	Fresh lychees.	189
Figure 5.4.	Durian.	190
Figure 5.5.	Mangosteens.	191
Figure 5.6.	Nuts of ginkgo.	197
Figure 5.7.	Preserved plums.	200
Figure 5.8.	*Paleta*.	203
Figure 6.1.	Edible insects.	207

Figure 6.2.	Cheese shop, Milan, Italy.	214
Figure 6.3.	Cheesemaker, Parma, Italy.	215
Figure 6.4.	Food shop window, Bologna, Italy.	219
Figure 6.5.	Carne asada.	232
Figure 6.6.	Meatballs.	235
Figure 6.7.	Lion's head meatball.	237
Figure 6.8.	*Felino*.	238
Figure 6.9.	*Menudo*.	242
Figure 6.10.	Stir-fry with chicken and vegetables.	244
Figure 6.11.	Pumpkin seed sauce.	247
Figure 6.12.	Classic hot and sour soup.	249
Figure 6.13.	Tea egg.	251
Figure 7.1.	Market, Lantau, Hong Kong.	262
Figure 7.2.	*Baccalà*.	265
Figure 7.3.	Jellyfish.	267
Figure 7.4.	Mackerel.	269
Figure 7.5.	Dim sum.	272
Figure 7.6.	Ceviche.	277
Figure 8.1.	Olive press.	288
Figure 8.2.	Rue.	293
Figure 8.3.	*Epazote*.	294
Figure 8.4.	Lily buds.	296
Figure 8.5.	*Annatto* (*achiote*).	300
Figure 8.6.	Market, Guizhou Province, China.	301
Figure 8.7.	Pods of the tamarind.	303
Figure 8.8.	*Piloncillo*.	304
Figure 8.9.	Bottles of black vinegar and rice wine.	306
Figure 9.1.	Cacao pods.	319
Figure 9.2.	Ingredients for medicinal Chinese tea.	321
Figure 9.3.	Traditional Chinese clay pot.	322
Figure 9.4.	*Pu er*.	323
Figure 9.5.	Pulquería, Mexico City.	327
Figure 9.6.	Roman woman pouring wine.	329
Figure 9.7.	Agave harvest.	332

List of Recipes

Recipes for Grain

Porridge

Polenta/Italy   127
Congee (Rice Porridge)/China   127
Panata (Bread Porridge)/Italy   128
Atole (Corn Porridge)/Mexico   128

Pasta

Basic Pasta Dough   128
Chinese Pulled Noodles   129
Pizzoccheri della Valtellina/Italy   130

Filled Dumplings and Dishes

Tortellini in Brodo/Italy   130
Chinese Dumplings   131
Apician Patina/Italy   131
Lasagne al Forno (Apicius 4.2.14)/Italy   132

Steamed, Grain-Based Foods

Tamales/Mexico   134
Baozi and Mantou/China   135
Xiaolongbao (Steamed Pork Buns from Shanghai)/China   135

Baked Bread

Grissini Torinesi (Breadsticks)/Italy   136
Pizza Margherita/Italy   137
Bolillos (Rolls)/Mexico   138
Bruschetta and Crostini/Italy   139
Pandoro (Golden Cake)/Italy   139
Cornetti (Pastries)/Italy   140

Grilled and Fried Grains

Tortillas/Mexico   141
Fried Crullers (Youtiao)/China   143
Zeppole (Fritters)/Italy   143
Churros/Mexico   144
Buñuelos de Jeringa/Mexico   145
Fried Rice (Yangzhou Chao Fan)/China   145
Risotto alla Milanese/Italy   146
Spanish Rice (Arroz Rojo)/Mexico   146

Sweets Made from Grain

Sbrisolona/Italy   146
Mexican Wedding Cookies (Pastelitos de Boda)   147
Chinese Peanut Cookies   147
Sesame Balls/China   147
Mooncakes/China   147

Starches

Potato Gnocchi/Italy   148
Camotes Fritos (Fried Sweet Potatoes)/Mexico   148

Recipes for Vegetables

Salads

Insalata Mista (Mixed Salad)/Italy   169
Stir-Fried Lettuce/China   169
Ensalada de Nopalitos (Cactus Salad)/Mexico   170

Mixed Vegetable Dishes

Caponata/Italy   170
Sautéed Chayote/Mexico   170
Casserole of Zucchini/Italy   170
Buddha's Delight (Luohan Quanzhai)/China   171
Braised Greens/Italy   171

Stuffed Vegetables

Chiles en Nogada/Mexico   172
Melanzane Ripieno (Stuffed Eggplant)/Italy   172
Fried Asparagus or Baby Artichokes/Italy   173
Scallion Pancakes/China   173

Beans

Frijoles Refritos (Refried Beans)/Mexico   173
Fried Tofu/China   174
Fresh Stewed Fava Beans/Italy   175

Vegetable Soups

Cold Avocado Soup/Mexico   175
Minestrone di Verdure (Vegetable Soup)/Italy   175
Ribollita (Bean Soup)/Italy   176

Recipes for Fruit

Fresh Fruits

Crostata di Frutta (Fruit Tart)/Italy   198
Pear Minestra (Anonimo Toscano, Fifteenth Century)/Italy   199
Pickled Peaches/China   200
Pickled Plums/China   201
Cherries in Conserve/Italy   201
Mostarda di Frutta (Fruit Mustard)/Italy   201
Fried Bananas/China   202
Tostones de Plátano (Fried Plantains)/Mexico   202
Guava Paste (Guayabate)/Mexico   202
Tang Hu Lu (Candied Bottle Gourd)/China   203

Frozen Fruits

Paletas (Popsicles)/Mexico   203
Gelato al Limone/Italy   204

Recipes for Meat

Raw Meat

Carpaccio (Raw Beef)/Italy   227

Boiled Meats

Bollito Misto /Italy   227
Vitello Tonnato (Veal with Tuna Sauce)/Italy   228

Fried Meats

Cotoletta alla Milanese (Veal Cutlets)/Italy   228
Peking Beef/China   229

Roasted Meats

Cochinita Pibil (Roasted Pig)/Mexico   230
Porchetta (Stuffed Roasted Pig)/Italy   230

Grilled Meats

Cantonese Roast Pork (Char Siu)/China   231
Carne Asada (Roasted Meat)/Mexico   231
Bistecca alla Fiorentina (Steak Florentine)/Italy   232

Braised Meats

Ossobuco (Braised Veal Shanks)/Italy   233
Mao's Favorite Red-Cooked Pork from Hunan (Hongshao Rou)/China   234

Steamed Meats

Birria (Stew)/Mexico   234
Steamed Beef/China   234

Meatballs

Polpette/Italy   235
Albóndigas/Mexico   236
Lion's Head Meatballs/China   236

Fermented Sausages

Salame Felino/Italy   236
Lap Cheong/China   239
Chorizo/Mexico   239

Organ Meat

Fegato alla Veneziana (Liver Venetian Style)/Italy   240
Lengua (Tongue)/Mexico   240
Blood Sausages/Mexico   240

Skin: Cicciole, Chicharrones, Crackling

Cotechino (Skin Sausage)/Italy   241
Menudo (Tripe Soup)/Mexico   242

Recipes for Poultry

Pollo Encacahuatado (Chicken in Peanut Sauce)/Mexico   243
Salt-Baked Chicken/China   243
Chicken Cacciatore/Italy   243
Chicken Stir-Fried with Vegetables/China   244
Chicken Burrito/Mexico   245
Uccelli Arrosto (Roast Birds)/Italy   245

Duck

Ragù di Anatra (Duck Sauce)/Italy   246
Duck in Pumpkin Seed Sauce/Mexico   247
Shanghai Braised Duck/China   248

Soups

Chicken Soup with Mushrooms/China   248
Hot and Sour Soup/China   248
Chicken Broth (Brodo)/Italy   249

Zuppa Pavese/Italy   250
Tortilla Soup/Mexico   250

Eggs

1,000-Year-Old Eggs/China   250
Tea Eggs/China   251
Frittata/Italy   252
Higaditos (Liver Omelet)/Mexico   252
Eggs and Tomatoes/China   252
Huevos Motuleños/Mexico   253
Savory Egg Custard/China   253
Flan/Mexico   253
Stracciatella alla Romana/Italy   254
Capirotada (Bread Pudding)/Mexico   254
A Common Capirotada (Diego Grando, 1599, pp. 77–78)/Mexico   255
Modern Capirotada/Mexico   255
Tiramisù and Zuppa Inglese/Italy   255

Dairy

Kumis (Fermented Milk)/China   256
Queso Fresco (Fresh Cheese)/Mexico   257
Mozzarella and Ricotta/Italy   257
Tres Leches Cake/Mexico   258
Empanada de Cajeta (Pastry with Milk Fudge)/Mexico   259
Panna Cotta (Cooked Cream)/Italy   259

Recipes for Fish and Shellfish

Raw

Ceviche/Mexico   276

Cold

Insalata di Mare (Seafood Salad)/Italy   277

Poached

West Lake Sour Fish/China   278
Peche Lesso (Poached Fish)/Italy   278

Stewed

Baccalà alla Vicentina (Cod Vicenza Style)/Italy   279

Fried

Calamari Fritti (Fried Squid)/Italy   279
Heng Yang Spicy Scallops/China   280
Coconut Shrimp/Mexico   280
The Best Stuffed Tench (Christoforo di Messisbugo, 1589, p .106)/Italy   280

Steamed

Steamed Fish Balls/China   281
Huachinango a la Veracruzana (Veracruzan Red Snapper)/Mexico   281

Recipes for Sauces

Pesto Genovese/Italy   309
Pico de Gallo (Tomato Salsa)/Mexico   310
Bagna Cauda (Hot Sauce for Vegetables)/Italy   310
Molé Verde with Tomatillos/Mexico   310
Guacamole/Mexico   310
XO Sauce/China   311
Tomato Sauce/Italy   311
Molé Poblano/Mexico   312
Chili Sauce/Mexico   313
Black Vinegar Sauce/China   313

Recipes for Drinks

Nonalcoholic Drinks

Horchata (Rice Drink)/Mexico   333
Aguas Frescas (Fruit Waters)/Mexico   334
Tea/China   334
Bubble Tea/China   334
Ginseng Tea/China   334
Italian Sodas   335

Alcoholic Drinks

Margarita/Mexico   335
Negroni/Italy   335
Bellini/Italy   336
Limoncello (Lemon Liqueur)/Italy   336
Nocino (Walnut Liqueur)/Italy   336
Chinese Cocktails   337

A Note on the Recipes

THE RECIPES found at the end of most chapters pull everything together—the technology, ingredients, and flavor combinations and preferences—and include some standard recipes of the three cuisines. These are more than simply recipes, however. Information about the social context, setting, mealtimes, and ideas about food will accompany each dish. So, too, will practical instruction for making these recipes, which we encourage as the experiential component of this text. Cooking and tasting are integral to the learning process here. Some recipes will be drawn from history, mainly to illustrate the roots of these cuisines, but most are contemporary and sometimes even drawn from descendants of the parent cuisines; that is, hybrids that exist outside of China, Mexico, and Italy. As this is a textbook on world cuisines, we must show how these recipes were adapted globally.

When thinking about the flavor combinations in these recipes, it will be apposite to recall what actually happens in the physical act of tasting. Naturally we first think of flavor on the tongue: the four primary flavors of salt, sour, sweet, and bitter. The fifth, *umami*, flavor might be included as well. It is essentially a flavor that intensifies other flavors. It is derived from the natural glutamates in foods such as mushrooms and Parmigiano cheese and monosodium glutamate. But rarely do we merely taste with the tongue. Aromas are equally important, and instead of merely listing them as a wine taster might—for example, the aroma of thyme, leather, or tobacco—they do fit into broader categories. Here we might list smoke; all herbaceous aromas; all floral aromas; all bright, citrusy aromas; and so forth, as primary flavor principles. Also essential is the "mouthfeel": Is a food tannic (that is, having the puckering effect of tea and other astringent foods)? Does it burn with spicy heat like a chili pepper, or is it pungent like horseradish, mustard, and garlic? It is creamy and unctuous, or slimy? Is it crunchy? All these are essential components of flavor.

More important, most recipes consciously make use of several of these flavor/aroma/mouthfeel elements, or even all of them in balance. Taste is comparable to music in this regard. Rarely is a single note presented, but rather various notes at different pitches in harmony, plus different textures, timbres, or tone colors, are heard. The same note from a saxophone sounds different from a flute. In cooking, sweet flavors are played against sour ones, crunchy textures are balanced with smooth and creamy ones, and piquant seasonings liven the mixture. For example, consider what makes an interesting barbecue sauce. The base may be tomato, but it is livened with acid such as vinegar. Aromatics like oregano or cumin may lie beneath the smoke. A hint of sweetness, perhaps from honey, rounds out the flavor. Garlic adds pungency. Chili pepper adds heat. Too much of any single element ruins the harmony. Take note in the recipes how flavors are balanced in this way.

Finally, perhaps the most important organ of taste is the brain. Before we register sensation in looking at the food, its color and arrangement, and before we smell it, before we hear it sizzling on a platter or resonating in our head as we chomp down on it, we assess it with reason. We recall past associations, we gauge what our companions will think about the dish, we imagine what eating this particular food says about us. Is the food considered sophisticated, homey, daring, healthy, reflective of our ethnic heritage? All these will have direct bearing on how the food tastes. Of course, these meanings shift over time, from place to place, and our taste preferences evolve. Some flavors we come to appreciate only later in life, as our taste buds dull or our associations change. Nonetheless, it is important to consider what these dishes mean to the people who eat them. The context is always critical. Only then can we begin to appreciate how they actually taste, keeping in mind that taste is both a historical and cultural construct and the product of our own particular social group, family, and personal psychology. As the saying goes, there is no accounting for taste, and, in the end, the only thing that matters is what someone likes.

The recipes included here will immediately strike the reader as unconventional. They do not follow the standard recipe format as used in most cookbooks in the past century or adhere to standards commonly used in professional kitchens that deal with enormous quantities of food and use expensive equipment that few students or home cooks can afford. The purpose of the directions offered here is purely pedagogical. Basic techniques and procedures will be explained as a starting point from which readers are encouraged to experiment, indeed to *play* with their food.

Cooking is largely an intuitive and creative process. You only learn by doing it. Practice will yield the finished dishes as you prefer them, but slavishly following someone else's recipes will not. The amount of seasoning you add, the cooking time, and the temperature will all depend on the equipment you happen to have at your disposal and the ingredients you can find from your local purveyors. Scientific precision, or the pretense of such, has become standard because cookbook authors understandably want to copyright their recipes, and for the same reason they give them their own individual twist to encourage sales. The recipes here are anything but unique. In fact, they are not really recipes but rather descriptions of cooking procedures, and they should not be followed meticulously. They merely offer instruction, so readers will become comfortable trying out the dishes from these three traditions. We must remember that tasting is an integral part of the learning process. Eventually you will learn what you and those you feed like best. This is especially important in seasoning, and as you will see I have usually left this entirely to the discretion of the cook. Only when measurements are critical, as with cakes, are they offered.

Equally important, these are not *classic* recipes according to some random, arbitrary canon or some objective standard of gastronomic truth. Unlike French haute cuisine, professional chefs never codified rigorously the basic repertoire of Italian, Chinese, and Mexican cuisines. These are procedures that were developed in home kitchens over generations. They necessarily change over time and may vary widely from region to region and from household to household. That is, they are living, constantly evolving traditions. It would be preposterous to claim that any of these directions constitutes the single *correct* way to cook a particular dish. Such a claim would inevitably meet with objection. Likewise, the descriptions of procedures are not intended to be definitive. They are, in fact, the way I have figured out how to make them in a standard American kitchen, though there has been no conscious effort to adapt them to American tastes. Undoubtedly they were made differently in the past in their country of origin and are made differently today by every cook. Again, think of these as the practicum, the pedagogical exercises that help you apply the principles learned in the text, much like the hypothetical problems one finds in an algebra text.

You will also note that the directions are suggestive rather than stringently prescriptive. For the vast majority of dishes, it will not make much of a difference if you cook it more or less or toss in one particular herb as opposed to another. For perfectly practical reasons, substitutions are suggested. There is no point in using substandard ingredients only because a recipe commands you to do so, or even worse, never to try a recipe simply because you lack a particular ingredient. In many cases, one fish easily stands in for another, the same technique will work on a wide variety of vegetables, or something may naturally be left out simply because you do not like it. As with all cooking, there are laborious, time-consuming ways to make things as well as shortcuts. I will leave it to you to decide which you prefer, though I will note that the slower procedures usually make things taste better and are almost always cheaper. For the purpose of learning, knowing the basics from scratch is liberating. It gives you the choice to cook the best food you can, whereas using too many industrially made *quick* and *convenient* ingredients makes you dependent on the people who manufacture them. It is not that such products are necessarily bad tasting or bad for you, but opening a jar of premade pasta sauce teaches you nothing. You are a mere consumer. Starting with fresh tomatoes, or even canned if they are not in season, will lead you to discover the basic principles that make a good sauce and then open up the possibility of creating something that suits your palate and will help you succeed in the kitchen.

Most important, I would like to emphasize that the academic study of food is far too often completely divorced from the bodily experience of cooking food and the sensory experience of eating it. Likewise the culinary arts often ignore the history, culture, and context of food that so interest food scholars. This book and these cooking instructions are an attempt to bridge that gulf.

Last, the recipes here are largely grouped by major ingredient family and technique, where relevant per chapter. Thus, you will find, as through the text itself, parallel coverage of the three cuisines side by side. This is intentional. The idea is to have you compare recipes across these three cultures in ways that would be very unlikely if you were to open up three separate cookbooks about each cuisine. Consider how each cuisine arrived at similar solutions to the same culinary challenges and why. Beneath the very different flavoring structures, there are some universal truths about cooking. There are some flavors and textures that transcend time and place and that might, without hesitation, be called *good*.

Source: CIA (2006).

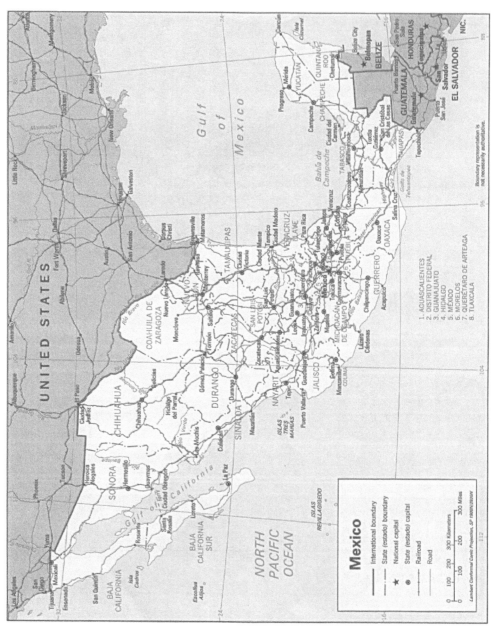

Source: CIA (1997).

Source: CIA (2001).

Introduction
A Theory of Gastronomy

Learning Objectives

- To introduce the basic outlines of the three oldest and most influential culinary traditions on earth.
- To enable students to think critically about the development of distinct culinary styles among the major world civilizations and how they have influenced cooking globally.
- To encourage students to think theoretically about the factors that enable great culinary traditions to arise and how they evolve.

Key Terms

Authenticity—the idea that certain ingredients and cooking methods are true to a given time and place and that authentic cooking involves no modern shortcuts or substitutions

Cuisine—a set of ingredients, techniques, and finished dishes in a standard repertoire, produced professionally and codified formally in gastronomic literature

Flavor Profiles—signature combinations of certain basic ingredients used primarily for seasoning, characteristic of and readily associated with a particular cuisine

Food Culture—a broad term intended to encompass the entire set of beliefs and practices surrounding food, agriculture, distribution, and consumption and any topic related to food

Food Ideology—a formal set of ideas that inform dietary practices that promises a transformation of the individual or society and usually encompasses an entire worldview, such as vegetarianism, kosher dietary laws, and Weight Watchers

Food Voice—the message intended to be communicated to others and one's self through the choice or presentation of food

Food Writing—ancillary texts that discuss food and consumption, encompassing diverse fields such as nutrition, agriculture, etiquette, economics and trade, religion, and fiction literature

Foodways—a term sometimes used synonymously with food culture but often focusing on patterns of food preparation and consumption

Gastronomy—writing about food from the perspective of taste, in the form of cookbooks, technical studies, restaurant reviews, and studies of what constitutes superior ingredients and cooking methods

Haute Cuisine—the tradition of French professional cooking as developed since the seventeenth and eighteenth centuries and codified in food writing

Terroir—the total combination of environmental factors, such as soil and climate, combined with human agency, such as processing techniques, that yield positive attributes characteristic of a specific place; used most often when referring to wines but increasingly refers also to plants and animals

This is a comparative overview of three major world cuisines: Italian, Chinese, and Mexican. Readers will learn how the cuisines developed over time, their internal structure and logic, and how they influenced the way we eat today. Covering the entire globe or even every major style of cooking would have been an impossibly huge undertaking. Moreover, the major national cuisines have already been thoroughly examined in specialized studies and numerous excellent cookbooks, and to replicate these in summary is not the object of this textbook. The goal is to analyze the building blocks that make up a cuisine, the material ingredients and technologies as well as attitudes and historical forces that have shaped taste preferences. These three major world cuisines were carefully considered. These are the oldest on the planet, with the longest span of evolution, and arguably these have had the greatest impact on the world today. They also best illustrate the crucial role of food in our cultural and social development.

Focusing on a single cuisine, as many works do, runs the risk of a narrow nationalist perspective. But nations themselves are often a product of recent history and haphazard political arrangements; borders are an artificial and sometimes misleading way to divide up cooking styles. Arguably, a recognizable Italian cuisine only emerged in the nineteenth century, as the nation of Italy itself was created out of diverse and varied cultures and as products and recipes came to be shared across the peninsula. To discuss Sicilian and Venetian cuisines in one lump category does injustice to the particulars of both.

A regional approach to cuisine comes closer to the practical reality of what takes place in a kitchen, and although it might be difficult to envision a unified American cuisine, regional patterns such as Southwest, Southern, and New England cookery have at least some commonly identifiable ingredients, techniques, and recipes. These patterns may of course be caricatured in the process of marketing brand names or selling dishes in restaurants, but they are nonetheless clearly recognizable and at least have some root in historical reality, even though such regional cuisines normally overlap, defying our hard and fast geographical distinctions. That is, you could find fried chicken virtually anywhere in the United States, although it is most often associated with the South.

However, a regional approach to world cuisine, given the sheer number of regions, apart from being impractically unwieldy would have to be superficial. Given the pace of globalization, it would probably be out of date within a decade as well. Broader regional groupings have been one solution for geographers and food theorists. Southeast Asia coheres as a unit of analysis, as does the Caribbean or Mediterranean, and these make sense in terms of ingredients, environment, and trade networks. But the number of such regions and the way they interconnect would also make an extraordinarily cumbersome and complex narrative. The Caribbean, for example, would have to be linked to Africa, Europe, and India, making the tidy regional category ultimately less valuable in describing the development of cuisine per se. Becoming bogged down in such detail would also lose sight of the topic: world cuisine and consistent themes in the development of the culinary arts on a global level.

Thus, this textbook ranges across three major world regions: the Mediterranean, East Asia, and Mesoamerica, all of which were cradles of civilization and the source of many of the most important plants and animals used as food historically. Focus within these regions is on three nations in particular: Italy, China, and Mexico, which not coincidentally happen to be three of the most popular cuisines around the world today. Although French cuisine has been a dominant influence in professional cooking for the past two centuries, it does not have the historical pedigree of the three regions covered. Nonetheless, the French tradition does provide important lessons for illustrating how culinary systems develop and are codified in terms of techniques, ingredients, and signature dishes.

There is also a conscious effort to avoid rehashing material one might easily find in cookbooks about these cuisines, so the approach here is purely comparative, exploring the three cuisines simultaneously as a way to identify larger patterns of global significance. That is, Italy, China, and Mexico have been chosen for points of comparison, and surprisingly each has traversed a similar historical path from the development of agriculture to industrialization thousands of years later, even though their cuisines are radically different. Most important, these three cuisines are of comparable antiquity, each having domesticated its own staple grain as the foundation of its culinary style. Each has progressed through a long and fascinating series of events, witnessed the growth of empires and invasions, and has interacted with its neighbors, particularly in trade for the acquisition of raw materials. And each, of course, endures to this day in the form of a highly complex and sophisticated culinary tradition. Each is also the epitome of a world cuisine, being linked to an enduring civilization and in turn having a major impact on cooking around the world.

The term *world cuisine* is used here in a special sense, referring not merely to any cuisine that has existed on the planet but to those that have had a broad historical influence on foodways across the globe. A world cuisine also tends to be dynamic and ever changing, easily adapting to and borrowing from neighbors and even farther-flung cultures in ever-widening geographic circles through the centuries. These may have originated as regional cuisines but have expanded on a national, continental, and then global scale. For example, the ingredients and techniques first developed by Mesoamerican cultures spread both northward and southward. Native Americans of North America adopted corn and squash. Likewise, ingredients and techniques of the Yangtze River region spread throughout what is today China as well as Southeast Asia. Both of these cultures also readily borrowed ingredients from the outside, especially as European plants and animals were introduced to Mexico in the sixteenth century and American ingredients were introduced to Asia. That is, world cuisines are never closed systems but evolve through the centuries and of course continue to do so, especially under the influence of industrialization and global trade.

Another key element in the spread of a world cuisine is its adaptability, if it can be exported and replicated despite the substitution of ingredients and cooking methods. First, the cuisine must include recognizable flavor combinations or flavor profiles and be a viable system for thinking about food despite inborn culinary prejudices encountered elsewhere. For example, many of the most important delicacies in Chinese cookery—shark's fin, abalone, bird's nest—were rarely exported abroad. But the cuisine as a system could be exported, albeit in an adapted form, because soy sauce, ginger, and garlic could be combined easily with other available ingredients. We will explore in detail exactly what makes some cuisines more exportable than others, but for the moment perhaps a brief illustration will highlight how this is a much more complicated matter than tracing trade routes or migration of peoples. Sometimes a cuisine becomes fashionable despite enormous distances and exorbitant costs. A perfect example of this is the cuisine of the Middle East, or we might say more precisely of medieval Persia. It is a sister cuisine to that of the Mediterranean, with a common lineage stemming from the development of agriculture in the Fertile Crescent about ten thousand years ago.

Persian cookery was important worldwide since the ninth century in the Abbasid caliphate centered in Baghdad and spread partly with the help of Islam. Its reach was also truly global; we have, for instance, noodles (*lockshen*) as far east as Indonesia and in Europe under the very same name in surprisingly, of all languages, Yiddish. Persian cuisine also spread with trade routes, specifically through the export of spices, so that items such as cinnamon and cloves and nutmeg, hailing all the way from the Moluccas, could be found in medieval European cookery. Along with these there were sauces made from ground nuts such as almonds and walnuts, dried fruits, and most importantly sugar, even in savory dishes, and often in tandem with acidic bases. The basic flavor combinations and textures were exported as a culinary system along with these ingredients themselves.

These recipes, with significant adaptations, were taken to India with the Mughal Empire in the sixteenth century, as well as to Mexico at the same time with the Spanish conquest of the New World and grafted onto Aztec cookery. As unlikely as it seems, the Mexican molé and forms of what has come to be known as curry from India are actually distant cousins with a common ancestor in Persia. These flavor combinations also later influenced China in the east and the rest of Latin America, and arguably much of the world. This is why Persian cookery qualifies as a world cuisine, even though it is rarely acknowledged as such. In the same way Italian, Chinese, and Mexican cuisines are of global importance, perhaps the three most important.

In choosing these three cultures, there is no intention to disparage or minimize the importance or even culinary artistry of other cultures. These three merely offer the longest history and are most useful in illustrating points of global significance. That is, this is a book about world cuisine, not three separate cuisines as distinct topics. There is also no laden value judgment in using the term *complex* to describe these cuisines. Rendering cassava root palatable, and not poisonous, is a long, arduous process for the Maya cook; beating seal blubber into a frothy "Eskimo ice cream" of just the right consistency is a labor of singular artistry and refinement. "Complex" is not meant to signify difficulty of execution but rather breadth of techniques, ingredients, and signature dishes, what we might call a detailed repertoire that is both bound by tradition and open to innovation.

Most important, these three are cuisines with a long and established tradition of criticism and commentary. Especially for China and Italy, there is a body of reflective writing that describes the historical development of cooking and that sets standards by which cooking is judged: a gastronomy, if you will. Such a tradition also existed in Mexico but was largely (and intentionally) destroyed by invading Europeans. Nonetheless, it can be partially reconstructed through archaeology and piecemeal descriptions left by colonial commentators. After the sixteenth century its development has been recorded, and the interaction of native and imported Spanish elements is easily traced.

First and foremost this is a book about cooking, meant for students in gastronomy programs and food studies, professional chefs, as well as home cooks who seek to gain some understanding of not only how to cook recipes from these three traditions but also who hope to discover more about why they took the form they have, their inner logic, and in particular the ideas that have driven cooks to make specific choices in devising classic techniques and flavor combinations. In other words, this is a multidisciplinary study involving history, anthropology, philosophy, sociology, religious studies, medical theory, and more. It is designed to unlock the inner workings of each culinary system in ways that will illustrate all cooking.

This is also a textbook about cooking, intended for use in cooking. The separation of the theoretical study of food from actual hands-on cooking is an unfortunate and largely accidental product of professional academia. At least in the Western intellectual tradition, objective study of a topic was considered to be an exclusively mental process, unsullied by the grime of the kitchen. To know a topic intellectually was radically divorced from understanding in a physical and sensory way. As every cook knows, a particular taste combination can never be adequately described in words, yet we "know" it. The "memory" of muscles performing the routine act of chopping is another kind of knowing, largely unconscious. That is, to know is not really only a process that takes place in our brains. Taste buds, fingers, and even our stomachs have an intellectual capacity. As thinking clearly about food is necessarily an embodied experience, it is absolutely requisite that one not only discuss food but also actually cook, taste, and digest it. This is why this book is both a product of what is known as food studies as well as a practical gastronomy text. Scholars have as much to learn from cooks as vice versa. In this case, the text is informed by recipes, and they in turn derive from the text, hopefully as one seamless whole.

One concrete illustrative example should suffice here. Think of the food at the epicenter of Mediterranean food culture: bread. We can describe the system of slavery on large-scale farms (*latifundia*) on which the wheat was grown in ancient times. We can explain the milling techniques and harnessing of animal power and eventually waterwheels to grind grain in enormous volumes. We can describe the complex trade networks through which flour and even water were distributed throughout the Roman Empire and the administrative infrastructure that made this possible. Large cities were fed through public charity as well as huge, mobile armies that did indeed "march on their stomachs." We can describe the bakeries and ovens in which vast amounts of bread were made every day. We can even explore the context of how bread was consumed in each household from the emperor to the lowliest craftsman, and how different kinds of bread were appreciated by different social classes, and how various foods were considered *companaticum* or things that go with bread, eaten by "companions." We can look at the bread sacrifices made to the goddess Ceres (from whom we get the word *cereal*) and the myths that reminded Romans of their utter dependence on bread. But to really know bread, this takes actually feeling the flour between one's fingers, judiciously adding water and yeast (or rather a starter) in ways that measurements can only vaguely approximate, kneading until just the right springy consistency and sheen is achieved. Then there is the patient raising of dough, the heating of the oven, the careful transfer of raw dough onto the fiery brick oven floor, which transforms it into a substance, a sustenance, without which the Roman Empire would not have existed. Last, and most important, there comes the tasting of bread, as close as one can get to the original experience if the ingredients and techniques have been faithfully reproduced. To begin to understand bread, both historically and today, requires a rational process as well as a physical kind of knowing through the muscles and senses. In this book, we hope to lead readers through both kinds of learning in commentary, recipes, and photographs that will explain three of the most important cuisines to have existed on our planet.

Food Culture and Food Ideology

Throughout this book the term *food culture* will be used. Specifically this refers to the total sum of activities and ideas surrounding food, including its preparation and consumption among a given group of any size—whether a nation, region, city, or even a family or small group of friends. It is a much broader category than cuisine, by which is meant a repertoire of recipes, how they are prepared and eaten. It is also meant to be broader than another common term, *foodways*, with which it is sometimes used synonymously. Here food culture would also include those ideas and pursuits that are related to food or contribute to the way people eat but that are not directly concerned with cooking. Thus, agriculture and the environment are part of food culture; so, too, are religious ideas and medical ideas that influence what people eat. Production and distribution, marketing, in fact every facet of what we now call the food industry would be included in the term *food culture* (in a way that it might not be encompassed by *foodways*). Food culture also includes manners and etiquette, rituals and celebrations surrounding food as well as phobias and taboos.

Usage of the term *food culture* will also extend to subjects such as tableware, cooking implements, and serving vessels. All of these have a profound effect on how and what people consume. A small bowl, for example, held close to the mouth is an integral part of Chinese food culture, which cannot be understood without it. So, too, is the standardized, uniform set of flat dishes in Western culture that shaped the entire direction of this cuisine. The vessels we eat from are often taken for granted but will naturally directly impact the form of prepared food.

This catchall phrase should be distinguished from another more weighty and equally useful term: *food ideology*. This is a set of beliefs specifically about eating that are motivated by political, religious, or social forces, which specifically intends to affect change. That is, they promise the consumer that if a set of usually explicit food rules are followed, one will become a better person—happier, healthier, more beautiful, whatever. Vegetarianism, for example, is not simply a cuisine; it is an ideology. It promises transformation of the individual into an ethically virtuous subject through avoiding the infliction of pain and suffering and prohibiting the using of animals merely as a means to an end (that is, killing them to eat and nourish yourself). Weight Watchers is also a food ideology; it promises a slim figure, happiness, perhaps better health and a better love life, if one eats as the diet prescribes. Structurally, a food ideology functions the same way as any ideology, whether political, religious, or social.

Food ideologies are transformative sets of food rules that may come ready-made, though they can be devised on an individual basis as well. The ideology may offer more than individual change, too; it can promise explicit social transformation. The temperance movement and Prohibition were largely motivated by a vision of a sober, industrious, and morally righteous society, but of course they directly changed the gastronomic culture of the United States, well beyond the years that Prohibition was officially law. The current ideology of local and sustainable food implicitly offers social benefits: the survival of the family farm and rural communities, environmental protection through less pollution caused by transport, sometimes animal welfare, and tastier fruits and vegetables. In other words, it is a total package intended to improve everyone's lives, but it still is intimately connected with the smaller category of cuisine.

The term is introduced here in particular because one specific food ideology will be at the core of every discussion. Gastronomy is essentially a food ideology. It promises sophistication, savoir faire, and "taste," but perhaps also access to social circles that money alone cannot buy. Of course, gastronomy is not only about social climbing; it is about living in the world with a keen appreciation for the pleasures of food, placing a priority on eating and having the knowledge and experience to be able to discern quality. The gastronome thus situates good food as an indispensible component of the good life. To eat well is to enjoy a higher, happier order of existence. The Slow Food Movement is merely one particular branch of the gastronomic ideology. This topic is clearly relevant to the entire subject of world cuisine, because each of these three cultures developed its own complex ideologies of fine dining. Individuals, as directed by gastronomic writings, adopted these ideologies.

In ancient Rome, for example, there were definite earmarks of the tasteful lifestyle—reclining on couches (*triclinia*) while eating, drinking fine wines, and eating complex foods seasoned with garum (fish sauce) and exotic ingredients. Being able to spend an exorbitant amount of money on gustatory pleasure was an indispensable part of social standing, and as such critics such as Juvenal and Martial scathingly satirized it. Importantly, a gastronomic literature developed that outlined for readers exactly how to pull it all off. This is no different from the gastronomic literature of the nineteenth century; they were arbiters of taste of their own day, the equivalent of the great early-nineteenth-century French gastronome Brillat-Savarin. For those who could entertain on a grand scale, these fashions provided a means for establishing social distinctions. Being "in the know" about fine dining could separate a person from the rude, plebeian masses and those with money but no taste.

However, such gastronomic ideologies need not be elitist. A food ideology can also stem from a group fostering hallowed ethnic traditions, perhaps with a food festival in which family recipes are lovingly prepared and passed on to future generations. It creates another kind of distinction, through group solidarity. Participating in the ritual, and literally consuming the markers of ethnic identity in the form of food, makes you a member of this group, unique and separate from outsiders. What it promises, as an ideology, is a sense of belonging, a way of being Italian, Jewish, and so on, and expressing it publicly. In a certain sense one also "performs" identity through food. Of course, outsiders are invited to participate, but it is in sense the "host" culture that thereby performs their identity for these visitors, who are necessary as a kind of audience. Though private expression of cultural identity also takes place in the home, at Christmas dinner, or Passover, for example, it is when outsiders visit that the message is broadcast.

The relevance of all this to world cuisines is that every culture discussed here also harbored various and often opposing food ideologies, each of which made their rival claims upon every individual. Each also informed in one way or another every food choice made by the individual. For example, in the West, a nutritional ideology may insist on skipping ice cream for dessert, the gastronomic ideology will say pass the chocolate fudge sauce, the organic will check the ingredients list, and the locavore, who only eats locally grown food, will source the cream within a hundred miles and go meet the dairy farmer. At no point need any individual pledge allegiance to any one ideology, and all of them can be involved at the same time, even in the simple act of eating ice cream. As another example, picture the devout Buddhist in China, who as a vegetarian would avoid meat, a medically knowledgeable person would balance the yin and yang flavors on the plate, and a gastronome might nonetheless seek out the rare and exotic ingredient that is perhaps poetically described and evocative of a particular place and season. Which ideology takes precedence depends entirely on the context—who comes to visit, the mood of participants, and so forth.

Every time we eat, and even when eating alone as a form of communication to ourselves, we perform an act of self-expression. We speak a certain message, as it has been called; we speak in a "food voice." As a means of communication, it is not merely a one-way expression of values or intentions. In a sense, we speak what we think our companions want to hear. We gauge our comments and actions, what and how we eat, based on our expectation of the perception and response of those around us. For example, a young woman on a date with a new friend sends a particular message by eating a salad. Culturally this communicates that this is an appropriate food for a weight-conscious female and indirectly sends the message to a potential lover that she intends to take care of her figure should the occasion arise for a long-term relationship. But if she orders a steak, it may communicate that she is in fact one of the guys and can keep up physically, as her hearty appetite suggests. There need not be any objective truth to these assertions; they are specific, culturally constructed messages, exclusive to time and place. These are stereotypes, intended merely to illustrate a point about food as communication. But there is a logical consistency in such choices. For example, ordering a searingly spicy meal will often denote an adventuresome, thrill-seeking spirit, as ordering an exotic oddity will suggest open-mindedness. Every act of eating is a performance of our identity, who we want to be or how we want to be perceived. The script, however, always keeps changing.

The relevance of this for the development of cuisine cannot be overestimated. What individuals make at home or what chefs concoct in professional kitchens is merely a vehicle for these modes of self-expression. A restaurant that does not offer the right script options to its clientele simply fails. A steak house in Mumbai is as bad a bet as a pork barbecue shack in Jerusalem because the ideologies and food associations differ from place to place, and even from individual to individual. (Though, of course, both of these might succeed catering to people with a penchant for transgression.) Cuisine not only offers what diners think "tastes good" or merely serves their food preferences as dictated by culture or even what is fashionable. Food also has to be "good to think" first, as anthropologists say, and then good to express to others. Macaroni and cheese would probably fail in an upscale restaurant, unless, of course, the message is home cooking, comfort, nostalgia, perhaps a dash of kitsch. Even still, it will never be from a box with a packet of powdered cheese sauce but something souped up and sophisticated, with deft handling and perhaps a twist. But the symbolic association remains intact nonetheless. In sum, because culinary traditions are so intimately influenced by food ideologies, it would be unthinkable not to take them very seriously. Cuisine is in fact nothing more than the intersection of ingredients and techniques with ideas about food. The cook is merely the conduit through which these interactions take place.

Among the more fascinating phenomena in the development of a cuisine is when the specifically ideologically driven recipe loses its original impetus and is eaten in an entirely different context than originally intended. It may become enshrined in the repertoire even though the original ideas are obsolete. For example, there was once a medical logic in combining fish with lemon juice. The acid was thought to cut through the gluey and cold phlegmatic humors of the fish, which tend to clog the system. This was eventually replicated out of habit and was a typical flavor combination long after humoral physiology was abandoned. It eventually seemed a "natural" combination but was, of course, entirely culturally derived. Another example would be the panoply of breakfast cereals, originally eaten exclusively as a wholesome, all-natural product designed for optimal health. With expert marketing, however, these became a convenience food, packaged for speed and efficiency, and ultimately they entirely changed breakfast as a meal. The health message was reintroduced in another guise, and there appeared new cereals, vitamin fortified, low in fat and cholesterol, high in fiber and antioxidants. But the primary association of cereal itself is no longer health but speed and convenience, which is an ideal combination for the twentieth century, as it turned out. This food happened to be ideally suited not only for the socioeconomic conditions that dictate such values but also for people's desire to eat quickly to get to school or work so they could succeed financially. It was cultural values that changed the meaning of breakfast cereal. Its early promoters merely seized upon the opportunity to give consumers what they wanted.

Slow-cooked oatmeal or homemade bread simply no longer fit the fast-paced ideology of twentieth-century industrial capitalism.

Examples such as these will abound throughout this text and serve to illustrate that what people think tastes good and the shape of any cuisine is in no way arbitrary or capricious. Like any art form, it caters to the desires and values of a civilization, and the two are inextricably bound. Cuisine is not a matter of culinary artists inventing new dishes that are hungrily snapped up by eager consumers. Cuisine is an integral part of culture as a whole, and to understand the former one simply must investigate the latter.

Authenticity

In popular parlance, this word is often bandied about, and people have a fairly clear idea of what it means in a culinary context: food as it is prepared and eaten in its original form in its native setting by people uncorrupted by exogenous or modernizing influences. It is usually understood in contrast to those cuisines in their so-called bastardized forms in the United States. Thus, one readily contrasts the mass-produced fast food taco with "real" Mexican food as one would find in Mexico, or chow mein in a can or adapted to American palates in a Chinese takeout as contrasted with what Chinese people really eat using indigenous ingredients and cooking methods as cooked by a traditional Chinese chef in China.

Yet the term *authentic* is fraught with problems that should be addressed from the outset in a text about world cuisines. Authenticity points to an imaginary tradition frozen in time and place that exists only as a point of contrast with more familiar and presumably corrupt permutations. It is also completely value laden. What do we really mean by the term? Cooking as done in a specific time and in a particular place, perhaps the real molé poblano as it was made in Puebla in 1937 by one particular chef? Surely it was differently prepared by another chef, and done differently in 1837 or 1737. That is, all cuisines are in a constant process of evolution. To claim that one version is correct makes a value judgment that in the sweep of history is illogical.

Saying there is an authentic molé is much like saying there is an authentic giraffe. Authentic when and where? As we find it in the Serengeti but not in a zoo? Authentic today but not ten million years ago, when it was a very different creature, perhaps with a shorter neck, different markings, and entirely different habits? Like a species, cuisines evolve, much more quickly of course, but nonetheless they are in a constant state of change, always in the process of adapting and becoming. To claim a single "correct" form of any dish is to suggest stopping evolution; it is to destroy the very process that brought the dish into being in the first place. In fact, if we consider our own species, since we have been mixing genetically for millions of years, claiming authenticity for a recipe might be considered as illogical as proposing an authentic or real form of human being. All people are the product of genetic interchange and random mutation over generations, and so too are all recipes. There are some outstanding ones created in this process, but more authentic ones?

Recipes are even more variable than species because they change every single time they are cooked, as well as when handed down from mother to daughter or from chef to apprentice or when transferred from place to place. We may be inclined to make a value judgment that one particular form of a dish excels, is the most complex or original, or is the classic as recorded in a cookbook. That is, it is perfectly appropriate to specify an individual creator, date, and place of origin of a widely recognized classic form of a recipe. This does have value insofar as recipes can be simplified or use inferior ingredients or techniques that yield less satisfying results. There are always inferior imitations and botched recipes. Such value judgments have a place in gastronomy, which is necessarily a subjective form of criticism.

However, the term *authentic* should go no further than this, primarily because of the wide, practical variations that are necessarily encountered in such an ephemeral art as cooking. Furthermore, the scant historical record rarely if ever records the origin of a dish, even if such a thing ever existed. Typically the claim of authenticity elicits counterclaims and pointless arguments of priority, which ultimately (cooking competitions notwithstanding) can never be resolved. Was there ever really a first Kentucky burgoo or Brunswick stew or cassoulet? Or were there merely many evolving, simultaneous adaptations of a basic idea? Only long after a particular dish has become well known and widely cooked by a significant body of people can it even be considered classic. But this has little to do with authenticity; it is merely a matter of custom and consensus.

Language usage provides a useful comparison. There have always been those who insist on proper grammar; these are usually teachers. This makes perfect sense, just as it is a good idea to learn how to make the classic crêpe suzette in cooking school. Both are classic forms, which in some contexts one may be expected to replicate accurately. The experts in these cases were the ones who set the canonical standard, and usually it is they who hope to see it preserved. Writers and chefs execute these forms so as to impress the experts, to "pass" when we go among them, literally and figuratively. There are undeniably certain places when proper grammar and the classic dish will be expected by a patron—on a job interview, in a certain kind of restaurant.

However, as everyone knows, real language usage constantly changes, and for effective communication one must necessarily adapt to it. The same is true of cooking. What tastes good always takes precedence over correct form or the proper formula. For example, would it make sense to use out-of-season peaches in a classic peach melba, or even worse, canned, because it is correct, when perfectly good strawberries are available? What would be the point of insisting on the first-person predicate nominative "It is I" when it fails to communicate as effectively as the colloquial "It's me"? Language and food must and do always change, whether we like it or not. The recognition that cuisine is always in a state of flux is actually liberating.

The question of authenticity is also of serious concern in historic cookery. Not only can one never know exactly what a particular food tasted like in the past but also remaining faithful to the original ingredients and techniques can be an overwhelming challenge. In this context, inauthenticity has a more specific meaning, without any intended value judgment. It merely means cutting corners, skipping steps, or substituting ingredients for the sake of convenience or fear. Authentic in this case means following a printed recipe as carefully as possible, recognizing that complete, utter fidelity is virtually impossible. When there is no printed recipe, then one must suspect that claims of authenticity are merely as one individual remembered it, on a particular occasion, and then sought to reconstruct the dish. This is itself a form of interpretation or adaptation, and the very idea of authenticity in this context must be considered spurious.

Even so, there are classic recipes as recognized by a body of tradition, critical acclaim, and established authority. It is also always useful to begin with reference to such traditions, as long as one is willing to concede that in practice cooking is an ever-changing and adapting art form. Even the classics themselves were once adaptations or hybrids, conceived though permutations of ingredients or procedures. Such is the classic pizza Margherita, an unlikely combination of native wheat and cheese with American tomatoes and Asian basil. Its descendants on American shores piled with all manner of toppings, at which the Italian flinches in disbelief, are merely further hybrids. Are they therefore unworthy of respect? Even the pineapple and ham pizza has become a classic in its own way. This is the nature of evolving art.

For cooks and culinary critics, education of the palate and awakening consumers' aesthetic sensibilities by appealing to tried-and-true flavor combinations are valid. Everything does not in fact "go" when it comes to cooking. There is also immense value, to which this book will hopefully attest, in tracing the lineage of and historical development of recipes. The pizza Margherita is a case in point, the pristine origin of an immense culinary family. But calling it authentic would only force us to examine its predecessors, stretching back to antiquity, and by these arbitrary standards, the predecessors would be more authentic. The word *authentic* will thus be jettisoned here as vague, capricious, and ultimately misleading in a book on the evolution of the culinary arts.

Toward a Theory of Cuisine

Though French cuisine is largely absent from this book, it has been necessary to use a French term to adequately describe the subject. *Cuisine* merely means "kitchen," from the Latin *culina*, hence the word *culinary* as an adjective. In English we adopted the term in its expanded sense, keeping the Germanic kitchen (*Küche*) for the physical space. Colloquially, the term merely means a style of cooking. One can say Hunan cuisine, as one can say "my cuisine"—the scale is irrelevant. The associations with the word in its fullest sense, however, are important, and there are very solid historical reasons why we use it in its French sense.

Cuisine generally means an entire codified set of cooking practices that are recognized by a group of people as unique. The practices are set down in words and, normally in order to understand and execute them, some form of training, either within a professional context or at the apron strings of a relative or mentor, is demanded. Those cuisines most easily recognized are often those with the greatest degree of professionalization and with the largest body of recipes in a standard repertoire that depends on signature flavor combinations, equipment, and techniques. Following the lead of the French tradition, we routinely speak of the cooking style of country, region, or people as a cuisine, and there is no real reason not to do so. All peoples on earth have a cuisine. Even the very idea of a cuisineless people is condescending and chauvinistic. All people cook, just as all people speak a language. Our brains are instinctively hardwired for both, as quintessentially human forms of communication. There are cuisines that we might call debased, perhaps those in which people's preferences are dictated by profit-seeking food industries. These, naturally, have little aesthetic value in the eyes of the culinary cognoscenti. But they are nonetheless cuisines, whether they take place in a large kitchen, a microwave oven, or even as street food. Many of the most refined culinary traditions emerged in this context. Think of sushi. Would we hesitate to call contemporary fast food a cuisine? The repertoire is well known and beloved by countless consumers. One might even call the menu classic. The equipment and cooking methods are rigorously codified, and even though the labor is mostly deskilled, a great deal of technical ingenuity goes into making the hamburger and fries consistent. That a cuisine exists should entail no value judgment. There are simple and complex cuisines. Complex cuisines are the subject here, recognizing that the complexity required to turn out fast food is mind-boggling.

Quality is another matter entirely and is almost completely subjective. It will have reference to perceived expectations, such as perhaps a judicious balance of flavors or efficient service that presents the food hot. Fast food may still qualify. What differentiates it from cuisines discussed here is not patronage either, but a critical body of literature such as cookbooks, restaurant reviews, magazine articles, and travel literature, as well as discursive writing on the pleasures of the cuisine, its health-promoting properties, the ideal forms of agriculture—indeed any kind of food writing. Last, there is the academic study of food. This literature tends to fix a canon, by which is meant a range of ingredients and recipes that "belong" in a given cuisine. These may be shared by other cuisines, in fact they almost always are, but the general patterns come to be so thoroughly recognized with the tradition that they become enshrined as the defining principles and delimiting parameters of the cuisine. It cannot be called Cajun without chili pepper.

The word *cuisine* also has more subtle connotations, particularly the association with the term *haute cuisine*—meaning extremely refined, expensive cooking that takes place in Michelin-starred restaurants in a tradition founded in France in the seventeenth and eighteenth centuries but stretching back to aristocratic dining considerably earlier. This culinary style was subsequently contrasted with cuisine bourgeoise, which was the cooking of city folk or middle classes, which differed not only in scale and cost but also ingredients. It depended more heavily on cheaper cuts of domestic animals and vegetables. This cuisine perforce had to be managed in a smaller kitchen with much fewer hands and less fuel and often on a small stovetop. This was, in turn, contrasted with rustic or provincial cookery, which was more closely tied to the farm and local produce, featured slowly cooked dishes based on regional preferences, and was thought to be passed down within families for generations. Thus, the term *cuisine* in France had implicit class associations, and when the term is used without a qualifier today it is usually taken to mean *haute cuisine*, or an elite cuisine of the court.

Court cuisines exist in practically every culture where the ruling elite has the wherewithal to live in a substantially different manner than the majority. The king of Thailand, the great Inka in what is today Peru, and the king of imperial Spain all patronized cooks who developed distinct cuisines based on their practically unlimited resources and staff at their disposal. As a general rule, where there is money, the culinary arts, like all others, will attain a high level of refinement and sophistication. These developments are dependent on a degree of social stratification; that is, many distinct layers of separate classes. In those societies in which the vast majority earn their living exactly the same way (usually farming), there will be only one class, even though there may be leaders, elders, priests, and medicine men. These people are so limited in number as not to constitute a separate class with their own distinct values and cooking styles. They probably have more food than others, but they basically eat the same food as everyone else, and in the same way.

Contrast this with a stratified society, in which there is a distinct and sizable number of elites and perhaps even a middle class, along with artisans and the poor. Rulers here draw from the resources of the populace, normally through taxes or tribute, and can amass large fortunes. With this wealth rulers will have access not only to very different ingredients, perhaps imported at great cost, but they will also be able to afford a permanent kitchen staff whose job it is to feed the entire household. Cookery then becomes one among many art forms that can be permanently patronized. Rulers, of course, also entertain other elites and thus spur competition and culinary innovation.

There may be a measure of social mobility here, perhaps lesser nobles trying to gain access to power, or wealthy merchants and businessmen in the cities. It is precisely these people, imitating their superiors, who cause fashions to evolve. The elite are forced to constantly "reinvent" themselves in ways that keep them distinct because their distinction may not merely be a matter of wealth. Sophistication and savoir faire, knowing how it should be done in matters culinary, sartorial, linguistic, and so forth, then become involved.

When historians and anthropologists speak of a cuisine it is often this kind of elite cuisine they refer to: a complex, self-conscious, reflective, and ultimately inventive cooking style. Stratified societies often become so through access to trade and the very possibility of accumulating great wealth. Through this outside contact they become cosmopolitan, worldly, and of course they learn about other traditions. Thus, the foreign and exotic often play a central role in how such cuisines constantly reinvent themselves, often by introducing rare ingredients. Spices in the European Middle Ages are the perfect example of this process, as wealthy people had increasing contact with the Middle East after about 1000 and adapted exotic recipes. Outside contact as well as social pressure from below are merely two of the factors that made such complex cuisines possible. Furthermore, since the culinary arts are so ephemeral, the impetus to record them for posterity, or merely to impress rivals, leads directly to cookbooks and related culinary literature.

What of those culinary traditions that go unrecorded? Are they unworthy of the name cuisine? One of the three great traditions discussed in this book falls in this category: the records of Mesoamerican civilizations were mostly destroyed or were not set down in forms recognizable to the Western tradition. Then there are traditions that are fairly static, in homogeneous societies and based on mostly indigenous ingredients with a more circumscribed repertoire. These are still cuisines, though they are perhaps less complex and less changing. It is difficult to avoid implicit value judgments in such comparisons. Perhaps an analogy from the plastic arts will shed light on this topic. In the Western tradition it has been conventional to speak of certain art forms, namely painting and sculpture, as "fine arts." This is partly because they have a theoretical underpinning and were highly professionalized. That is, one could practice them full time, and there were formal channels of apprenticeship and education, even academies. This only happened because there was a significant body of people willing to patronize artists. One could make a living as a painter. Often these arts were appreciated solely for their aesthetic value without reference to function. One never asks what a painting is for, or what it is used for. It exists merely for pleasure—for art's sake.

In contrast to these, there are so-called crafts whose form is dictated primarily by function. Their practitioners, although professional, enjoy lesser social status, and they usually get paid much less for their work. The materials they use are less expensive, and their clients pay less per object. They often produce work in bulk as well, rather than individual and unique works of art. The scale of their work is often much smaller, say a pot or a chair as compared with a bronze statue, although paintings and sculpture can be on an intimate scale, too. Since the Renaissance, fine arts such as painting, sculpture, and often architecture were kept distinct from pottery, metalsmithing, glass making, cabinetry, weaving, and so forth, which were all considered lesser crafts. This is of course a completely arbitrary separation and has nothing to do with inherent value or artistic worth, but it is important historically.

It is also of great importance in understanding the word *cuisine*. Just as with other media, cuisine has usually been separated consciously from mere cooking. Chefs were considered distinct from cooks. The former worked in professional kitchens on a grander scale and wore uniforms as marks of their status. In addition, they often had formal training or a degree. The uniform (today chef's whites and the toque) were specifically designed to keep out pretenders to the art and to remind the public of their legitimacy. There was also an explicit statement of qualitative difference if a dish was described as part of a cuisine rather than cookery. The latter took place at home for a family, or in a small establishment, used simpler ingredients on a smaller scale, and cost much less. Cooking used simpler implements and fewer techniques. There was also thought to be less artistry and invention involved. Its product was less unique, less a statement of individual creativity and genius.

This book by no means wishes to perpetuate these distinctions, but they are important in understanding the historical significance of the word *cuisine* and also why elite cuisines developed. But it is equally important to remember that there has always been a constant and crucial flow between these two supposedly distinct categories. Haute cuisine is always informed by home cooking, often in recurring cycles, and equally home cooking borrows and adapts elite recipes. There is both grassroots upsurge and trickle down. This is arguably as central to the development of the three traditions discussed here as it is to our own vibrant culinary culture. Restaurants must serve food people will eat, and even though their cuisine may legitimately push the boundaries, they must have patrons. Moreover, a culinary style that is completely cut off from the public, perhaps a closed court or one with no social mobility or access to power, may have a complex repertoire, but it is usually stagnant and bound strictly by tradition. That is, without social pressure from below, it never does reinvent itself. It never needs to create distinction with exotic ingredients or bold, new dishes. Nor does it ever seek to rediscover rustic, simple home cookery. It is essentially cut off from outside pressure to evolve. There have been many examples of court cuisines like this. Revealingly, the three complex traditions discussed here benefited precisely because of social mobility, external contacts, and the fruitful interchange of haute cuisine and home cooking.

In fact, it is possible to discern regular patterns, a kind of oscillation between extravagant cooking and simple homey cooking in the evolution of cuisines. The initial trigger may simply be economic, as it becomes unseemly to eat luxurious meals while the masses are starving. Empathy for common people then translates into appreciation for fresh, local ingredients used in simple, unadorned dishes served without complication. We see this time and again when chefs insist that a food should "taste of itself" rather than being disguised or elaborately stacked, fussed over, or garnished wildly. After a few decades, in an effort to satisfy patrons, cooking once again becomes elaborate, uses exotic ingredients and techniques, and is designed to showcase the creative genius of the chef. It also becomes expensive, as a way to create distance from those without money or not sophisticated enough to appreciate its subtleties. This constant tug of war between these two opposed aesthetics, while neither completely dominates, will serve to explain how culinary fashions change in the short term and even within the cooking of a single chef throughout her career. Moreover, such changes happen in generations, spanning perhaps a few decades. This tension between sophistication and simplicity, as in all the arts, has always been a catalyst in the evolution of cooking, as numerous examples throughout this text will attest.

The question of why complex cuisines develop in the first place involves many more factors, though. It is never accidental, and the three examples of East Asia, Mediterranean, and Mesoamerica illustrate this process best. First, gathering and hunting societies can use a vast range of plants and animals on which they subsist throughout the year. Ironically, it is usually a much greater variety than the so-called civilized societies, because they are completely dependent on the seasons, cannot accumulate significant surpluses, and eat foods that are considered less palatable to sedentary cultures. But these societies rarely develop what we would call a complex cuisine. Their technology is necessarily simpler and more portable and the techniques fewer. Despite the variety of foods from season to season, the diet is still monotonous. Their methods of preservation may be very sophisticated; consider the pemmican (dried meat pounded with fat and berries) among Native Americans of the Plains or preserved salmon in the Pacific Northwest. In general, these cuisines are stable and unchanging, given the limitations of the environment and assuming there is little outside contact.

Furthermore, there must be a need to change, such as either population pressure or dwindling resources or outside contact. Aboriginal societies in Australia subsisted on essentially the same diet for millennia. This turned out to be a diverse cuisine with a huge repertoire of ingredients. There were ancient oral traditions regarding which plants and animals are best to eat and how to cook them. The cuisine never effected global change though, and never became a world cuisine, partly because this was never an imperial culture that needed to expand and they were not forced to find new resources. They struck a balance with the environment that held up until outside invaders appeared.

The typical reason a gathering and hunting society changed was growing population triggered by climate change, the diminishing viability of gathering and the nomadic lifestyle, or simply the accidental discovery of flora and fauna that could be domesticated. We should say as an accident of geography some places simply lucked out. They had available animals that could be herded and plants that would yield to cultivation. The Fertile Crescent, what is today Iraq, Syria, Lebanon, Israel, and eastern Turkey, is a prime example. Ruminants such as sheep and goats, native to the area, are fairly tame and will instinctively follow a leader, animal, or otherwise into a fold. Barley and the simple ancestors of wheat can easily be gathered, planted, and stored. The process of domestication was not necessarily intentional, nor was selective breeding, but over time farmers and shepherds chose the biggest seeds to replant and the largest, most docile animals to breed. Ultimately they changed the species in the process. Wheat and barley in the Middle East, corn in Mesoamerica, and millet and rice in Asia literally made civilization possible.

In some places there simply were not viable species to domesticate. Zebras and gazelles cannot be tamed or bred in captivity. In other cases, there was never the need to domesticate food. In California, an abundant supply of acorns and plenty of game and fish combined with a relatively low population density meant that there never was a need to adopt agriculture.

Another factor may be the East-West axis of transport, which was perhaps key to the spread of domesticated crops. Traveling north or south the latitude, weather, and hours of sunlight change, and it is usually very difficult to move plants along this axis, say from North to Central Africa. However, it must be admitted that corn did move along this axis, although this entailed centuries of selective breeding. Overall, it was much easier to move plants between east and west along a relatively stable climate zone—from the Middle East to the Mediterranean or to India. This will be very important for the story of exchange, even across long distances, that follows. Corn, for example, was easily grown in Spain and northern Italy as well as in Turkey.

Apart from domestication, some regions benefited from having a wide variety of ingredients, in essence a complex biosystem, from which to draw. The variety of indigenous ingredients certainly plays a major role in how cuisines develop over time, and Italy, Mexico, and China are exemplary in this regard. It is also crucial that although located in temperate zones, they have easy access to subtropical regions that held different crops that they sought out in trade. For example, the Aztecs acquired cacao and vanilla from the hotter south. China acquired fruits and spices from Southeast Asia, and the Mediterranean procured dates, nuts, and sugar from the Middle East. In other words, proximity to other regions is a boon and geographical isolation is a positive hindrance to the development of a world cuisine. However, important crops were distributed from isolated places over vast distances. Cloves and nutmeg from the Moluccas in what is now Indonesia are the best examples.

It should also be pointed out that our three civilizations were not the only ones to domesticate their own staple crops. New Guinea had yam and taro and Ethiopia teff. But these were never distributed worldwide, although taro certainly is one of the most important crops globally today. More important, these cultures never developed the complex social structures, centralized governments, and trade routes that are requisite for the development of a world cuisine.

Added to this complex picture is another accidental set of resources: first clay and the technology to fire it and later metallurgy. This topic will be covered fully in the chapter on cooking implements, but without a doubt, as with transportation and war, agriculture and irrigation, as well as the manpower to support these endeavors, raw materials and the need to develop technologies that exploit them are central in the elaboration of complex cuisines. The stewpot, the wok, and the *comal* (griddle) are as important to our three cultures as their staple crops.

Another factor is the succession of civilizations. None of these civilizations had an unbroken lineage from ancient times. They have invaded outside civilizations, have borrowed from previously existing cultures, and were also conquered by invading armies. The ability to draw from previous cultures, as did the invading Aztecs in Mexico or the Ming in China, is also key to this story. Unmitigated destruction does not pass on civilization, culinary or otherwise. Yet some relatively uncivilized nomadic peoples have gained the full trappings of civilization in one fell swoop, merely by successful invasion. Logically, among the key ingredients contributing to a world cuisine is the development of an empire with extensive trading routes and an elite court with open access by means of civil service or through rising up in the ranks within the military. Such middling ranks are essential especially for spreading court culture down the social ladder. A closed court is rarely a dynamic one.

Equally important, and perhaps surprising in this context, is the structure of the family. Recipes can easily be handed down among elites in written or oral form because there are always apprentices and a succession of chefs. But in the home kitchen the presence of elderly generations in teaching cooking cannot be overestimated. Where an extended family household including several generations is the rule, younger daughters tend to learn cooking from grandmothers or elderly aunts. In most economies, where both men and women work outside the home, it is these elder women who care for the children and pass on culinary wisdom. It is very clear that a break in this mode of transmission, as has happened in our own times, can completely alter culinary traditions. It is not that young men and women are completely helpless in the kitchen, as many suspect, but they cook very differently from their ancestors, except when the extended household is still the rule, as it often has been in Chinese and Mexican homes and was once in Italian homes.

It cannot be coincidental that it is precisely in these three culinary traditions that the reverence for elders, the appreciation for holidays and celebrations, and also for religious piety go hand in hand with the survival of traditional cookery. This is important as a central thesis of this book: It is never merely elites who create vibrant and dynamic cuisines. It is also the fruitful interaction between high and low, elite and peasant, and bourgeois and aristocratic that spurs the culinary arts to excel. Without the crucial caretaking of age-old traditions among people in their homes, much of the innovation at the dazzling courts would have been literally a flash in the pan.

Innovation and Tradition

That the three culinary cultures discussed here survived through the centuries was also due to the ability to strike a careful balance between tradition and innovation, which is a hallmark of truly great cuisines. Successful culinary art rests somewhere on a continuum between the familiar and new, usually altering some elements such as form, ingredients, and presentation, but rarely all of them. That is, the consumer must have some reference point, even if all the elements are thrown into disarray, as happens with creations like bacon and egg ice cream. This has two disparate, but recognizable, reference points conflated together. With no context or reference, a dish will likely fail. The consumer cannot get it or will not eat it.

This balance is characteristic of all great cuisines historically. They are never completely locked into tradition, or they stagnate. Neither can they be totally and utterly new. Even the great codifiers of the Western culinary tradition—Apicius, Bartolomeo Scappi, Carême, Escoffier—built upon earlier traditions while introducing bold new ideas. Then they in turn were built upon by followers, and their rules were later broken. This constant process of rule making and rule breaking is what keeps all arts alive.

There is also the broader tendency for some historical periods in general to be rule making and others rule breaking. What triggers this is difficult to discern, but it is probably a combination of economic factors, political climate, and the nature of the marketplace and patronage. It is true, for example, that in difficult economic times people will turn to more traditional and comforting foods. Naturally, bountiful patronage among a broad range of consumers will encourage serious innovation. This probably accounts for the rise in scientifically inspired cooking and so-called molecular gastronomy and may explain its demise with the very recent economic downturn. It may also be that all fashions eventually wear out, and to keep ahead at critical junctures, innovation is simply necessary.

Exportability

Alongside this rhythm of tradition and innovation is what might be called the exportability factor: How easily can a cuisine and its basic flavor profiles be replicated elsewhere? It may be that certain ingredients and cooking styles were taken with immigrants, or the basic outline of the cuisine traveled with exported goods, which were normally imperishable spices or condiments. Later, of course, cuisines were consciously exported as an indispensable part of culture, along with empire. Hence we find baguettes in Vietnam, a former French colony. However the basic culinary building blocks arrived, some cuisines are inherently easy to copy. The details are often lost in translation, but it seems that in the case of these world cuisines, the fact that their basic flavor profiles could be easily recognized and replicated meant that they had an easier time traveling and going mainstream elsewhere. Admittedly, these are often shallow stereotypes of each cuisine: soy, ginger, and garlic can turn anything into Chinese food; tomato sauce and mozzarella are the epitome of Italian; chilies, salsa, and guacamole are instant Mexican. These ingredients are so firmly associated with these places that mozzarella in Chinese, guacamole in Italian, or soy in Mexican food would be an affront to proper taste. These extremely simplistic associations enable a cuisine or some vague simulacrum of it to travel easily, precisely because they had become formulaic.

In many cases it was items of international commerce that heralded the arrival of a new cuisine. Tea and porcelain from which to sip it spurred the fashion for Chinese things in Europe, as did chocolate for Mexican flavors. In Europe, these were accompanied by an interest in exotic flavorings; soy and ketchup, for example, arrived in England in the seventeenth century, as did mango pickles from India, and before that tobacco from America. These products were all fostered by a nascent consumerist economy and a significant number of people willing to pay for them. Most are essentially superfluous luxury items rather than nutritious foods, but they brought with them the idea of the new and exotic and an ability to literally taste the other. Naturally, there eventually followed the fashion for consuming foods from these far-flung places, which at first were brought back with returning expatriates on duty throughout the empire. Eventually, steamships with refrigerated containers supplied the mother countries in Europe and the United States. This is how exotic tropical fruits first appeared on the global market. It is only within the context of nineteenth-century imperial policies that such massive movements of products make sense. That is, it took more than entrepreneurial importers finding new markets for exotic goods; it also required a conscious government effort to foster trade with the colonies by protecting shipping, offering what amount to monopolies to investors, and even giving tax incentives. This is how the United Fruit Company was able to start huge banana plantations in Central America and how countless other products—coffee, pineapples, chocolate, sugar—came to be grown throughout the world to supply the luxury trade. Often the products moved with immigrants within the empire, at first with limited interest among the majority population, but later gaining a certain fashionable mystique. Thus, eventually it was not merely products but whole cooking styles that had been imported.

Interestingly, some cuisines made the colonial transition easily and others did not. It appears that a clear internal logic and a recognizable flavor profile are essential for understanding this exportability factor. Expectedly, the transfer of recipes gave rise to hybrid cuisines. Italian, Mexican, and Chinese food are all enormously popular in the United States, and as they are prepared, they must be considered legitimate as distant descendants of the classic culinary traditions. They have been dramatically transformed in the process of transmission, as have the people who brought them, but they are sometimes just as interesting in translation.

Although this book will focus on the historical roots and current practice of these cuisines in their homeland, which will naturally look quite different from what an American audience has come to expect, there will be some homage to these hybrid descendants and the recipes created by emigrants. In many cases the hybrids are actually important for understanding cooking in the original context. Take, for example, once again, pizza, once mostly found in Naples. After its enormous popularity among Italian Americans, and then all Americans, it was in a sense reintroduced to Italy, becoming a more national dish, even though that claim had been made for the pizza Margherita in the nineteenth century. Most pizza in the United States is very different from what is normally served in Italy, yet its very success was dependent on its adaptability, its ability to hold diverse new toppings, be divided into slices and carried away, with a sturdier crust to make that possible. All these are the ultimate reason for its success, and ironically as Italian eating habits come to more closely resemble Americans', the portable slice is increasingly found in little corner *pizzerie* there, too. Adaptability is the key to this and many other dishes in world cuisines.

None of this would have been possible without restaurants to promulgate these new cuisines. The restaurants often substituted ingredients, played down regional differences, and watered down more pronounced flavors for an audience with a less adventuresome palate. But it was this very process of adaptation that enabled these to become popular "ethnic" cuisines in the first place. It is precisely because they were not tossed into the great "melting pot" but kept their basic features and evolved to suit local preferences. These cuisines were caricatured and even bastardized if one thinks of the travesties sold in cans or at fast food outlets, but these, too, played a crucial role in popularization, which in turn led people to seek out what they considered to be more "authentic" forms.

In a hybrid cuisine, Tex-Mex is a good example. Certain key features became rote, or expected, by customers. When ordering tacos and burritos, customers expect a plate lathered with sauce and cheese, refried beans, "Spanish" rice, tortilla chips, and salsa. These can be found in some form in Mexico, at least ancestral versions of these, but that should not mean that the hybrids are necessarily corrupt or inferior. In fact, in this case, some of these dishes developed while Texas and the American Southwest *was* part of Mexico. The point here is that all cuisines are to some extent hybrids, and all have involved the importation of ingredients and the migration of peoples or movement of borders. All cuisines have been adapted over time; this is nothing new in the modern era. The very fact that we find tomatoes on our pizza, chilies in Szechuan cooking, and cinnamon in a molé is ample evidence that cuisine is by its very nature an ever-evolving hybrid, especially those here designated as world cuisines.

Good Food

Finally, it is difficult to avoid subjective value judgments in a text largely about taste. All people have their own taste preferences that are clearly based on current trends or even their own quirky personal favorites. Because this is a textbook rather than a cookbook, there is no need to cater to popular taste. We will not shy away from describing dishes that have been enjoyed within these traditions that might strike the modern reader as strange or even offensive. Algae, for example, was integral to Aztec cookery, as were insects and huitlacoche, a fungus that grows on corn. Small rodents such as dormice were eaten in Rome, and the Chinese tradition has had practically no food taboos, except for pregnant or nursing women; otherwise, nearly everything is "fair game." We will not insist that the reader hunt these ingredients down, even if that were possible, but an open mind to bygone and even current ingredients and flavor combinations will be absolutely necessary.

However, a totally relativistic and value-free discussion of food that assumes everything is good merely because someone enjoys it would be a mistake for a work rooted in gastronomy, the art of good taste. Thus, we have framed some fairly objective criteria for evaluating dishes deemed to be "good food." In some ways these criteria overlap what is called today "slow food," but we have avoided that term partly because of its proprietary nature and its manifesto designed in reaction to fast-paced industrial civilization. The Industrial Era has produced good food. Most of what we eat today, in fact, is a product of that system. More important, some very good food has traditionally been fast—both in the household and on the street. Speed plays an important part in many culinary cultures, not only our own, and it would be a disservice to excise this entire variety of cookery merely because of the pace of preparation and consumption. It would be like leaving drawing out of the history of Western art or calligraphy from the Asian tradition. Sometimes these capture the essence of everything positive in art and can be its purest form of expression. Think of a freshly caught sardine, quickly grilled and gulped down on the beach in Sicily. There is no way this could be construed as slow, but it is nonetheless very good.

What then do we mean by good food? First, it is food made with integrity, from ingredients that are fresh if possible and by hands using native techniques in recipes that are recognized and bound by long tradition. They may be innovative nonetheless and even use new ingredients. For example, when corn was first introduced to Italy in the sixteenth century, if Italians had set out to make *masa* dough and tortillas, we would rightly say that they had imported Mexican cuisine rather than a new dish integral to Italian cookery. Instead they dried the kernels, ground them, and boiled the meal in the form of polenta, in exactly the same way they had been processing barley and millet for centuries. Despite the new ingredient, polenta was and still is an Italian dish with integrity, a permanent part of that tradition.

Good food should also ideally speak of the soil, climate, and people who prepare it in a unique way. The term *terroir* has been used to describe these and many other combined factors. The idea of terroir was first seized upon by the wine industry, and increasingly growers of fruits and produce use it as a marketing tool. Think of Vidalia or Walla Walla onions. The idea is that certain plots of land, their soil chemistry, sunlight, and water quality, in tandem with human intervention, methods of manufacture, and traditions, yield products that express the unique characteristics of the land, which is deemed of exceptional quality.

There is undeniably some validity to the concept, even though it is difficult to quantify or analyze these various elements with any scientific analytical precision. It is true, though, that the same grape will taste very different planted in different places and especially when made into wine. The same tomato that flourishes in New Jersey often languishes elsewhere, and of course subtle differences of flavor come directly from the soil, apart from the obvious effects of climate and latitude. Equally it would be pointless to deny that in some places and with some methods of agriculture and processing, a superior product is obtained. Often these are rightfully associated with a specific place: Parmigiano-Reggiano, Roquefort, Appenzeller, all of which names are protected by law.

But are these determinations of quality at some level also merely matters of taste? Some people undoubtedly prefer the taste of Parmesan from a cardboard can, or at least have been convinced by exposure, affordability, and ubiquity that this is what Parmesan cheese should be, and they honestly like it.

It would be apposite here to make an important distinction between pleasure as an index of value and another set of gastronomic criteria. Rarely in this book will we have recourse to use the word *pleasure*, and certainly not in terms of food. Pleasure bears little direct correlation to the idea of good food or to the topic of cuisine, or at least it is impossible to assess within the parameters of a discussion about cuisine. That is, the aesthetic reception of food in the mouth of each consumer is more properly the subject of physiology, psychology, and philosophy. Why people like what they do is the product of individual and multifarious associations. It is influenced by culture, and we can make generalizations about what whole groups like, but when it comes to sensory reception of each individual, the locus of pleasure, we would necessarily be at a complete loss to adequately describe why it happens as it does. For example, a three-star Michelin-rated restaurant will mean very different things to different people. To one it can be a sublimely pleasurable experience, to another a complete waste of money. These individuals' expectations largely determine what happens in their mouths; it preconditions the pleasure, or lack thereof, for each. This holds true for individual foods as well. Expectations can even override something that does not taste very good and yet yield pleasure nonetheless. Think of the first taste of caviar, coffee, or whisky.

The gastronome will insist that this is merely a matter of educating the palate, learning to like a food and learning about it. For those who have crossed that threshold, it seems unthinkable to refuse foie gras, raw oysters, stinky cheese, and so forth because of some particular prejudice, informed by "infantile" taste, or worse, a politically driven agenda. Yet the fact remains that some people recoil in horror from these foods. The durian (a large, spiked fruit from Southeast Asia with a ferociously aggressive aroma) is an interesting example for many Westerners, who liken it to rotting carrion, fecal matter, and sweaty socks. Sometimes these deep-seated prejudices can never be effaced. A recent experiment of mine in a classroom evenly divided between students of European and Asian descent revealed almost without exception that only the Asians enjoyed the durian and finished it, and conversely only the Europeans enjoyed the Point Reyes blue cheese. Their cultural prejudices and habits completely determined their level of pleasure. We can discuss taste preferences, but there is no disputing them.

Moreover, gustatory pleasure is a completely subjective notion, which is not directly related to the art of cookery. One individual will have a jaded palate and require strong seasonings to enjoy a dish; another is famished and even the simplest of foods will taste good; another has an aversion to garlic; another had a bad experience with white food as a youth and can no longer bear the sight of it. Pleasure is clearly a different and in many ways a more complicated topic than cuisine.

Still, what possible objective criteria could be applied to food, over long periods of time, among different civilizations, as universally good attributes of food? The use of the term *sustainability* may seem very strange in this context because it is currently such a buzzword. This is because we have only recently begun to lose sight of what has always been a universal and unbreakable rule of civilizations: they must strike an equitable balance between population and resources. Furthermore, they must not degrade those resources in ways that precipitate the collapse of civilization even when population pressure is not an issue. Arguably, both factors were critical in the disappearance of ancient Sumer, as the long and slow salinization of the soil through irrigation made agriculture less viable and less able to support a large population and ultimately left southern Iraq a desert.

Such collapses are just as often triggered by climatic change; a shift in temperature, even a few degrees hotter or colder, can lead to crop failure and devastation. A few centuries of warming can produce a great surplus of resources and dramatic population growth. A simple rule of all civilizations is that more food equals more people. This may in turn force emigration or whole new ways to exploit resources and whole new economic systems. These forces have all been critical in the cultures discussed here and have shaped their cuisines in profound ways.

Sustainability refers not only to a balanced use of resources in ways that make them renewable to future generations. For many civilizations an abundance of space meant that they could practice slash-and-burn agriculture and move periodically. But for the sedentary societies with limited space, they had to farm, fish, and hunt in ways that would not deplete resources permanently or drive species to extinction. Those that did ultimately perished, or scant survivors were absorbed into other civilizations. In the case of the Maya, the great urban centers were abandoned, and survivors moved into the periphery.

Sustainability in this context also means devising a diet that can support a population in health without disastrous attrition. Obviously there have been cultures around the world that have lived very well on diets that by modern standards would be considered nutritionally inadequate or imbalanced. In addition, there is no doubt that nutritionally related diseases such as rickets, beri beri, and scurvy have been endemic in many places. Yet the very fact that these civilizations survived is proof that they were able to supply the basic nutrients necessary for life and reproduction. That is not to say everyone ate well, and for most of human history subsistence crises were a regular occurrence, especially for those close to what biologists call the "carrying capacity"; that is, the number of people a given environment can support given a certain level of technology. The slightest change in weather or human fertility can trigger what has been called, after the early-nineteenth-century philosopher and economist Thomas Malthus, the "Malthusian Scissors." That is the tendency for a population to naturally outstrip its resources, only to be cut off by what he considered natural checks: famine, disease, even war.

Perhaps the best example of this would be Europe in the fourteenth century, with a teeming population and just on the brink of disaster after a few serious crop failures early in the century. The so-called scissors that intervened to rebalance the population was the bubonic plague of 1348, which cut off about one-third of all Europeans. Thereafter survivors enjoyed unprecedented prosperity, meat entered the diet as it never had before, and this had direct consequences for the development of European culture as well as its cuisine. Cookbooks of the late fourteenth and fifteenth century are filled with new recipes for meat, not because it was a status symbol, but because it was cheap and affordable.

All this is merely to assert that demographics have a direct impact on the evolution of cuisine, and vice versa. The population boom of China in the early modern period has been attributed in part to the arrival of the sweet potato. In the most basic way population density determines what kinds of resources will be available to what kinds of people. It will influence the dramatic distinctions between diets of the rich and poor, and of course the experience of hunger has an immediate impact on culinary culture. Think of the frugality and resourcefulness of those who lived through the Depression of the 1930s. Their recipes look and taste differently because they were keenly aware of "keeping the wolf away from the door," as food writer M. F. K. Fisher put it.

Though it may seem odd to discuss economics, demographics, and climate change as part of cookery, these are perhaps the most important and unrecognized structures that drive long-term change. They are just as important and directly connected to the development of great trade routes, overseas exploration, and new technologies. That is, when we think of the Silk Road, the "discovery" of America, or the Industrial Revolution, it is important to keep in mind that none of these would have taken place without fundamental changes in the economy that made them necessary or highly desirable. Without these we would never have found tomatoes in Italy or food in cans. As German philosopher Ludwig Feuerbach (1804–1872) put it, *Der Mensch ist was Er esst* (Man is what he eats). Equally, what he eats is the result of myriad underlying processes that are environmental, demographic, economic, and of course cultural. The form and development of all cuisines are as much a product of these factors as are individual creativity and ingenuity.

How to Use This Textbook

This textbook is meant to be read from cover to cover over the course of a semester or at one's leisure. Country maps of Italy, China, and Mexico are included in the frontmatter, as are a List of Figures and a List of Recipes. After an historical overview in chapter 1 (which features a time line) and a discussion of technology, utensils, and techniques in chapter 2, the text is organized around the main components of the cuisines: grains and starches; vegetables; fruits and nuts; meat, poultry, and dairy products; fish and shellfish; fats and flavorings; and beverages. These chapters are packed with information on ingredients and techniques, and the reader is encouraged to return to previous chapters to review information on these when mentioned later in other contexts. By the end of the text, the goal is to give the reader a good sense of how and why these cuisines developed as they did and how they influenced each other and subsequently the entire world.

Importantly, most chapters include recipes that are the practical exercises to accompany the narrative. As a rule, these recipes are grouped by technique or major ingredient so readers can see that each cuisine comes up with its own solutions to basic culinary challenges, yet some basic forms of food are universal. Thus, the meatballs are grouped together, as are the noodles, the breads, and so forth. For a full appreciation of the information presented here, readers are encouraged not merely to read these recipes but to cook them, taste them, and, naturally, share them. There are fascinating parallels not only in the development of these cuisines but also in the recipes themselves. Approaching the material in this way will make the reader not only more informed about the internal logic of each cuisine but also should yield more creative and confident cooks, both in the home and in professional kitchens.

All chapters begin with Learning Objectives and end with Study Questions for contemplation, class discussion, and paper ideas. A Glossary of important terms that crop up several times in the textbook is included in the back of the book. Most terms are defined when they appear, however. A Bibliography provides the sources for further reading and research. Copious photos, which are bolded in the index for easy access, enhance the narrative and recipes.

Study Questions

1. In what ways is food used as a means of communication and an expression of identity?
2. What role does the social structure and imitating one's superiors play as a catalyst for changing fashions?
3. How does family structure influence the preservation of culinary traditions?
4. How can we define the term *authenticity* in cooking?
5. How do economics and material prosperity influence taste preferences?
6. Why do some cuisines become complex and invested with a standard repertoire and professional status for cooks?
7. How do factors such as population, environment, and resources contribute to the evolution of cooking?
8. Explain the ways trade influences cuisine. Use concrete examples.
9. Provide examples of the ways elaborate cooking gives way to simple homey cooking and vice versa.
10. How does connoisseurship and food writing encourage the development of sophisticated cuisines?

❋{ 1 }❋
Historical Background

Learning Objectives

- To familiarize the reader with the historical roots of the three world cuisines in a coherent, chronological narrative.

- To appreciate the complexity of how the environment, movement of peoples, and political fortunes affect what people eat.

- To understand why these three cultures developed in parallel ways and with similar consequences throughout history.

- To recognize that although some flavor combinations and ingredients may be universally appreciated, taste does indeed change over time.

- To understand how social emulation, external trade, and political structure all influence and serve as a catalyst in the evolution of gastronomic trends.

TIME LINE

ITALY	CHINA	MEXICO
30000 BC	------- Hunting and Gathering -------	
	Upper Paleolithic Period	
8000–5000 BC	------- Agricultural Revolutions -------	
	Neolithic Period	
	------- Early Civilizations -------	
3000 BC		
	Xia Dynasty	
	1994–1766 BC	
	Shang Dynasty	
	1766–1027 BC	
1500 BC		
	Zhou Dynasty	Olmec
Etruscans	1122–256 BC	1200 BC–AD 300
700 BC		
Roman Republic		
508–44 BC		
300 BC		
	Qin Dynasty	
	221–206 BC	
Roman Empire	Han Dynasty	
0 44 BC–AD 478	206 BC–AD 209	
	(Xin Dynasty)	
	AD 9–24	
AD 200	Three Kingdoms	Maya
	220–280	AD 300–c. 1500
400		
Ostrogoths	Sui Dynasty	
600	589–618	
	Tang Dynasty	
	618–907	Teotihuacan
		c. 700

800	Franks		Toltec
		Song Dynasty	800–1000
1000		969–1279	
1200	High Middle Ages		Aztec
			c. 1300–1500s
		Yuan Dynasty	
	Bubonic Plague ------------	1279–1368 -------------	
	Renaissance		
		Ming Dynasty	
1400		1368–1644	
1500	-------------------	Columbian Exchange	-------------------------------
	Habsburg-Valois Wars		Conquest by Spain
			1519–1521
1600		Qing Dynasty	
		1644–1912	
1700			
1800	------------------	Industrial Revolution	-------------------------------
			Independence
			1821
			Mexican-American
	Unification		War 1846–1848
	1860		Emperor Maximilian
			1863
			Porfirio Díaz
1900		Republic of China	1877–1911
	WW I		Mexican Revolution
	1912		1910–1920
	WW II	People's Republic	
		1949	
		Cultural Revolution	
		1966–1976	
		economic reforms	
		1978	
2000			

IN THIS CHAPTER we will explore the development of the three major world cuisines—Italian, Mexican, and Chinese—in tandem, historical epoch by epoch. While comparing events separated geographically and culturally is to some extent artificial, there are significant parallels in the way each of these societies progressed and similarities in the solutions they found for feeding a growing populace. The intention here is to provide the reader with a basic overview illustrating the various roots of culinary traditions, which although completely divergent in their respective paths nonetheless highlight the ingenuity and resourcefulness native to each society. For example, although the agricultural revolution took place independently in each of these regions and was based on the domestication of entirely different flora and fauna, the consequences of the shift to agriculture bear fascinating comparison. Equally, the introduction of global trade and industrial-scale agriculture had comparable effects on the way people ate, even though the details of how these events took place in each region were completely different. The ultimate purpose of this chapter, apart from offering a chronological context for the entire book, is to show how each cuisine is the result of conscious human choices in how to grow, process, and serve food.

Hunting and Gathering

About thirty thousand years ago, our subspecies *Homo sapiens sapiens* was the last remaining hominid on the planet. By that point we had been around nearly one hundred thousand years, much of that time sharing space with our close relatives, the Neanderthals. Like them, we were hunters of big game and used a wide variety of tools and weapons made to a great extent from stone, hence the Stone Age, and as archaeological evidence suggests, we both cooked our food. Philosophers have long tried to define the essence of humanity, but the eighteenth-century English biographer James Boswell succinctly quipped that "beasts have memory, judgement and faculties and passions of the mind, but no beast is a cook." Despite the ability to hunt and cook with fire, which renders food more palatable, nutritious, and safer as it kills pathogens, the Neanderthals died out about thirty thousand years ago, leaving us as the sole heirs to some ten million years of hominid evolution. As for the importance of cooking in our evolution, although the details and timing are widely contested, there is no doubt that cooking food offers greater caloric value and nutrients and may have given hominids who cook an evolutionary advantage. They were able to reproduce in greater numbers and eventually dominate. It may even have provided the extra calories to facilitate the development of large brains, which are characteristic of modern humans.

Although our species had originated in Africa, we had spread throughout the Eurasian continent thirty thousand years ago as well. Although still hotly debated, the first waves of migration arrived in the Americas not long after. It had been thought that humans trekked across the frozen Bering Strait some twelve thousand years ago at the very end of the last Ice Age, but increasing evidence—genetic, linguistic, and archaeological—may push that date back to thirty thousand years or more. Thus, we can begin the story with humans ensconced in our three areas of focus roughly twenty thousand or thirty thousand years in the past, although arrival in what is today Mexico probably occurred significantly after this. All humans at this point earned their sustenance by hunting and gathering, or as some contend, more properly gathering and hunting, as most calories and expended labor are likely to have come from gathered plants rather than the boon large kill. First, although we may harbor romantic notions of humans living in close harmony with nature, colored by our knowledge of Native American hunter/gatherers, our prehistoric forebears hunted animals to extinction, burned forests, and even polluted. Given the necessarily low population density, the damage they inflicted might be considered negligible in comparison to the havoc we have wrought on the planet, but nonetheless, early humans did whatever was necessary to survive, sometimes at the cost of their own long-term success. By 30000 BC, there were perhaps as few as three million humans on earth.

There are good reasons for this low population density: most humans were nomadic, following the herds and fresh plant gathering sites. Although a necessarily rough existence, in many ways the hunter/gatherers enjoyed a better life than their agricultural and pastoral descendants. First, they suffered fewer diseases because they were always on the move, leaving their refuse and perhaps the sick behind when they left. Never occupying a place for too long and never living in close proximity to animals, they remained fairly healthy, eating an extremely varied diet of practically everything and anything that could be found or captured. Ironically, they were also extremely well fed considering their energy expended in relation to calories consumed. By one estimate, this economy provides about ten thousand to fifteen thousand calories for every hour of labor. That means a good hunt for an hour can feed a person for five days or support a family of five for a day. In sheer terms of leisure time, they obviously understood something we have long forgotten.

Hunting large game, however, is not exactly easy, and there is much evidence of sophisticated strategies for taking large game animals, such as running herds over the edge of cliffs, as at Solutré in France. This took a great measure of cooperation as well as sophisticated technology. Apart from well-fashioned spearheads and arrowheads, hunter/gatherers used nets, weirs for fishing, and rudimentary traps. Communal dismembering and cooking of such animals inevitably led to more complex social gatherings, storytelling by the fire, and the development of complex rituals. In other words, processing and cooking food were indispensable ingredients in the proliferation of larger organized social units. The same might be said for processing wild grains, roots, and other sources of carbohydrates such as acorns and nuts. Quite simply, those groups that managed to cooperate were better fed and survived at a greater rate than others.

At this point, cooking technology was simple. Food could be simply thrown into the fire, or suspended over or near it, held by a simple stick. This is the origin of roasting. But more complex methods were also apparently used. A flat stone could be placed over glowing coals to become a rudimentary griddle, and food could be wrapped in leaves and placed in a fire or buried in a fire pit and covered over. This prevents burning and can also flavor the food. A stone-lined pit or tightly woven basket can be filled with water into which hot rocks are thrown, creating a simple boiling vessel. Although it sounds implausible, a skin stretched over a fire will hold liquid in which food can be cooked, and the skin remains intact. Equally, entrails provide a handy container into which grains or vegetables can be stuffed before cooking, which is the ancestor of today's sausages and haggis. All these methods show that cooking can take place with practically no permanent vessels whatsoever, keeping in mind that nomads would rarely have been able to carry heavy vessels anyway. Other indispensable cooking implements include rocks fashioned as simple mortars and pestles or grinding stones in which to break up grains or tough, fibrous plants. Stone-cutting implements, and especially sharp-edged flints, were used to skin animals and cut meat. Clearly, the absolute basic cooking technologies necessary for any kitchen were used by hunter/gatherers.

The specific foods hunted and gathered in this stage of our culinary history would be too long to recount in any meaningful detail, but suffice to say both the Eurasian continent and the Americas abounded in many plant and animal species sufficient to maintain life and good health for humans, albeit with a low population density. Through Europe and Asia there were numerous wild sheep and mountain goats, wild horses, boar and aurochs (the ancestor of the modern cow, but now extinct), deer species, and even saber-toothed tigers. The Americas had bison on the plains, as well as deer and countless smaller mammals such as beavers, raccoons, wolves, foxes, and rabbits, while to the south there were llamas, armadillos, ducks, and turkeys. Birds naturally proliferated everywhere for those wily enough to capture them. Any people near the coast, rivers, or lakes made use of fish and shellfish, and huge prehistoric shell middens (refuse heaps) attest to the importance of seafood in the coastal diet.

Perhaps even more important were gathered wild grains and legumes. The ancestors of wheat and barley could be found in Europe, millet was vitally important in Asia, and the probable wild ancestor of corn, teosinte, is native to America, as is the sweet potato and various other tubers like yucca. We should also mention the hordes of teeming, crawling insects, fungi, nuts, and berries—all of which provided a fairly well-balanced diet and changed from season to season. Each region also contained manifold wild leafy greens and vegetables as essential sources of vitamins and minerals, and by many accounts of equal importance to hunted game.

What could possess our ancestors to trade this life of plenty for one of unmitigated toil and a monotonous diet based on cultivated plants and a narrow range of domesticated animals? Most likely they were forced to do so by environmental factors, which initiated the next great food revolution after the advent of cooking.

The Agricultural Revolutions

There is much debate whether permanent settlements were a precondition for the development of agriculture or its consequence. In most likelihood, there were a wide range of variables that, over the course of several thousand years, led people from the hunting/gathering way of life to one of planting food and raising domesticated animals. There was also probably a long transitional period during which people would plant crops and then return months later to harvest them, what one might call a seminomadic mixed economy. That is, the Agricultural Revolution did not take place overnight but rather over many centuries. It was not a single event, but rather it occurred independently throughout the world, at various times in history. Most importantly for our story, it did occur in Italy, China, and Mexico at a very early date and roughly contemporaneously.

The oldest archaeologically verified sites for these events, also called the Neolithic Revolution, occurred in the Fertile Crescent some ten thousand years ago. This region encompasses a broad arc of what is today Iraq, Syria, Eastern Turkey, Lebanon, Jordan, and Israel. The dry climate of this region has obviously been an advantage to archeologists, who quite naturally focused their efforts here. These were also the sites of the earliest civilizations. However, elsewhere around the world, findings are challenging the primacy of the Fertile Crescent, and someday it may be discovered that equally old sites exist in both China and Mexico. For example, a wide range of domesticated chili peppers have recently been discovered whose residues date back six thousand years, and it is claimed that squash seeds in the Andes are ten thousand years old. In any case, it is undeniable that these events occurred independently in perhaps as many as ten distinct regions, including the Mediterranean, China, and Mexico, around the globe.

The question remains why the Agricultural Revolution occurred at all. The initial discovery that wild plant seed could be transported and grown elsewhere may have been an accident; the seeds were perhaps stored in a pit, accidentally getting wet and sprouting. Likewise, animal domestication may simply have been the result of wild dogs learning to live in proximity with humans, exchanging a few handouts for the task of barking at intruders. But domestication is a more complex process than merely living next to plants and animals. It also involves the conscious selection of certain desired traits deemed to be improvements, over successive generations, which ultimately make the organism decidedly different from its wild progenitor. Or put more simply, the Paleolithic farmer each season chooses the largest seeds from plants bearing the greatest number of grains to save as seed for the next year. Year after year, those larger seeds pass on their unique genetic makeup to successive generations, gradually making the domesticated grain larger, sturdier, and better yielding than the wild relative. The same goes for domesticated goats and sheep. The biggest animals, or those that yield the most milk, are kept as breeders, and over the years the animals are thus selectively bred to provide the traits the farmer prizes most. The "domesticated" species—ones that can literally be brought into the house (*domus*)—have sometimes dramatically changed in the process.

The Fertile Crescent was extremely fortunate in that many species that can be domesticated are native. Wild grains, which easily cross-pollinate, accidentally gave rise to the ancestors of modern wheat. Barley, too, is native. Ruminants, which are herding animals that are easily tamed and will gladly follow any leader, were important because they feed on grass, which otherwise is of no use to humans. They are, in a sense, factories for converting grass into protein, either in the form of meat or milk. Carnivorous animals, apart from being likely to eat you, need food that could otherwise feed humans, and thus are a comparatively inefficient and unlikely candidate for domestication. Some animals simply refuse to reproduce in captivity, the timid deer being a good example—at least until the advent very recently of venison farms.

Again, all this answers the question of how it may have happened, not why. A single satisfying scenario explaining the event throughout the globe has not yet been postulated. But for the Fertile Crescent it has been suggested that at the end of last Ice Age the great glaciers began to melt and recede. This created natural irrigation and a warmer climate and ultimately more food, which may in turn have stimulated population growth, which in turn necessitated the need to find more reliable and settled modes of food acquisition. The opening of vast plains that were once frozen may have also made tracking herds of animals more difficult, as they dispersed northward across the Eurasian continent. Something similar may have occurred in the Americas, but this scenario has been repeatedly criticized and is surely too simple to explain the myriad ways and reasons various peoples adopted agriculture. Most likely, population pressure played an important role because everywhere plants and animals were domesticated, there was a population boom, perhaps not as a precondition, but certainly as a consequence. It is a simple rule of nature: more food equals more people. And even though the agricultural and pastoral way of life requires more labor and offers less calories and constitutes a far more narrow diet based on grains, it is nonetheless more stable. This is for one simple reason—grains can be easily stored—especially in waterproof, fired clay pots. Settled people can also preserve food through smoking, drying, and pickling, and they can store food in much greater quantities than would ever have been possible for nomadic peoples. Fermentation was not only an important preservative but also made possible risen dough and bread, as well as fermented grain in the form of beer and grape juice in the form of wine. All this not only changed the human diet completely but also made life less precarious. People were better able to stave off famine when there was drought or crop failure.

Better-fed people were ultimately able to support cities and civilization. Where these developments never occurred, conditions never made this switch necessary. For example, in what is today California, the indigenous peoples were still enjoying a largely hunting/gathering economy before contact with Europeans. Or simply, there were not just the right species that could be domesticated, as in parts of sub-Saharan Africa and Australia.

Focusing now on our three world cuisines, it is clear that the major staples that continue to be at the center of the human diet were introduced during the Agricultural Revolution, or shortly thereafter, though in the case of China, rice was a relative latecomer. Wheat, which is still the single most important ingredient in the Mediterranean diet, was introduced from the Fertile Crescent to Italy before 5000 BC, or just a few thousand years after initial domestication. At the same time, and perhaps much earlier, corn and beans were domesticated in Mexico. Millet was being cultivated in China, and within another thousand years rice was as well. Wheat arrived there later, from the Middle East.

Although these grain staples are the foundation upon which these civilizations were built, it should be emphasized again that the health of the populace deteriorated as a result of a narrowing range of nutrients obtained from this starchy diet. There was consequently a high infant mortality rate. Malnutrition as well as infectious diseases spread though contact with human feces and living in proximity to animals, many of whose diseases easily mutate into human pathogens. There was also prevalence of dental caries (cavities) for the first time, as a direct result of starches converting into sugar and rotting the teeth. Even the stature of the average human diminished. Though they may have been barely surviving, they did indeed survive. In each of these civilizations, the grain staple also took on mythical importance, being honored with grain gods such as Ceres (from which we get the word *cereal*) and worked into sacred rituals and fertility festivals. People clearly understood that their survival depended entirely on grain, with its high protein content.

To mitigate their suffering, these early civilizations also learned to ferment grain into beer and other sweet liquids, including honey, into alcoholic beverages. These, too, were often imbued with mythical significance. Think of wine among the Bacchic cults of the ancient world, *pulque* in Mexico, or rice wine in Asia. Intoxication, in the right time and place, was considered a form of ecstatic worship, when one would be removed from one's place (ex-stasis) to commune directly with the gods. The development of civilization and intoxicants were closely intertwined.

Also important in these early culinary cultures were other fermented products: cheese in particular in the Mediterranean and pickles created with brine and natural bacterial fermentation elsewhere, as with soy products in China. These preservative processes, of course, extend the shelf life of food, but they also make many foods more nutritious, in a sense, by predigesting them and making available more nutrients. The same could be said of the process of nixtamalization of corn, in which the dry kernels are soaked in alkaline ash-laden water or lime (calcium oxide), which causes them to swell. They are then not only easier to pound into dough but also the process makes available more nutrients such as niacin and calcium. Thus, what might seem to have been merely a culinary procedure directly facilitated nutritional health and survival. Although the origins of nixtamalization are obscure, there is evidence of it being used by 1500 BC in Guatemala, and the great ancient civilizations of Mexico definitely depended on it.

We should also note the revolutionary impact of ceramics on all these developments. Pottery enabled the storing of foods and liquids and of course the serving of them with greater refinement. Earthenware vessels also make excellent cooking vessels because they require little fuel, especially when compared with roasting. Because water is an excellent conductor of heat and clay helps retain that heat, boiling, or more likely slow simmering food in the form of soups, stews, and starchy gruels, was given a definitive boost. With the invention of the pottery wheel, cooking and serving vessels could be made more rapidly and in greater numbers, ultimately making them more affordable for the average household. Bronze was equally important for cooking vessels, being far more durable than clay and able to withstand much higher temperatures. By the time of the Shang dynasty (1766–1112 BC), the Chinese were able to cast huge bronze cooking vessels. Stone or brick ovens are another prehistoric invention crucial to baking bread and can also be used to cook other foods because the walls of the oven retain heat for hours.

At this point we are entering the historic phase of human history, when written records provide evidence of agricultural techniques, cattle rearing, property sales, taxation, and so forth. Inhabitants of the Fertile Crescent have left us the earliest written scripts dating back some five thousand years, including the earliest recipes in the form of three cuneiform tablets written in Akkadian dated to about 1600 BC. Etruscans introduced writing systems to Italy around 700 BC, who adopted a form of the Greek alphabet, which in turn was an adaptation of the Phoenician. China's ancient pictographic writing system originated around 1500 BC during the Shang dynasty; it was carved into oracle bones and was an ancestor of modern scripts. The recent discovery of inscription on the so-called Cascajal block in Veracruz, Mexico, dated to about 900 BC, suggests that the Olmec (the earliest advanced Mesoamerican civilization, c. 1500–400 BC) also had a writing system not long after those of the Old World. The existence of writing systems in these three regions points to something far more important: the development of cities and a much larger scale of organization, complex political systems, and a brisk economy, in a word, *civilization*. With this came job specialization, a greater proportion of people who could earn a living doing something other than farming, as well as patronage of the arts. It is here that we must look for the development of cuisine: cooking as a refined art undertaken by specialists and eventually recorded in writing.

The Classical Era: Romans, Han Dynasty, Olmec

The great efflorescence of cuisine in our three regions occurred in roughly contemporaneous empires of the ancient world, which in many ways experienced parallel developments. The culture of the Romans has left us the most extensive written evidence, but that of the Qin dynasty in China is also particularly rich in food-related texts, agricultural treatises, and dietary works. Evidence of Olmec cuisine is almost entirely archaeological, but there is no doubt that the roots of Mexican cuisine should be traced here.

The earliest cookbook written in Italy is actually of Greek provenance, as Greeks colonized the southern half of the peninsula and Sicily. Near the wealthy and thriving city of Syracuse, from a town called Gela, came one Archestratus, whose works survive in fragments as preserved by the third-century AD compiler Athenaeus in his *Deipnosophistae*. Archestratus himself originally composed the work around 330 BC, and most of what survives is about fish. Interestingly, the work is in verse and was probably intended to be read at a symposium (drinking party). Above all, Archestratus was a connoisseur rather than a chef. He writes with conviction about where the best of every species comes from, and he had evidently traveled throughout the Greek world sampling food. For example, he waxes rhapsodic about tuna from Byzantium and finely bolted white barley bread from Lesbos. He is most interested in the provenance of food, the finest ingredients unsullied by fancy nonsense, such as cheese with fish, which was apparently a common combination. The bonito (a fish), he believed, is best wrapped in fig leaves with a little oregano and cooked quickly in hot ashes. His commentaries and recipes suggest that while there was a highly advanced level of culinary sophistication enjoyed among the Greeks, truly good food is presented with respect for the ingredients and simple preparations that accentuate their qualities.

The Romans, however, were cut from a different cloth. One gets an excellent idea of their stern values in the republican period from the agricultural text of Cato the Elder (234–149 BC), *De Agri Cultura*. In it he describes the operation of a typical Roman farm, at the point when agriculture was shifting from cereal-based self-sufficient farms to the large-scale commercial ventures supplying growing cities with luxury items such as wine, olive oil, and fruit. His text is designed for those seeking investment opportunities, whether elites considering how to spend their inheritances or retiring soldiers after years of loyal service seeking to settle down. These farms, which were extensive operations run by slaves and which normally contracted out larger tasks such as harvesting, were meant to turn a profit but could also serve as places of rustic retirement and leisure. Cato offers advice for bringing in and pressing the olive harvest, storing wine, planting asparagus beds, and curing hams, and even a few recipes for cakes and pastries. Most of these contain cheese and honey; others are fried and are ancestors of fritters and funnel cakes. Cato praised the simple and rustic life as an antidote to what he considered the effeminate influences of Greek civilization, but in the end it would be Greek culture and cuisine that won out among the Romans.

In the Imperial Era (27 BC to AD 476), the Romans experienced dramatic cultural changes as Greece, Egypt, and the Near East were conquered and incorporated into the empire. Greeks were often brought to Rome as slaves, household servants, and even cooks. Eastern luxuries and spices were imported along trade routes stretching all the way to India and China. This was also a society in which fortunes could be easily made or lost, and sometimes even slaves could buy their freedom and become successful businessmen. While there was an established patrician class, their position in society was constantly threatened by wealthy upstarts, who may have had the money to entertain on a grand scale but were certainly not to the manner born. Sometimes no doubt they made complete fools of themselves trying to impress invited guests with their supposed sophistication.

It is in this context that we can appreciate the marvelous story of Trimalchio's Feast as told by Petronius (c. AD 27–66) in his novel *Satyricon*. Petronius was of the old moneyed class, and the historian Tacitus wrote this of him: "He was not considered a glutton and a spendthrift, but rather a refined pleasure seeker who was brought into the circle of the emperor's favorite few and became the arbiter of good taste in Nero's reign, the person who amid such wealth determined what was beautiful and delicate." For some perceived insult he was also ordered to commit suicide; some scholars have suggested that Trimalchio was intended to be a satire of Nero. Whatever the case, the story unfolds as two patricians are invited to a feast thrown by Trimalchio, a former slave who made a fortune speculating in food. The feast is designed in indescribably bad taste. For example, a bronze donkey with dishes slung from its mouth offers olives and dormice dipped in honey and sprinkled with poppy seeds. The Romans were especially fond of dormice and had specially made *gliraria*, little perforated ceramic pots, in which to fatten them. When Trimalchio arrives he proceeds to ignore his guests, and a wooden hen appears that lays half-pound eggs, which are actually pastries containing whole fig peckers (a tiny bird) in a spicy, yolk-based sauce. The main course consists of twelve dishes, one for each sign of the zodiac, pairs of beefsteak for Taurus, bull's eye for Sagittarius, kidneys and testicles for Gemini, sow's womb for Virgo, and so forth (figure 1.1). The piece de resistance is a roast boar wearing a hat, whose gut when sliced open spills forth

Figure 1.1. Scene from the banquet in Federico Fellini's 1969 film, *Satyricon*. United Artists/Photofest. © United Artists.

blood sausages. Although such perversities might fit in well at some cutting-edge restaurant today, elegant Romans would have taken this to be nothing more than the pathetic attempt of some nouveau riche poseur trying far too hard to impress his guests and making a complete idiot of himself. Petronius's text is, of course, a warning to anyone who tries to do the same. It is snobbery, of the sort that would only appear in a setting of great social mobility, when the position of aristocrats is threatened.

It is perhaps in this light that we should consider the cookbook attributed to Marcus Gavius Apicius, *De re Coquinaria*. There was a historical figure Apicius in the first century AD, renowned in his own day as the most extravagant gourmand. Stories circulated about his taking his own life when he realized his fortunes were dwindling and he would no longer be able to live lavishly. But this cookbook was compiled several centuries later and was probably just labeled *Apicius*, which by that time was merely a byword for luxurious dining. Although there are some rather simple and by any standard appealing recipes in *Apicius*, many use exotic ingredients such as ostrich, flamingo, and dormice. Moreover, most of the recipes include a wide array of seasonings, such as cumin, silphium (an aromatic resin from Northern Africa, now extinct), the ubiquitous garum (a kind of fish sauce), honey, dried fruits, and plenty of pepper. The complex jumble of flavors not only suggests that Romans had seriously jaded palates but also that they were trying very hard to impress with exotic ingredients and fantastic combinations of flavors. How else can one make sense of his rose patina, a combination of rose petals, brains, fish sauce, pepper, eggs, and sweet raisin wine, cooked in a kind of steamer? Or equally the eponymous Apician Stew (*minutal*), which contains oil, garum, leek, mint, small fish, meatballs, rooster testicles, sweetbreads of a suckling pig, plus pepper, lovage, coriander, wine, and honey? There is nothing inherently unpleasant about these dishes if one remains objective, but it would be a mistake to consider this ordinary, everyday Roman fare. This is extremely fancy food, meant to titillate, of exactly the sort we might find Trimalchio serving.

It should also be taken as a rule that when a civilization reaches this level of sophistication, and the ability to eat on such a grand scale is available to a significant proportion of the population, there will also be a demand for physicians to lay down dietary rules, either to cure the many aliments brought on by culinary excess or merely for those conscious of their health. That is, nutritional theory might be considered a rival food ideology that details a specific way of eating diametrically opposed to the lavish style of elites. The Greeks here, too, devised the dietary system inherited by the Romans, most importantly in the figure of Galen of Pergamum, who served as personal physician to Marcus Aurelius and several successive emperors (figure 1.2). Galen was not only the most prolific medical writer of the ancient Mediterranean, and in fact more of his works survive than any other classical author, but also he codified a dietary system that influenced the eating patterns of Europeans for the next millennia and a half after his death around AD 200.

GALIEN

Figure 1.2. Portrait of Galen, nineteenth century. © Stefano Bianchetti/CORBIS.

The system of humoral physiology posits that the human body is regulated by four essential humors: blood, phlegm, choler, and black bile or melancholy. When in proportional balance the body was said to be in a state of health (*eukrasia*). Should one of the humors be in excess, then distemperature or sickness result. For example, too much of the hot and dry humor choler leads to fevers and other hot ailments. An excess of the cold and moist humor phlegm leads to colds, coughs, and the like. Foods, too, were categorized according to their predominant humor, so that one could describe pepper as hot and dry or a cucumber as cold as moist. Thus, individuals were recommended to eat foods that could counteract any humoral balance to return their bodies to a state of balance. Foods were considered a kind of medicine; although not as powerful as compound drugs, they could nonetheless be used to maintain health. When regulated in concert with other external factors such as climate, amount of exercise and sleep, and emotions, all of which also alter the body's humoral makeup, individuals could devise a regimen suited to their own particular "complexion" or temperament. A balanced diet, albeit very different from our own concept, was another style of eating that a Roman might adopt, either on a temporary basis or more or less permanently. One can, of course, eat right following a binge or in response to sickness.

There was yet another Roman food ideology, also with an attendant culinary style, whose prime goal was not explicitly health but rather simplicity and lack of adornment. It, too, was designed as an antidote to the culinary riot of the wealthy, and it looked to basic foods prepared without fuss as a way to strengthen the body and spirit. One might even call this style Stoic, and authors such as Seneca (4 BC–AD 65) promoted a lifestyle with as few needs as possible, because luxury, he thought, only makes a person weak and dependent. It is better to live simply so one can withstand the vagaries of fortune. In his epistle 95, Seneca directly attacks Roman fine dining: "What a price we pay today for pleasures coveted beyond reason. No wonder there are innumerable diseases: count the cooks. . . . Let alone the tribe of pastry makers, and servers who once the sign is given rush forth with dishes. Good Gods! How men work to satisfy a single belly . . . I recall once a story about a noble dish, into which every delicacy was heaped by a caterer bound for ruin—with mussels and trimmed oysters, mixed with sea urchins, all surrounded by boned and cut up mullets. These days people spurn single foods, but mix every flavor together." Seneca obviously had a pretty decent idea of Apician cookery.

Among some authors this call for simplicity was gastronomically driven, the ability to taste unsullied flavors fresh from the countryside offering a greater pleasure than the mixed and jumbled dishes of professional cooks. In Juvenal's Eleventh Satire, written in the early second century AD, the author invites a friend not for rare and expensive luxuries like the largest mullet but true delicacies one cannot find in the market. From his own farm there will be a fat kid unweaned from milk, mountain asparagus, fresh eggs taken directly from the nest, as well as local grapes, pears, and apples—good, simple, rustic fare. There will be nothing normally found at cook shops—sow's womb, rose-laden dishes, boars, pheasants, flamingoes, and gazelles. "No such nonsense in my humble abode." It almost seems as if he were directly responding to Apicius here. All this is convincing evidence that while some Romans splurged, others were content with the pristine flavors of the countryside.

Last, there was yet another important set of food ideologies among the Romans, those deriving from religion. Apart from the significant proportion of Jews in the empire, who had their own unique kosher food laws, Christianity also presented a radically new way of thinking about food. Although the early Christians decidedly rejected the food prohibitions of the Jews, considering no food unclean, their own food restrictions gradually crept in. For one, there was a strong undercurrent of asceticism in Christianity. The logic was that to deny the body's urges strengthens the soul; to suffer for one's faith is meritorious in the eyes of God. Early monastic communities inflicted upon themselves heroic bouts of abstinence. This would leave a permanent streak of food guilt in European food culture, as well as an explicit idea of gluttony as a deadly sin. Fasting and penitence, while having roots in Jewish practice, eventually took a unique form in the first centuries AD as Christian dogma and the church itself were coalescing. Eventually the fast came to mean abstinence from meat and meat products such as eggs and cheese a day or two every week, during holy fasts scattered throughout the calendar, and finally for all of Lent, the forty-day period between Ash Wednesday and Easter, not counting Sundays. This rhythm of fast and feast had important consequences for the culinary culture of Italy throughout the Middle Ages, a topic that will be discussed at length here.

If Ancient Rome developed multifarious modes of thinking about food and rival culinary styles, the same must be said of China at the same time. In many ways, China during the Han dynasty (206 BC–AD 220) is comparable to the Roman Empire, and of course they did have indirect contact through trade. The elaboration of Chinese cuisine in this period was dependent on several factors, the most important of which appear to be the reverence for the elders and extended households in which they could pass on culinary skills and the vibrant court culture to which access could be gained through bureaucratic service to the state. This in turn spread elite culinary fashions beyond the confines of the imperial palace walls, as imperial officials took up their posts throughout the Empire.

Central to the culinary culture of China, even if indirectly, is the thought of Kongzi, or, as he is known in the West, Confucius (551–479 BC). In brief, Confucius believed that proper behavior, understanding your place in the social hierarchy, and always obeying your superiors were essential not only for social harmony but also even cosmic harmony and the stability of the state. Thus, ritually prescribed ways to address superiors and for treating inferiors prevent misunderstanding and discord. Individual desires are subordinated to the good of the whole, and proper respect for superiors (and elders) ensured the smooth functioning of daily life. This translated into a carefully codified set of manners, a way to avoid potential conflict, including at the table. It is no coincidence that the Chinese developed chopsticks and banished sharp knives from the table, a sign of barbarity and lack of morals. The success of the Chinese state was accomplished in large measure by the monopolization of violence, and this translated directly into courtly manners and carefully observed protocols.

Another thinker influenced a culinary aesthetic counter to the elaborate court cuisine in China: Laozi (sixth century BC), the father of Daoism. He and later Daoist thinkers praised the beauty of nature in its simple, unaffected state. His importance to culinary culture is the espousal of a simple diet tied closely to ingredients directly from the soil, treated without artifice. Among some sages, this simple diet became a form of asceticism, especially to promote longevity, though this never became a mainstream practice and Chinese culture was generally devoid of the kind of food guilt that developed in the Western tradition. Nonetheless, an appreciation for food in its natural state appears to be an indirect consequence of Daoist thought.

Last, Buddhism, as imported from India, along with a form of monastic life, supplied the third great Chinese food tradition. Siddhartha Gautama, who upon enlightenment became Buddha around 500 BC, taught that the world is filled with suffering, and the way to overcome this is merely to deny all desire, recognizing that the self is an illusion. Neither asceticism nor luxury can free us from an obsession with the self, only a "middle way," eating enough to survive but not paying undue attention to food—either its pleasures or denial. With this philosophy came the central tenet of nonviolence as a way to actively banish the suffering of humans and other creatures. This translated into practical vegetarianism and dishes specifically designed as meat substitutes. Although not all Buddhists practiced vegetarianism in the centuries following the introduction of Buddhism to China, it remained an ideal, especially for monks and the devout. Buddhist monasteries also served as places of refuge, medical dispensaries, and centers of vegetarian cuisine. The importance of nonmeat foods such as tofu (soybean curd) can be indirectly related to Buddhist nonviolence.

Recognizing that these three traditions were by no means exclusive, one could practice elements of each or indeed all at the same time. A famous painting called *The Vinegar Tasters* nicely illustrates the importance of the three currents of thought. In it a Confucian scholar tastes the vinegar and finds it sour. The vinegar is a metaphor for life, and the Confucian scholar finds life with its strict protocol and deference unpleasant; presumably he is always being offended in some way and is angry because people fail to follow the rules. The Buddhist finds the vinegar bitter, for all of life is suffering, and he tries not to let the taste linger. Only the Daoist tastes with a smile on his face, accepting the vinegar for what it is in its natural state.

The roots of classical Chinese cuisine can be traced to the brief Qin dynasty (221–207 BC) and the long and prosperous Han dynasty, which extended from 202 BC to AD 220, with only one interruption. In this period Confucianism became the dominant state policy, and access to bureaucratic service was available by passing a state examination. The importance of this avenue for social mobility is that it spread the culinary achievements of the imperial court both to the provinces and down the social scale. That is, emperors enjoyed a dazzling court cuisine in the confines of the palace, but there was an effective means of transmitting these skills elsewhere as state functionaries took their posts throughout the empire and hired professional chefs who may not have been able to replicate the scale of court cookery, but certainly its ideals. Most important, the Han emperors also consciously promoted agricultural improvements throughout China, partly by offering incentives for small independent farmers and low taxes, but also through what might be considered a kind of agricultural extension service. Farming textbooks and encyclopedias were written and distributed, and scholars were sent out to instruct farmers and to introduce new crops and farming techniques. Huge irrigation projects were initiated, as well as a form of public relief in times of dearth. One classic agricultural text by Shengzhi of the first century BC describes multiple cropping, use of mulch, new methods of irrigation using circulating water to keep rice fields warm in the spring and cool in the summer, and a detailed explanation of the use of fertilizers.

Perhaps the greatest technological innovation of the Han dynasty was the perfection of iron farm implements and the signature cooking vessel of Chinese cuisine—the hammered iron wok. Using relatively little fuel and cooking small pieces of food rapidly, the wok would have a permanent impact on cooking in China, especially when sesame and its oil were introduced in Han times. The Han also invented porcelain, a form of very high fired, light and strong white ceramic that could when thrown with deft hands achieve translucence. Porcelain bowls would become a standard serving utensil. They were highly sought out in China and abroad in a brisk trade that was developing between East and West. In addition, new milling techniques and the ability to finely ground and bolt grains led to the invention of noodles. In fact, an intact bowl of noodles made from millet flour was recently discovered at the Lajia site on the Yellow River dating back some four thousand years. Noodles were certainly being enjoyed in China long before they were in Italy.

As in the West, China also devised a complex dietary system, which survives in practice today, in what is probably the oldest living medical system on earth. Its origins are traced to a text called the *Yellow Emperor's Classic of Internal Medicine*, written by the "celestial emperor" Huangdi (c. 2697–2597 BC). The written form in which it survived was composed in the Han dynasty, roughly contemporary with Galen. In this system health is also conceived of as a state of balance between the two primordial forces of the universe—yin and yang. Yin is the female principle and corresponds to the qualities of darkness, cold, and softness. Yang is the male principle whose qualities are light, warmth, and firmness. Illness derives from an imbalance in either force, and as in the Western allopathic (conventional) system, foods and medicines can be used to bring the body back to homeostasis (equilibrium). There is also the concept of *qi*, a kind of energy force that flows through the body, keeping us alive and fending off sickness. Certain foods are said to increase good orthogenic *qi*, or prevent pathogenic *qi* from entering the body from the outside. In practice, this is an extraordinarily complex system, but suffice to say the Chinese take careful account of the state of the body and how various foods will interact with it, and this had a profound effect on the culinary culture as well.

Next we come to the third ancient culture, the Olmec, about whom the least is known. Nonetheless, it has become increasingly clear to archeologists that Mexican cuisine as practiced in later empires had its roots in these early times, especially as the key ingredients were already being extensively used. The Olmec occupied what is today south central Mexico around Veracruz and thrived between 1200 and 400 BC. The height of their civilization corresponds to those in Italy and China but ends significantly before the rise of classical civilization in the Old World. Like elsewhere, the Olmec cultivated a domesticated grain staple, in this case corn. With this there were beans, squash, sweet potatoes, and manioc root. As the civilization expanded and its population grew, it has been suggested that corn became even more important in the diet, just as rice became crucial to China and wheat to Italy. The Olmec also cultivated several different varieties of chili to flavor food. They obtained cacao from the south, where they also found jade and obsidian. Avocados, another important ingredient in Mexican cuisine, were also enjoyed by the Olmec. There were abundant wild mammals and fowl native to this area, including deer and peccaries (a distant relative of the pig), but the Olmec had also domesticated dogs, which apparently was the most widely used meat. The region they occupied is also fertile and well irrigated, but they also practiced slash-and-burn agriculture to clear forests. This necessitated periodic relocation, and it may be that the collapse of this civilization was brought on by environmental disaster.

The Middle Ages: Maya, Tang, Franks

Much more is known about the Maya, who occupied an area south and east of the Olmec in the Yucatan Peninsula and southward to Guatemala, Belize, Honduras, and El Salvador. The height of "classical" Maya culture occurred from AD 250 to 900 and thus corresponds to the end of the Classical Era and Early Middle Ages in Europe. This was a highly advanced civilization with a fully developed system of writing, complex mathematics and astronomy, and the resources to build extensive cities and pyramids.

As elsewhere in Mexico, corn was the staple grain but was supplemented by amaranth, jicama, manioc (*Manihot esculenta*), sweet potatoes, and another root vegetable called macal (*Xanthosoma nigrum*). There is some debate whether the Maya formed their ground corn into tortillas or mainly ate a kind of corn porridge or steamed corn dough in the form of tamales. Flat griddles for cooking tortillas have not been found at Maya sites, though that is hardly conclusive evidence, and flat corn cakes can also be cooked on a stone or directly in hot ashes. Corn was the most important food here, and the *Popol Vuh*, a sort of Maya bible or creation myth, states that the gods created humans from corn dough. Then as now, black beans cooked with *epazote* (an aromatic herb) and spiked with chili supplied protein. Although direct accounts of Maya foodways date only from the colonial period, in accounts written by Spaniards such as Diego de Landa, we can by inference get an indirect idea of what the Maya ate in the classical period. They definitely enjoyed chili pepper, sprinkled into their *posole*, a corn drink. They also drank chocolate, and in fact the word for *chocolate* was a key in deciphering the Mayan language, as a pot still containing residues was inscribed with the word *ka-ka-w*, from which we get the word *cacao*. The Maya also had a range of fermented alcoholic corn drinks such as *atolli*, sweetened with native honey, though this is the Nahuatl term of the Aztecs.

People, primarily the elite, also enjoyed an abundance of wild animals such as deer and dogs as well as the peccary, iguana, armadillo, and monkeys. Most important, at least as concerns modern foodways, were the turkeys, though these were most likely the ocelated turkeys, captured in the wild and fattened, rather than the domesticated species (*Meleagris gallopavo*) familiar to us. Along the coasts and in the marshy lagoons the Maya also made extensive use of seafood, turtles, and snails.

The major sites of Maya civilization in the south began to be abandoned after about 900 AD, but the people themselves by no means disappeared. Their descendants exist to this day. But the reason for the collapse has been widely debated and was perhaps the result of overpopulation, depletion of the soil, or some environmental disaster such as drought. Despite this collapse, culinary advancements of the Maya were handed down to subsequent civilizations.

At the same time the Maya civilization was at its height before the collapse, on the other side of the globe there flourished the Tang dynasty in China (618–907), a period in which China was politically unified. This dynasty is also considered the height of a system of large-scale landholding comparable to feudalism in Europe, in which an aristocracy dominated government positions. Although technically still open to anyone who could pass the state examination, the old aristocratic families developed an elite culture in which skills in calligraphy and poetry, refined manners, and musical skills set this class apart. Not surprisingly, a refined culinary culture, open to exotic new delicacies, was also cultivated. There was also a rich food literature, including extensive dietaries and pharmacopeias, or, as they were called, "Food Canons" (*Shi Jing*). Among the earliest of these was composed by Meng Shen in the seventh century. The *Xin Tang shu* (New Tang History) is especially detailed in describing the exotic plants grown in imperial gardens, many of which hailed from western Asia and Persia.

From their bustling metropolis of Chang'an, the Tang became a remarkably cosmopolitan civilization, open to foreign traders and influences from the outside. This period is often considered the greatest efflorescence of Chinese culture and the arts, and it is equally important in the history of cuisine. It was during this period that China had contact with Persia along the great Silk Road, and while silks and luxuries traveled westward, foodstuffs and new plant species traveled to the east, including spinach, lettuce, almonds and pistachios, figs, and dates.

The staple grain continued to be millet, with rice more important in the south. In the north wheat was increasingly important, and various forms of dumplings and noodles were already widely known as well as flat breads introduced from Central Asia. Surprisingly, other products of the West became fashionable—dairy products such as cheese, butter, and yogurt, as well as wine made from grapes—although their long-term impact on Chinese cuisine was negligible. It was the Tang who established tea as the premier beverage, being imported from Burma and northern India and eventually grown extensively in China, so much so that its botanical name is today *Camellia sinensis* (Chinese camellia). A form of tea connoisseurship was already in evidence as well, to which the *Book of Tea* by Lu Yu of the eighth century attests.

Although Europe generally languished culturally and economically in the wake of the collapse of the Roman Empire after the fifth century, there was one brief period of vibrancy in the empire created by the Frankish king Charlemagne. It was, for example, in this period that the two surviving manuscripts of Apicius were copied in the monastery at Fulda. The Franks were originally one of a number of barbarian tribes who dismembered classical civilization and set up their own kingdoms throughout Europe: the Visigoths in Spain, the Ostrogoths in Italy, the Vandals in northern Africa, and the Franks in what is now France, Belgium, and Germany and eventually northern Italy, after they conquered the kingdom of the Lombards. The Franks eventually rose to dominate Western Europe after Charlemagne was crowned emperor by the pope in 800. In terms of cuisine, these Germanic peoples' foodways differed radically from those of the earlier Roman civilization. Whereas the Romans primarily valued neatly plowed fields of wheat, orderly orchards, vineyards, and well-managed herds, the Germanic tribes depended far more heavily on hunted wild game, especially pigs, which were allowed to roam through the forests feeding on acorns and beechnuts. And, in fact, much of Europe reverted to forest as extensive agriculture and trade networks collapsed and dependence on wild food became a necessity. The economy also shrank dramatically to a more local subsistence economy.

In these years, the roots of what would later become serfdom also developed. In return for military service, the Frankish kings granted large blocks of land to their followers, or vassals. These people eventually became a hereditary landed aristocracy with virtually independent power in their own counties and dukedoms. To support their position, the rural populace was reduced, not exactly to slavery as they could not be bought and sold, but nonetheless they were bound permanently to the soil and forced to work on the lord's own plot of land (*demesne*). These serfs would also be forced to pay various fees—to marry, pass on land to children, and importantly for various services provided exclusively by the lord, such as milling grain. The lord would also often have exclusive fishing and hunting rights, violation of which for the serfs was a punishable offense called poaching. The long-term effect of these arrangements is that peasant agriculture became by necessity rudimentary, and villages were forced to cooperate, perhaps owning a single plow among them, but certainly timing the season of agricultural tasks as a group. Thus, agriculture became not only bound by tradition and very conservative but also was less likely to be able to afford innovations. Commercial activity was also difficult beyond the immediate vicinity. Agriculture under serfdom was reduced to a precarious hand-to-mouth existence, repeatedly subject to famines and subsistence crises.

The very fact that Charlemagne would have to issue orders throughout his realm, known as *Capitularies*, encouraging the planting of various crops and the keeping of hens for eggs so his soldiers could be fed while on campaign, suggests the dismal state to which food production had deteriorated. Charlemagne's own personal eating habits, as recorded by his biographer Einhard, also offer a fascinating glimpse into the culinary culture of this period, in a sense descended from three distinct traditions. On the one hand, he was a Germanic warrior-king, and as such would be expected to consume great quantities of roasted flesh and drink beer as a sign of virility. A visitor to his court was seen to gnaw on a hunk of flesh, crack open the bone, and suck out the marrow. Charlemagne commented that he must be the son of a king to eat with such gusto. On the other hand, Charlemagne was a Christian and was required to fast periodically and eat in a penitential fashion, which he did, especially when his physicians warned him to cut back on meat consumption. Last, he also inherited the classical tradition, styling himself a new Roman emperor, and accordingly should eat moderately throughout the year.

That these three ideals could never be melded is indicative of the interesting hybrid culture of the European Middle Ages, in which it was customary to swing from the riotous excess of holy feasts to austere fasts throughout the year.

Around the year 1000 a global shift in climate had a dramatic impact on human history. Although we are concerned with global warming in our own day because we are contributing to it, such changes have always occurred. In the Middle Ages, grain surpluses and the ability to extend cultivation to higher latitudes and to marginally less productive areas all led directly to population growth. Again, more food equals more people. The population of Europe roughly doubled between 1000 and 1300, from thirty-eight to seventy-four million. Warmer weather was compounded with the invention of the mould-board plow that could be used on heavier soils and that could turn the earth over into furrows, aerating it. The invention of a new harness, which fit comfortably around the collarbone of the horse, provided a faster form of traction, and new nitrogen-fixing crops such as clover and alfalfa were used as animal fodder, sometimes worked in rotation with other plants. There were also land reclamation projects, greater use of the waterwheel to mill grain, and an incentive to provide more food to supply the teeming urban centers. In a nutshell, Europe experienced a minor agricultural revolution.

Moreover, the reopening of trade routes to the East, stimulated by an aristocracy increasingly bereft of its military functions at home and eager to plunder other countries, led to direct contact with a flourishing Islamic civilization. As the Iberian Peninsula was wrested from its Moorish rulers, as Muslims were pushed out of Sicily, the armies of Christendom also conquered and temporarily ruled the Holy Land in a series of military forays known as the Crusades. All these exposed Europeans to a culture far more complex and sophisticated, and one that was not only materially wealthy but also boasted a complex cuisine incorporating exotic spices, dried fruits, citrus, nuts, new vegetables such as spinach and eggplant, and sugar. The impact of Islamic cuisine on that of the Middle Ages, at least among elite diners, cannot be overstated.

Europeans did not merely borrow Middle Eastern recipes, however; in many ways, they innovated with the new ingredients. The almond is a case in point: when pounded and mixed with water it was made into almond milk, a substitute for cow's milk during Lent. It was incorporated into spicy sauces thickened with breadcrumbs and soured with verjuice (the juice of unripe grapes). The earliest medieval cookbook, *Libellus de arte Coquinaria*, is known to us only through thirteenth-century translations, likely from Latin, into, surprisingly, Danish, Icelandic, and Low German. It reveals that an elite culinary culture was shared throughout Europe. The most extensive cookbook of this period, the *Viandier*, was perhaps composed as early as the late thirteenth century, though attributed to one Guillaume de Tirel, nicknamed Taillevent (which means "wind slicer"). Taillevent enjoyed immense popularity and became chef to King Charles V of France. Although written in French, the book nonetheless reflects the elite international cuisine of the High Middle Ages, which definitely includes Italy. In it we find dazzling dishes using the full panoply of spices imported from the East, including cinnamon, nutmeg, cloves, and ginger, as well as the ubiquitous black pepper. In addition, there were other less familiar spices to us today, such as grains of paradise (melegueta pepper) from the west coast of Africa and cubebs, long pepper, cassia buds, and spikenard, which were all exotic Asian imports. Medieval diners also enjoyed brightly colored foods daubed with saffron yellow or colored red with sandalwood or purple with alkanet (a powder derived from a plant root).

These spices were naturally extremely expensive, imported from as far away as the Moluccas in what is today Indonesia, and serving them was the most conspicuous form of consumption, in this case literal. But Europeans also paid another extraordinarily high price for spices, albeit indirectly. Trade with the East also brought pathogens with which Europeans had not had contact for several centuries, most notably the bubonic plague. It struck in 1347 to 1348 with remarkable ferocity and reduced the entire population by about a third. Ironically, however, though devastating for those inflicted and for the economy in the short turn, survivors were able to reap the benefits of a greater demand for labor and consequent high wages. Feudalism as it had been practiced fell apart, and peasants could negotiate much better rents for their land. Most important, higher wages meant more people could afford meat, and the appearance of cookbooks catering to what might now be called emergent middle classes are filled with recipes for pork, beef, and fowl that would have probably been rare exceptions in the diet only a century before.

The effect of this redistribution of wealth can be seen in a number of anonymous cookbooks composed in various Italian city-states in the Late Middle Ages. The *Libro della cocina* by the "Anonimo Toscano" (Anonymous Tuscan) dating perhaps to the late fourteenth century contains not only many dishes for capon and chicken, veal, kid, and lamb, as well as many fish dishes, but also certain preparations that are distinctively Italian. For example, there is lasagne made of fine white flour dough, rolled out into a thin sheet and cooked in capon broth and covered in grated cheese. Likewise, there is Genovese *tria*, a thin spaghetti boiled in almond milk and given to invalids. Strikingly, vegetable dishes proliferate, which is rare for medieval cookbooks. There are recipes for cabbage and fennel cooked in a variety of ways, as well as asparagus, chickpeas and fava beans, leeks, and turnips. The *Libro di cucina* of "Anonimo Veneziano" (Anonymous Venetian), composed about the same time, includes various herb tarts, gelatin aspics, sausages such as mortadella, and even the original "polenta," made here with millet, and various types of "ravioli," which translates literally as "little turnips," some encased in pasta and others fried "nude," as one would say today.

Late Middle Ages: Italian Communes, Aztecs

The greatest single cookbook of the Late Middle Ages, *Libro de arte coquinaria* (Book of the Art of Cooking), was composed by Martino of Como and was incidentally also the first printed cookbook, as its recipes were used in *De honesta voluptate* (On Honest

Pleasure) of Bartolomeo Sacchi, called Platina, published around 1470 in Rome. Although significantly influenced by Catalan cuisine and heavily indebted to earlier cookbooks, Martino presents the height of Italian cookery in this period. We find veal head, various roasted fowl, and classic standards like blancmange—a smooth, pounded chicken dish with ground almonds, sugar, ginger, and redolent of rosewater. He also describes how to make several types of macaroni, ravioli stuffed with cheese and ground veal or pork, and a slew of typical medieval sauces based on garlic, fruits, or mustard. A whole range of pies and fritters including every possible ingredient; fourteen different ways to make eggs, including roasted on a spit; as well as unique ways to prepare every fish that could possibly be caught, grace his pages.

At the same time as the Italian city-states were flourishing in the Renaissance, there was a powerful and sophisticated empire ruling much of Mexico. The Aztecs were relative latecomers to Mexico. They claimed to have come from a place called Aztlan to the north, and indeed linguistically they are related to the Apache and Shoshone. They arrived about 1325 to the shores of Lake Texcoco in central Mexico long after the collapse of the Toltec state around 1150. At the time there were many small, autonomous regions whose rulers intermarried and just as often fought with each other. The Aztecs, as fierce warriors, essentially filled a power vacuum and over the years conquered almost all of the region, building a formidable empire. Their capital at Tenochtitlan (now Mexico City) was a sprawling metropolis covering roughly five square miles. It was constructed in the middle of the lake, connected to the mainland by long causeways and crisscrossed by canals. Bernal Díaz, who accompanied the conquistador Hernán Cortés, wrote that it reminded him of Venice. This was a highly stratified society, with the emperor at the apex, a class of warrior nobles, and beneath them a large group of merchants who sold wares in the city's extensive markets, which were strictly regulated by the state.

The Aztecs adopted much of the civilization of those peoples they had conquered, including their foodways. Corn (maize) as elsewhere was the staple, and it was nixtamalized, ground on a *metate* (a saddle-shaped stone with a stone rolling pin) into a *masa* (dough), and then flattened into disks and cooked on a *comal* (flat griddle), creating tortillas. It could also be steamed in a corn husk, popped, or made into a fermented corn drink called *attolli*. Amaranth, relatively unimportant today, also played an important role, though Spanish conquerors consciously hindered its cultivation when they learned of its importance to the Aztec religion. The Aztecs also had sweet potatoes and manioc root and made a kind of thick, gelatinous gruel from the ground seeds of the chia plant. Beans, especially in combination with corn, were also a crucial source of protein in the diet, and various squashes as well as their seeds were commonly eaten. Among the more interesting foods eaten by the Aztecs was spirulina, a very high-protein algae collected from lakes.

The lakes and waterways were a significant source of seafood for the Aztecs, and they also played another important role. *Chinampas* (floating islands) were built on the water, covered with dirt, and used to plant rows of corn, which could produce four crops a year. They were anchored with willow trees to prevent erosion. The Aztecs also had remarkably complex systems of terracing and irrigation.

To dispel the notion that the Aztecs were perennially short of animal protein because they had no domesticated cattle, rabbits, dogs, and turkeys were all domesticated. There were also abundant wild animals such as ducks, capybara, coati, and the armadillo. There was also ritual use of human flesh as part of a sacrifice to feed the sun god Huitzilopochtli, but this was definitely not a regular ingredient or a response to a shortage of meat as the population grew. Typically the victims were captured in war or offered as tribute to Aztec overlords by conquered peoples.

The chili pepper was the most ubiquitous flavoring ingredient, and reputedly the Aztecs ate nothing without them. In fact, during penitential fasts they would abstain from salt and chilies, which was considered a hardship, comparable to the prohibition on meat during Catholic Lent. Chilies were used fresh and also dried, soaked, and ground up to form the bases of sauces, exactly as they are today. We should also mention here one of the greatest gifts of Mexico to world cuisines: the tomato. Early accounts scarcely distinguish between the tomato and the tomatillo, but both were certainly used in Aztec cuisine. Tomatoes are actually native to South America, but by this point they had been adopted in the north, and our word comes directly from the Nahuatl word *tomatl*. They were normally coarsely pounded, mixed with chilies, and served in little bowls into which diners dipped other foods—the direct ancestor of modern salsa. Avocados, too, were made into a pounded sauce called *ahuaca-mulli*, a food that has scarcely changed since Aztec times, although it is now called guacamole.

The first full account of an Aztec meal comes from Spanish conquistador Bernal Díaz, and it is a banquet thrown by Aztec emperor Montezuma, which included more than thirty different foods kept warm on charcoal braziers, served in three hundred dishes. Díaz heard it said that human flesh was included, though he could not make out the contents of the many different dishes. But he could recognize pheasants, partridge, quail, ducks, peccary, doves, and rabbits. He also described an elaborate hand-washing ceremony, fascinatingly similar to those at use in European courts at the same time. There were also golden bowls of chocolate, of which the emperor was said to be extraordinarily fond (he was said to drink forty cups a day), and fruits, and to end the meal some pipes of tobacco, another gift of the Americas (figure 1.3). Apart from the tobacco, the banquet sounds strikingly similar to those given in Europe at the same time, and either Díaz was merely interpreting the scene in terms he thought his readers would understand or there are very important parallels between these two disparate cultures.

For one thing, the Aztecs were by no means less wealthy or sophisticated. Equally, their society was similarly stratified, which meant that those down the social scale could imitate the feasts of their superiors. In fact, Bernard Sahagun, essentially an ethnographer hired in the sixteenth century by King Philip II of Spain to record Aztec customs, describes a banquet thrown by a merchant. In it there are similar ceremonies to that enjoyed by the emperor: flat tortillas are served, along with stewed dishes, as well as chocolate, and there was smoking. What this means is that culinary customs circulated among various social strata, exactly as they did from European courts. What is striking in both these accounts, though, is the observation that participants ate very moderately,

Figure 1.3. Mexican Indian preparing chocolate. From the *Codex Tuleda*, 1553 (vellum), Museo de America, Madrid, Spain. The Bridgeman Art Library, Mexico School.

and they severely restricted the use of their prime alcoholic beverage, *pulque*. Interestingly, when the Aztecs were introduced to pork, beef, and wine, they attributed their rapid decline in population to these. In reality, the decline was primarily a result of disease. On this note, we turn to the greatest single event in the globalization of world cuisines—the Columbian Exchange.

Columbian Exchange: Europe, Mexico, and the Ming

As we have seen, plants and animals used for food were shared throughout the Eurasian continent and Africa in several successive waves—in prehistoric times, in the Classical

Era, and then again in the Middle Ages. None of these had as dramatic an impact on world cuisines as the rapid globalization that began in the early modern period, not only with the conquest of the Americas by Europeans but also with direct trade contact between Europe and Asia. The story actually begins before 1492 in Portugal. Prompted by their king, Henry the Navigator, the Portuguese began to sail southward along the west coast of Africa in the early fifteenth century. With their sturdy, oceangoing vessels, new navigation techniques, and a strong mapmaking tradition, they successfully launched trading voyages to secure African gold, ivory, and slaves. By venturing away from the coast and catching the counterclockwise currents south of the equator, they were eventually able to reach the southernmost tip of Africa, the Cape of Good Hope, in 1488 under the leadership of Bartolomeu Dias. Before long they ventured into the Indian Ocean and established direct and permanent trade contacts with India by setting up fortified garrisons in areas conquered by Portuguese explorer Vasco da Gama. Although pepper and cinnamon were available there, they also knew that other spices hailed from farther east, and this led to further forays to what is today Indonesia, then to China and Japan.

The Portuguese spice trade did not immediately displace the long-established Venetian trade routes through the Mediterranean. But it did mean that spices arrived in Europe in much greater quantities and were available to an increasing number of consumers. In the long run, spices would eventually go out of fashion as markers of social status or would be marginalized largely to desserts in the following century.

When the Genoese merchant Christopher Columbus first conceived of the idea to reach Asia by sailing westward, it was only natural that he would first seek the aid of the Portuguese, but by the late fifteenth century they were already fairly certain of securing the route around Africa. Only after seeking help from other wealthy nation states such as France and England (who had their own plans to find a Northwest Passage, spearheaded by John Cabot, another Italian merchant) did Columbus turn to Spain. The Spanish were preoccupied with their own domestic problems—the conquest of the last Muslim stronghold on Iberian soil, in Granada. When that had been accomplished in 1492, and the Jews expelled in the same year, they then considered financing Columbus's voyage. Columbus was trying to reach China, and to his dying day he believed what he had discovered was land immediately adjacent to Asia. Nevertheless, those who followed in his wake, including Florentine Amerigo Vespucci, for whom this half of the earth is named, began to realize that this was indeed a New World.

The significance of these events for the history of food is that Old World plants and animals were brought to the Americas—the horse, cow, pig, and chickens to name a few, plus wheat and practically every other vegetable familiar in Europe. Likewise, New World plants were brought to Europe. Not all were immediately or enthusiastically adopted. Corn was grown principally in Spain and northern Italy and, in fact, anywhere where eating polenta (porridge) had been the custom. Interestingly, Europeans did not adopt the process of nixtamalization; they merely ground the dry kernels into meal and boiled it, which ultimately led to nutritional deficiencies like pellagra among those who relied on corn. Potatoes, from South America, did not catch on for several centuries or remained a botanical curiosity, or food relegated to the Irish, who were considered half barbaric. Tomatoes likewise were difficult to fit into the European culinary schema, and they are not found in Italian cookbooks until the very end of the seventeenth century. But some species did quickly find a niche—New World beans were immediately grown and consumed, as they were not even recognized as different species. The turkey was rapidly adopted, as it was similar to guinea fowl or other large birds. Chili peppers (*capsicum*) were grown in some places, in Spain and southern Italy, but they never became widely used throughout Europe. Ironically they spread within a few decades to places where we could not even conceive of the cuisine without them: Hungary, Southeast Asia, and the Sichuan region of China.

This exchange was thus not merely between Europe and the Americas. With the growth of the slave trade, African plants like okra were planted around the world. Peanuts, originally from South America, were in turn brought to Africa, and from there to North America, and at the same time to Asia. When the Spanish conquered the Philippines in the late sixteenth century, they established a Pacific trade route from Mexico to Manila, which further dispersed plants globally. It is difficult to imagine sweet potatoes being grown extensively in China, but in fact they have played an extremely important role since the sixteenth century. With the exchange of foodstuffs, there was also, of course, the exchange of peoples as well as pathogens they brought with them. Diseases like measles and smallpox, to which Native Americans had never been exposed, absolutely devastated the native population. For example, the population of Mexico, estimated at twenty-five million before European contact, was reduced to about one million within a century.

The culinary cultures of our three world cuisines were irrevocably changed by the arrival of new ingredients from around the world. But other social, political, and demographic factors also played a decisive role. We will begin with China in the Ming dynasty because it neatly overlaps the period of exchange, extending from 1368 to 1644. China experienced demographic fluctuations that were replicated across the Eurasian continent. Through the Middle Ages as in Europe, there was population increase due to improved agricultural techniques, including the introduction of early ripening rice, which could allow double cropping, as well as terracing projects, draining of marshes, and building of canals. By the early thirteenth century the population stood at about 150 million, due in part to a more secure food supply. This population increase was cut off by Mongol invasions, civil wars, and plague, the same epidemic that hit Europe, so that the population was reduced to about one hundred million. This gradually began to increase, particularly in the sixteenth century, so that by the end of the Ming period (1644), thanks in large measure to political stability and intensification of agriculture, the population had recovered to 150 million. Products, particularly corn, sweet potatoes, and peanuts, from the Americas played a significant role in this growth, and all were adopted enthusiastically.

It was during the Ming dynasty that rice emerged as the principal food in practically all of China, with the only exception being in the north where wheat and millet were still very important. Soy remained as the most important adjunct to the principal starch along with a variety of fresh vegetables, so that by comparison with the rest of the world, during the Ming dynasty people were relatively well fed. Pork and poultry and a wide variety of salt and freshwater fish provided much of the protein, though normally as an adjunct to rice, which was considered the central feature of every meal.

The Ming emperors, following a long period of Mongol rule, were anxious to revive the ritual forms of Chinese etiquette, including those of the table. From their new capital at Beijing in the fifteenth century, a vast household staff served the court in banquets easily surpassing those of contemporary Europe and the Aztecs. Chefs were recruited from throughout the empire, and in 1435 some five thousand servants were reported as catering to the imperial table. The staff included cooks and servants but also a staff of palace eunuchs attending to the emperor personally and preparing food for sacrifice. There was also an entire retinue hired to manufacture wine, mill flour, maintain herds, and grow vegetables. There was also within the Forbidden City a huge contingent, numbering several thousand female servants. Altogether these constituted a large-scale city merely to attend to the emperor and his family. The Ming also held highly ritualized state banquets, which were at once a means of legitimizing their rule and a form of propaganda, but also a way the elaborate court cuisine influenced the rest of the country. These banquets were also orchestrated to emphasize the guests' status. Society was rigidly hierarchical, with the necessary subordination of some individuals.

For a brief period the Ming were also actively engaged in voyages of exploration. The eunuch Zheng He led seven voyages between 1405 and 1433 to Southeast Asia and India and through the Indian Ocean and as far as the east coast of Africa. On his first voyage there were 317 ships and twenty-eight thousand men. The ships were stacked with silks and porcelain, which were the very goods Westerners desired. It was only an accident of history that the Chinese did not round the Cape of Good Hope and reach Europe before the Portuguese reached Asia. With a new emperor and change of policy, further voyages were banned. This shows that the domination of world trade by Europeans was in no way inevitable or based on decidedly superior navigational technology.

The Early Modern Era

The globalization of trade compounded with political developments had a permanent effect on the three world cuisines. A closer look at the food culture and culinary literature of each region shows that the modern form of Italian, Mexican, and Chinese cuisine took definitive shape in the centuries following the Columbian Exchange, namely between the sixteenth and eighteenth centuries.

Colonial Mexican cuisine was to a large extent a hybrid of medieval Spanish and Aztec foodways, but it is easy to overestimate the extent of mutual influence. The idea of

Mestisaje, or mixture of peoples, cultures, and cuisines, was the official government policy after independence in the nineteenth century. It is important to remember, however, that elite Spanish colonists preferred their own European foods (especially wine, wheat, and olive oil) and sought to import them for several centuries after the conquest. Meanwhile, the indigenous population maintained their own corn-based cuisine. Nonetheless, there were in the long run significant exchanges of ingredients and techniques. While the basic ingredients remained by and large American, new sources of animal protein, a second grain staple, as well as various flavorings that ultimately hark back to medieval Persian cuisine, were added. A perfect example of this is the *molé poblano*. Although many legends surround its invention, supposedly by a Spanish nun in Puebla, there are dozens of variants, and most likely no individual "invented" it. The word *molé* itself comes directly from the Nahuatl *molli*, meaning "sauce." Its base, of various types of chilies such as mulatos, anchos, and pasillas, is native. Heating them in lard, derived from pork, is a Spanish import. The spices added, such as cinnamon, cloves, and black pepper, as well as nuts like almonds and sesame seeds, were imported from the Middle East. Raisins also hail from the Mediterranean, as do the garlic, aniseed, and coriander. Pumpkin seeds and cacao beans are native. The thickening agent can be a stale tortilla, bread, or both. In the end, the basic form of this molé, especially when served over turkey, is native Mexican, but the flavors are distantly related, oddly enough, to the Indian (that is, East Indian) cuisine that was imported there by Mughal emperors and had the same roots as medieval Spanish cookery. This is only one example of how Mexican cuisine is not only a hybrid of Spanish and Aztec but also truly of global influences.

With the introduction of Christianity, bread and wine were necessary products for the celebration of the Eucharist. At first these were largely imported from Europe. There were other direct imports, such as rice and beef and its associated dairy products and cheese. Cheese was especially important in the north where cattle ranching came to dominate. Elsewhere, pork became the predominant meat. Chickens and their eggs were also introduced, and kid (*cabrito*). In some cases, typical Spanish dishes were merely adapted to native flavorings, as in the case of meatballs (*albóndigas*). In others a familiar Spanish word was totally changed in the new context, as with the tortilla, which in Spain is an egg-based "little tart" but in Mexico became the new word for a flat, corn-dough round.

At the same time that colonial Mexican cuisine was forming, Italy achieved what was undeniably a high point of culinary sophistication in the late Renaissance. This happened partly as a result of the wealth accumulated from trade, money lending, and manufactures among the small Italian courts. Ironically it coincided with domination by larger nation-states such as Spain and France, which were struggling for domination of the peninsula. For the first half of the sixteenth century, Italy would become the battleground for rival imperial ambitions, and although this did thwart any possibility of achieving nationhood, small city-states such as Florence, Venice, and Rome, as well as smaller courts like Ferrara and Mantua became the models of refined European cuisine, just as they were in painting, architecture, and music.

We have seen that Italy produced the very first printed cookbook, but there followed in the sixteenth and seventeenth centuries a veritable profusion of culinary literature: cookbooks, banqueting guides, carving manuals, as well as a rich dietary literature. The cookbooks of Christoforo di Messisbugo in Ferrara, Domenico Romoli in Florence, and Bartolomeo Scappi in Rome are all indicative of this era. Scappi's *Opera* (Works) of 1570 might be called the first truly encyclopedic culinary text, including every possible known ingredient, a buying guide, menus, and even illustrations of kitchen and cooking procedures. Its recipes are precise and clearly worded, and rather than dictate procedures that must be followed to the letter, it offers multiple variants and substitute ingredients and in a very real way teaches the reader how to cook. The full range of cooking procedures is also in evidence: baking, roasting, boiling, frying, braising, and so forth. It also describes the origin of the finest ingredients and when they are in season, and features regional specialties from Lombardy and Milan, fish dishes from Venice, a herb tart from Bologna, macaroni from Genoa, spinach *alla Fiorentina*, Neapolitan pies and pizza. In other words, although there was not yet a state of Italy, Scappi highlighted the best dishes from throughout the peninsula and described regional cheeses, salami, and cured hams. He was also willing to go farther afield, as with a recipe for couscous, or as he calls it *succussu*, as well as recipes for the newly arrived turkey. One could justifiably call this the very first modern cookbook.

The style of Italian cuisine, although in many respects very modern, still retains elements of medieval cookery. There are still spices used in a wide variety of dishes, and sugar is used as a practically universal flavoring. The emphasis on color lingers to some extent as well. What is new, however, is an attention to garnishes and presentation, a new range of dishes incorporating butter and other dairy, and a variety of pastries and pies never before seen. Nonetheless, this is a distinctively Italian cookbook, and many of his dishes would scarcely seem out of place on an Italian table today. Take, for example, a recipe for *brisavoli*, which is made from lean veal, pounded thin and splashed with wine, vinegar, and garlic and sprinkled with fennel pollen, or coriander, salt and pepper, and weighted down in a press for an hour. The thin pieces of meat are then either floured and fried and served with a sauce of sugar, cinnamon, and orange juice or a sauce of vinegar, sugar, cinnamon, cloves, and nutmeg. They can also be grilled with a strip of pork fat for each. Apart from the spices, the procedure remains the same today.

As we have seen in all cultures with a refined sense of gastronomy, there was also an accompanying dietary literature, and sixteenth-century Italy was no different. Based primarily on ancient Greek dietary theories, a number of textbooks informed readers of the precise nutritional qualities of every ingredient and how they should be combined to form dishes both appropriate for an individual's specific temperament and humorally balanced dishes whose cooking procedure and combination of flavors made them more healthy. Typical of these is Baldassare Pisanelli's *Trattato della nature de' cibi et del bere* (Tract on the Nature of Foods and Drinks) of 1586. In it, he advises, for example, that lemons are a food cold and dry in the second degree. Thus, they are useful for all hot ailments, such as fevers, but also for cutting through thick humors. The Genovese have a custom of chopping lemon finely and adding salt and rosewater, which accompanies meats. The lemon cuts through the fibrous flesh, making it easier to digest. In this way, humoral medicine informed culinary practice, or at least was used to rationalize an established custom.

In the centuries that followed, the aesthetic of Italian cuisine closely matches that of the Baroque in other arts. Coming into play are a certain lightness and delicacy, a profusion of garnishes for ornamental effect, as well as a penchant to perfume food with exotica such as ambergris and musk. Bartolomeo Stefani's *L'arte di ben cucinare* (The Art of Good Cooking) of 1662 is typical. In it he recommends that sturgeon, one of the most highly prized foods, can be cooked with wine, mastic (an aromatic resin), salt, bay leaves, and butter *in bianco* (in white sauce). Or it can be cooked on a grill with salt and cloves, basted with oil, and served with a sauce of *tarantello* (cured tuna belly). Sturgeon can be made into little delicate pastries garnished with anchovies or shrimp, or even meatballs laden with aromatic herbs with marzipan, pine nuts, and raisins. Although these flavor combinations have gone out of style, the basic preparations remain in the elite repertoire of Italian cuisine.

By the eighteenth century, we encounter a cuisine that has become thoroughly and familiarly Italian, most notably in the cookbooks of Vincenzo Corrado of Naples. Products from the Americas such as tomatoes, peppers, and squash are now featured, although tomatoes made their first appearance in a kind of salsa in the works of Antonio Latini, another Neapolitan, at the end of the seventeenth century. In Corrado's book, they are used in a variety of new ways—as in a frittata (an Italian omelet) made of peeled and seeded tomatoes cooked with butter, spices, and marjoram, flavored with pancetta (unsmoked, rolled bacon), and mixed with ricotta and eggs. There is also a true tomato sauce with garlic, red pepper, herbs, oil, and broth to be used with lamb. The cuisine of the nineteenth century, however, at least as recorded in cookbooks, largely reflects the fashion for French cuisine, and we must wait until the work of Pellegrino Artusi in the late nineteenth century to find a detailed record of home cooking in Italy.

Meanwhile, Chinese cuisine in the Qing dynasty after 1644 warrants examination and must be set against an unprecedented surge in population growth. Although crops from the Americas such as the sweet potato had been introduced earlier, it was in this period that they led to a population that could no longer be reliably supported by its own resources, resulting in widespread famine. If there were 150 million people at the start of the Qing dynasty, there were 450 million by the mid-nineteenth century. In part this was stimulated by conscious government policies that promoted planting of sweet potatoes, corn, and regular potatoes, all ingredients we rarely associate with China but which became a major food source for the poor. Visitors described the disparity between the living standards of the rich and the poor as the most glaring on earth.

There was, for those who could afford it, an elegant form of dining and gastronomy. The cookbook of Li Yu of the seventeenth century cultivated a refined aesthetic, combining rusticity with a fastidious attention to rare, light, and elusive flavors. He rejected garlic because it left the breath foul, and he would not discuss certain animals, including cattle and dogs, because of their usefulness to humans. He was criticized by the mid-eighteenth-century cookbook author Yuan Mei, an accomplished poet and painter, whose *Sui Yuan Shi Tan* (Recipes from Sui Garden) was a kind of introduction to basic gastronomic principles. He stressed the importance of choosing the finest ingredients from specific locales, carefully adding only the choicest condiments with the purest flavors. But cost alone should not be a criterion of good taste, and he complains of having been served a completely flavorless, plainly boiled swallow's nest once. It would have been better to fill the bowl with pearls if the only intention was to impress with extravagance. Yuan was also remarkably attentive to cleanliness in the kitchen, and he described kitchen implements and cooking procedures that bring out the best flavors of foods. Most important, he was concerned with finding the right balance of flavors in every dish, not overpowering certain ingredients or combining too many flavors into an indistinguishable muddle. His approach to food was a level of connoisseurship on the same level with Scappi's in Italy.

The failure of Ming China to modernize has long been a question, particularly when Japan so rapidly caught up with Europe in the nineteenth century. Certainly the impact of the foreign domination of trade and the introduction of debilitating drugs like opium played a role. But this was not merely a matter of the Chinese being drugged or inherently backward or hopelessly traditional. It was merely the effect of having a huge, teeming population living at the margins of existence with little possibility or incentive to innovate because most people were living fundamentally from hand to mouth, eking a bare living out by cultivating smaller and smaller plots more intensively. Furthermore, a large family is necessary to farm this way, ultimately causing further population growth. Thus, there was little hope of limiting family size for fear of not having enough hands to work the land, combined with little ready money to invest in new crops, fertilizers, or improved systems of farming. Despite centuries as one of the most powerful nations on earth, the advent of industrialization in the nineteenth century would largely leave China behind.

The Industrial Era

The Industrial Revolution was the single most important event of the modern era, and it ultimately and irrevocably altered the diet of most people on the planet. Industrialization was made possible only with the precondition of what historians have called an Agricultural Revolution, or the rationalization of farming, which increasingly became a capital-intensive and large-scale enterprise. Although the picture of immediate transformation of the landscape has been revised in recent decades, there is no doubt that Britain, followed by other Western European nations, shifted from primarily agricultural to primarily manufacturing economies. In a nutshell, food production ceased to be the occupation for the majority of people, who increasingly moved into teeming cities to find work. Among the innovations introduced in the eighteenth and nineteenth centuries were new, improved systems of crop rotation, which made it possible to plant fodder crops to keep larger herds alive through the winter. There was also a massive effort at improving plant yields and productivity of dairy and meat cattle through selective breeding. Last, machines were gradually introduced, including the seed drill by Jethro Tull, the McCormick reaper, and other devices that presaged the eventual mechanization of agriculture and that increasingly made human labor obsolete. That is, it took many fewer hands to farm than ever before in human history.

The connection of these transformations to industrialization is that the surplus labor, abetted by a population boom, was eventually absorbed by factories where mass-produced and inexpensive goods were churned out. People, who for the first time were enmeshed in a consumer-oriented society, bought these goods, including mass-produced foods. For example, rather than bake bread at home using stone-ground flour, people increasingly purchased bread that was made from flour milled under intense pressure with steel rollers. Whiter and lighter bread could now be afforded by the masses, though the milling process and chemical additives made it a less nutritious and sometimes dangerous product. However, we should not over-romanticize the life of the preindustrial peasant. In many ways life would be, in the long run, improved and the food supply made much more secure, which in turn fostered demographic growth. Moreover, in cities people had always purchased their bread and many other manufactured foods. But the law would not enforce food purity legislation and production standards, as well as minimum wages and safe working conditions, for many years. In general, it is safe to say that the diet of the working classes deteriorated, especially as sugar was consumed in increasing quantities to sweeten coffee or tea or in the form of mass-produced candy and chocolates, the nineteenth-century junk food. The diet of elites, however, became opulent, particularly in restaurants where newfound wealth could be flaunted.

Obviously none of this happened overnight, and our three regions were relative latecomers to industrialization. It was only in the nineteenth century and with unification, for example, that northern Italy began to modernize, leaving southern Italy rural and comparatively impoverished, in many places to this day. Even in the north the system of *mezzadria* (sharecropping) left many families at the threshold of subsistence until the mid-twentieth century. This was only exacerbated by wartime rationing and dictator Benito Mussolini's (1883–1945) policy of autarky or self-sufficiency, when Italy attempted to restrict imports and supply all its own food. Ironically, the extremely simple and healthy "Mediterranean diet" was to a great extent the result of failed government policies and wartime deprivation rather than a long-established culinary aesthetic of simplicity.

Not everyone suffered deprivation, though. A crucial component of this new economy was the entrepreneurial middle classes or bourgeoisie, which maintained a cultural and culinary aesthetic distinct from the working classes and rural populace. These were largely wealthy urban and educated people, who could naturally patronize the arts, buy cookbooks, and hire servants to cook for them. While they aspired on grand occasions to emulate the cuisine of the court, their cooking was on a smaller scale and used fewer exotic ingredients and less complicated procedures that could be carried out without a large, well-staffed kitchen. While French influences, and its bourgeois cooking traditions, continued to influence Italy and all of Europe, a spirit of consciously cultivating and preserving what was construed as a national cuisine spread throughout middle-class kitchens. The most representative cookbook of this era is Pellegrino Artusi's *La scienza in cucina e l'arte di mangiar bene* (Science in the Kitchen and the Art of Eating Well) of 1891. Artusi himself was a banker rather than a cook, but he did systematically collect recipes from throughout Italy. Ironically, rather than foster a national cuisine, the book is evidence of persistent regional variations. His lighthearted style and diverting stories made this a persistent best seller, which also had the effect of spreading local preferences throughout the peninsula. For example, industrially produced dried pasta, once the mainstay of the south, would become a first-course staple for all Italians.

Mexico experienced pockets of industrialization, particularly around urban centers, but here, too, there remained large stretches of subsistence agriculture. In a sense, the process that began in Britain is still under way in Mexico, as the effects of the so-called Green Revolution of the twentieth century are still radically altering the lives of the rural populace as mechanized, large-scale farms depending on chemical fertilizers and pesticides are pushing out the small family farm and home garden plot (*milpa*). In the nineteenth century, and motivated largely by a nationalism similar to that of Italy, cookbooks were produced that were meant to reflect the unique blend of Spanish and indigenous traditions, bereft of French influence. *El cocinero Mexicano* (The Mexican Cook) of 1831 was merely the first of these, celebrating a unique Mexican cuisine. Nonetheless, there remained strong class biases in dietary patterns, based still on racial associations as a legacy of the colonial era. For example, white bread was still the preferred grain staple of the upper classes, while corn tortillas were considered a lower, and less nutritious, staple best left to the Indians. By the turn of the century, scientists and politicians even went so far as to insist that if Mexico would take a place among the modern nations, the indigenous population had to learn to eat wheat, which they supposed was the key to the greater intellectual advancement of Europeans and Americans. In the end, a truly mixed cuisine did develop, but under the aegis of mass production; that is, tortillas and *masa* flour produced industrially.

The history of Chinese cuisine in the nineteenth century follows a completely different pattern. China did not industrialize and remained an imperial nation until the overthrow of the last Qing emperor in 1912. However, in many respects the capital at Beijing had already established a fertile ground for culinary traditions from throughout China to mix and influence each other. While distinct regional patterns existed and remain to this day, there was also a broadly national style emanating from the aristocratic classes. By this time, European powers, intent on maintaining brisk trade for tea, porcelain, and silk, essentially muscled their way into dominating the empire and were granted trading outposts, such as Hong Kong to the British. In two Opium Wars (1839–1842, 1856–1860), intended to force the sale of opium from British India, the Chinese were soundly defeated when faced with technologically superior Western weapons. Despite largely failed attempts to modernize, China remained politically controlled by Western powers and was everything but directly colonized. With the declaration of the Republic of China in 1911, the country fell under the sway of warlords and suffered ongoing civil strife. It was these events that marked the decisive end, not only of the two thousand years of imperial tradition, but of China as a superpower. Combined with invasions of the Japanese, the civil wars only came to an end with the ascendancy of the communist People's Republic of China under Mao Zedong in 1949. Culinary traditions can scarcely be said to flourish during violent upheaval and famine, but it was really the Communist Party that almost completely eradicated Chinese haute cuisine. However, there were published cookbooks and even cooking schools that taught approved dishes. Most important was the state control of restaurants, whose menus were stripped of all bourgeois and imperial dishes, or they were renamed to reflect appropriate revolutionary values. Henceforth only the simplest plain fare appropriate for the working classes would be served. Mao himself, to set an example, ate only rustic dishes from his native Hunan Province. It is only since the 1980s under liberalization and with extensive foreign investment that restaurant culture has returned to China. This restaurant cooking is largely recovered from home cooking. Some of the first private businesses to open after 1979 were in fact restaurants.

We will pause here with the twentieth century, as the remainder of this book focuses on the contemporary scene, stretching back a generation or two. We will begin with the material foundations of the three world cuisines: cooking implements, technology, and techniques.

Study Questions

1. What role does climate change play in the rise and fall of civilizations?
2. How does domestication of plants and animals change human existence?
3. Compare and contrast the experience of invaders and how they both inherit and change the culinary customs of those they conquer.
4. How and why does the state sponsor agricultural innovation, and what specifically do they foster in terms of new crops and techniques? How do new agricultural techniques influence the development of cuisine?
5. Why are some periods open to foreign influences and readily adopt foreign imports, while others do not?
6. How is food used as a marker of social status, and how does this influence the popularity of specific items?
7. Why do complex dietary systems develop in each of the three world cuisines?
8. Identify the origin of the following plants and animals: peanuts, pineapple, corn, okra, tomatoes, turkey, potatoes, rice, artichokes, chili peppers.
9. Explain how the history of food is a process of increasing globalization from prehistoric times to the present.
10. How is feasting used as an instrument of power?

{ 2 }
Technology, Techniques, and Utensils

Learning Objectives

- Understand how and why different cultures reach similar solutions to basic culinary challenges in the kitchen.

- Appreciate how the size and shape of serving utensils influence cooking methods and ways of eating.

- Consider why the same basic ingredients are processed in such radically different ways across the globe.

- Understand how various cooking methods affect the flavor, color, and texture of ingredients.

- Identify antiquated cooking utensils and the advantages that have been lost with their replacement by machines.

- Appreciate the wide variety of materials used in making cooking vessels and utensils and the particular properties of each.

- Develop a critical vocabulary for describing subtle differences in cooking techniques among these three world cuisines.

THIS CHAPTER introduces the most important technologies, cooking techniques, and implements commonly used in the three world cuisines, both traditional and modern. All cuisines are fundamentally shaped by the tools used in the kitchen and by vessels in which food is cooked and served. In turn, the aesthetic values, available natural resources, and the social context of dining in our three regions have ultimately determined the types of implements most favored by professional and home cooks.

Cuisine should be considered not merely a collection of recipes, but an entire attitude toward eating, an expression of the interaction among ingredients, technologies that transform them, and formal codes of consumption and manners that determine how people eat. For example, the use of a bowl and chopsticks in Chinese culture is the direct result of rice being the basic staple, but it also dictates the predominant chopping and cooking techniques, as well as the interaction of diners at the table in a system of etiquette. Moreover, recipes themselves and the texture and consistency of the food are a product of the vessels in which or on which the food is served. A steak is practically impossible to eat from a bowl, just as finely chopped vegetables and rice are difficult to eat from a plate set on the table. To properly understand the spirit of a cuisine, therefore, one must explore the connections between the physical properties of prepared foods and the manner of eating them as a unified artistic expression.

Energy Sources

The first and most simple cooking technology, older than our species, is quite simply **fire**. Although there are examples of cooking in natural hot springs and volcanic fissures, and flame can be gathered so to speak from natural fires, for the most part flame is something that must be created by humans using wood, charcoal, or other fuel source such as dung or peat. Fire can be started with friction, the prototypical rubbing together of two sticks, or with more complicated bowing techniques, in which a wooden rod is entwined with a bowstring. When the bow is drawn back and forth, the stick spins rapidly on a wooden board, with luck igniting it or normally proximate pieces of kindling. Flints struck with metal have also been used to generate sparks that can set alight bark or dry tinder. Paleolithic remains of such fire-starting tools are prevalent in each of our world cuisines, and although there are easier ways to start a fire today, direct cooking over flame remains the preferred method for many dishes: little roasted fowl (*uccelli*) in Italy, a roast kid (*cabrito*) in northern Mexico, or a roast suckling pig for New Year's in China.

It is only really in the Industrial Era that fuel sources other than easily found natural materials began to proliferate. First among these is **coal**, mined in large quantities since the eighteenth century. Coal-burning stoves, although they create black, sooty smoke, provide an intense heat, and when food is prepared in covered iron pots, it remains unscathed. There are still many places in China where coal is used as fuel in an iron stove, and China remains the largest consumer of coal worldwide (although today largely for generating electricity).

Gas has been another fuel source since the Industrial Revolution, either natural gas (mostly methane, CH_4) or a chemically distinct form of gas such as propane, butane, or ethane. The use of gas for fuel has radically transformed the ability to control cooking temperature and to a great extent made possible the professional kitchen of the modern era and the precisely prepared and timed recipes associated with it. The energy output of gas flames is measured in British thermal units (BTUs). One BTU is the amount of gas necessary to increase the temperature of a pound of water by one degree at normal atmospheric pressure. A typical home range will put out about 10,000 BTUs, although commercial ranges can approach twice that number. This is especially important to consider in recipes that require intense searing heat, or when using a wok, as it works best in a professional kitchen with a volcanic blast of flame, often firing up at ten times or more the power available at home. Many Chinese recipes can only be prepared with a professional range. It is also important to remember, though, that a professional Chinese wok is much larger and is used to cook much more food at a time than the household wok. The gas jets found in a typical Italian household range are hot enough to sear food and also make possible extremely fine gradations of heat, often necessary when cooking something like tomato sauce in a pot.

Electricity has been used for cooking since the turn of the last century. In many ways it is the least effective way to cook because the coils on an electric range are slow to heat up and cool down and the cook has much less control of temperature. For ovens, however, electricity is often the least expensive form of energy. The solar oven is a recent addition to cooking technologies and can be as simple as a reflective metal surface or box lined with aluminum foil and covered with plastic. It is dependable only in hot places with regular, year-round sunshine.

A Taxonomy of Cooking Techniques

A discussion of kitchen tools must be foregrounded by a definition of cooking techniques, and although many vessels can be used for a variety of methods using various fuel sources and cooking media, nonetheless a working terminology will be useful in the descriptions of pots and pans that follow and how they influence cuisine. Many of these terms have been drawn from classical French cuisine, and they have become universal in all gastronomic theory and thus have been applied sometimes at the cost of precision. For example, although **braising** is a fairly uniform technique in the European repertoire, Chinese cuisine makes further and more subtle distinctions based on the braising liquid, be it water, broth, or so-called red-cooked dishes based on soy. In Italian cuisine there is sometimes scarce distinction between braising and what might be called a slowly cooked stew (*in umido*), which also contains a large piece of meat that is removed and served separately. Therefore, these categories should not be considered rigid, but rather ideal forms of each technique for the purpose of analysis, while in practice they often overlap. Furthermore, many recipes in the traditions discussed here demand several techniques in succession: steaming then frying, poaching then baking, and so on. The finished dish may thus defy definition by a single technique.

Cooking is here defined as the transfer of heat to raw food. There are many other ways to transform ingredients, such as fermentation, application of acids or salts, drying and freezing, and a number of modern chemical and physical processes including the use of alkali such as lye, which is an ancient technique. These will be discussed in the chapters that follow. Here we discuss only changes made by means of heat. Many traditional cooking methods are defined by what scientists call **conduction**, from the Latin *conducere*, "to lead along with." This simply means the direct transfer of heat from a source such as a flame to a relatively cooler space, such as a pan. Historically, most cooking in China, Italy, and Mexico has been by means of an open flame applied directly to a vessel, thus necessitating and accounting for the dominance of oil as a cooking medium.

Convection, in contrast, circulates heat through another medium serving as the vector, such as boiling water, or heated air circulating around the food, as in a convection oven. Although they will not concern us much here, it is worth mentioning modern methods of heating such as radiation by means of microwaves that directly agitate the molecules of water in the ingredients, heating them up, and the even more recently invented **induction**, which uses electromagnetic waves that transfer heat to an iron-based vessel by means of high-frequency alternating current, as in modern induction stovetops.

A further practical distinction can be made between those methods that use a liquid medium and those that use ambient heated air, either directly or indirectly. The direct methods apply heat from the source without an intervening barrier—as in **roasting**, which uses air as the vector, whereas **sautéeing** posits a pan between the food and the heat source and is thus an indirect method. We can also distinguish between methods on the basis of heat intensity. Thus, **boiling** differs from **poaching** only by degrees, just as gentle baking in a moderate oven differs greatly from food cooked in the same space at 800 degrees—as in a wood-fired oven. Furthermore, duration of contact with heat is a significant variable. Quick immersion in boiling water is called **blanching**. Food cooked in a lot of fat is deep fried, but if quickly submersed it is called velveting in China.

The liquid-based cooking methods can be classified as follows in three basic categories: water, fat, and other liquids. There are techniques that use water as the basic medium, including direct immersion or boiling, poaching, and blanching. Indirect methods would include boiling in an enclosed space, as with food stuffed in a sausage casing like an Italian mortadella (a large sausage made from finely ground pork with large cubes of fat and peppercorns), an enclosed vessel such as an egg coddler, or *sous-vide* technology that cooks food very gently in a vacuum-sealed plastic bag. Pressure is yet another variable, which further agitates molecules and speeds the cooking process, accomplished in a pressure cooker. It is a technique used in the world cuisines discussed here only recently but is used widely for cooking beans and stews.

Another subclassification of water-based cooking is **steaming**, wherein food is placed over steaming water held in a basket, the signature technique for dim sum. Food can also be steamed indirectly in an enclosed space, either wrapped in leaves or corn husks as with Mexican tamales; in a large, hot, stone-lined pit in which food is placed and covered (as pork is often cooked in the Yucatan); or in a steam oven that may circulate steam around a covered tray of food. **Slow cooking** in a vessel held over hot water, as when making zabaglione (a custard made with egg yolks and marsala wine), or heated in a bain marie (water bath) is also a type of heat transfer by means of simmering water and steam plus ambient heated air. Cooking rice might also be considered a form of steaming—although it begins in boiling water, much of the cooking takes place as the water is converted into steam in the pot.

Braising is a technique in which a large piece of food is cooked slowly in an enclosed vessel with water or other liquid that comes halfway up the ingredients and which may be periodically replenished. Sometimes the food is seared first, and then the slow cooking is intended to reduce and concentrate the flavors. All three world cuisines use this technique, especially for tougher cuts of meat. Thus, it is a technique that combines immersion in liquid and steaming but should be distinguished from stewing, in which the ingredients are fully immersed in a liquid, either covered or uncovered, and cooked slowly.

Fats are the second major family of liquid-based cooking, within which there are subtle distinctions among various methods. The most straightforward is direct immersion into hot fat, whether vegetable oil or animal based. Food may be deep fried when held under the surface of the fat with a tool or frying basket, as with fried wonton skins, or it may cook on the surface of the fat, as with *frittelle* and *zeppole* (Italian fritters). Conversely one can fry in a few inches of fat in a shallow skillet, as is often done with cutlets or battered food. In these methods only the lower surface of the food is in direct contact, and thus the cooking food must be turned over. Frying is a small amount of fat is comparable but must be considered a hybrid method, because the food is both cooked by direct contact with the medium and also by the heat generated by the cooking surface. The fat is used here both for flavoring and to prevent the food from sticking, but it does not transfer most of the heat into the food. The same should be said of closely related methods such as sautéeing, in which the food is constantly flipped through wrist action in the pan over high heat, as well as **stir-frying**, in which a comparable effect is achieved in a wok (a large, hemispherical cooking vessel with two short or one long handle) with the aid of a long-handled, spatulalike tool.

Third among the liquid media are broths, wine, milk, or a liquid of vegetable origin such as coconut milk. When cooked in a large amount of liquid for a long time, the result can be called a **stew**. When cooked gently in a small amount, it is referred to as a **braise**, which is done either on the stovetop or in an enclosed vessel in an oven.

The next major classification of methods is foods cooked with heated air, either directly in contact with the heat source or indirectly as in an oven. The simplest method is roasting, and while this term has lost its specificity today, it traditionally only meant foods cooked in proximity to flame. This was accomplished with birds or a whole animal threaded on a spit set beside a roaring fire, so that drippings could be captured in a pan, but it could also be done above hot coals, as when a spit is mounted with ingredients directly over the hot embers. In China, a duck, for example, may also be roasted slowly, hung vertically so the fat drips down (figure 2.1) The term *roasting* today is also often used to describe meats, vegetables, or potatoes that are daubed with fat and cooked in a very hot oven.

Grilling is a technique similar to true roasting but uses a metal barrier to support the food while it cooks. This can be anything from a perforated sheet of metal to a grill or iron bars. Significantly, fat and drippings fall from the food into the coals, creating smoke and flavoring the ingredients. Smoking is another method, although it can be used merely to flavor and preserve food, as in a cold smoke used to preserve hamlike speck in the Alto Adige region of Italy, when the ingredients are technically still raw. Hot smoking and a traditional barbecue apply heat as well as smoke, and the product is indeed cooked; in the case of barbecue it is also steamed slowly with moist heat. There are cooking methods in all three world cuisines comparable to the traditional slow barbecue, usually for cooking tougher and fattier cuts of pork.

Toasting is also a method of direct heat transfer through air and can be accomplished next to a fire, in a toaster with heated metal coils, or beneath a gas broiler. The

Figure 2.1. Draining Peking ducks. © Dean Conger/CORBIS.

term can be applied beyond bread and is used, for example, with spices heated in a pan without liquid or to chili peppers toasted in a dry *comal* (griddle). Ultimately, the term means quick heating in order to achieve caramelization of proteins on the surface; that is, **browning**, or as it is called in professional circles, the Maillard reaction. **Broiling** is another modern term, referring to powerful heat applied from above, normally in a modern gas or electric oven. Similar effect can be achieved, however, with a hot iron or salamander (a flat wedge) positioned above the food, which quickly heats the upper surface only. Similar effects can be achieved with a portable gas jet designed to quickly brown the surface of food. **Charring** is among the oldest forms of this cooking method, as food is passed quickly over a flame.

Baking involves placing food in a heated chamber either in direct contact with heated air or in an enclosed casserole or similar vessel. Again, some forms of baking must be considered hybrid cooking methods. Some oven forms, for example, use burning wood inside the main chamber and are thus a form of roasting in proximity to flame and baking, with a measure of smoking involved as well. Pizza in a traditional pizza oven or foods cooked in a Mexican *horno* (oven) are not simply baking. Only when the hot embers are raked out of the oven and the food cooks solely through retained heat is it quintessential baking, though here heat is also transferred directly through the hot oven floor or a dish or tray that conducts heat. Strictly speaking, a modern oven with exposed gas jets also radiates heat from the bottom, though normally not directly onto the food. Most important, what is often colloquially called roasting in an oven is really baking. The same must be said for so-called **pan roasting**, which is merely cooking food in a pan without a liquid cooking medium. Akin to baking, though not using an oven, is the technique of cooking food under hot coals, either directly or wrapped in leaves, as well as the ancient Roman method of cooking *sub testu*, when the food is covered with a ceramic dome and hot coals are heaped on top, which was an early way of making breads and cakes without an oven. The iron Dutch oven is a comparable cooking method, in which food is "baked" enclosed and over hot coals, though this term referred to several quite different devices in its early history.

Implements for Food Preparation and Cooking

None of the cooking techniques mentioned here can be accomplished without implements to contain or hold food and tools for processing it. Although tools used in the three world cuisines vary widely, their solutions to the same basic physical challenges are comparable. Food must be cut; hence there are knives, though they come in many forms and shapes. Vessels must hold food or food and water, though the material they are made from will not only directly influence the cooking procedure but also the texture and consistency of the final dish. In thinking of cooking in general, it is useful to keep in mind that the intersection of fuel source, technique, and tools ultimately explains how ingredients are transformed and is the basic structure upon which cuisines are built. It is also important to keep in mind that not all food is cooked per se; it can be eaten raw or fermented, chemically altered, or processed in other important ways that require no heat.

Cutting Tools

The most important and practically indispensable tool in the kitchen is the **knife**. Knives were from prehistoric times fashioned from chipped flint or obsidian. Both are forms of quartz, which can be carefully struck with another rock, called knapping, to form a razor-sharp edge, suitable for arrowheads, spear points, and hand-held knives. These were used both for skinning and dismembering hunted animals and for other culinary tasks involving cutting.

Regardless of the material from which knives are made, they can take hundreds of forms. They can range from huge, single-edged, swordlike carving knives to tiny paring knives. Their edges can be straight or tapered, serrated or beveled, and their bodies flexible, as in boning knives, or rigid. There are specialized cheese knives, including the wedge used to break up Parmigiano, melon ballers, and zesters for citrus rinds, and even can openers are a kind of cutting tool. A full catalogue of every knife form would take an entire book, but there are certain distinctive shapes used in our three cuisines that warrant attention.

Although commonly called Chinese cleavers, the rectangular-bladed implement most commonly used in Asian cookery is really a knife, as only the heaviest and thickest are used for cleaving food; that is, cutting through joints and lighter bones. The shape of the blade is important in determining both knife technique as well as the final shape of food to be cooked. The blade is not rocked over a cutting surface as with Western knives but is rather used to chop in quick, downward strokes. It can also be used for precision cutting with the food held in the left hand and rotated around the blade held horizontally in the right or by using diagonal or horizontal cuts for certain ingredients. For example, a breast of chicken may be sliced horizontally beneath the hand, dividing it into two thin cutlets. The flat edge of the blade is also used for flattening food, either to make the final pieces even in diameter or as a tenderizing technique. Most important, the blade also serves as a surface whereupon cut ingredients are transferred from the cutting board to a bowl or cooking vessel. These knives come in a variety of sizes as well, some designed for chopping large cuts of meat or hard vegetables, others with narrower blades for more delicate tasks. A Chinese chef may also use two cleavers to finely mince ingredients, the sound of which is best compared to a horse gently galloping, and the steady rhythm is said to reduce the ingredients to uniform consistency.

Italian knives have traditionally come in a wide variety of specialized shapes, many of which are illustrated in the *Opera* (Work) of the famous Italian chef Bartolomeo Scappi published in 1570. There are short oyster knives with curved blades, broad-bladed knives for cutting and serving tarts, knives for cutting pasta, and others for specialized butchering tasks. There is even illustrated a handy carrying case, worn at the side suspended from the belt. Specific knives for pastry making are also shown as well

Metallurgy and Knife Technology

Metal cutting edges were used as early as 6000 BC by the Egyptians, who smelted, cast, and worked copper. These simple copper implements were created by melting copper (at 1083°C or 1981°F) in a kiln, but on its own copper is fairly soft and does not keep an edge well. Likewise, gold and other soft metals, though easily hammered, cannot be made into working knives. Alloys of copper and tin were introduced in the Bronze Age in the Fertile Crescent, roughly 3500 BC, around the same time that the earliest civilizations were forming. China discovered bronze technology several centuries later, and inhabitants of the Andean region in what is now South America discovered bronze around 1000 BC. Deposits of copper and tin rarely occur together, though, and it has been suggested that this was a primary stimulus for trade in the ancient world. Cyprus, for example, became a major source of copper, and the word *copper* is the origin of the island's name; Britain was a source for tin, even for ancient Mediterranean cultures. As copper and tin melt at a fairly low temperature, the resulting bronze alloy can be easily cast and worked with cold hammers, and thus bronze was most important in making cooking vessels. Some of the oldest and most beautifully wrought are from China.

The Iron Age occurred around the twelfth century BC in the Middle East and several centuries thereafter in southern Europe and China. The Americas received iron technology only with the arrival of Europeans. It has been argued that the collapse of trade routes in the Mediterranean ultimately hastened the switch to iron, which is not only more difficult to process but also weaker as a metal than bronze. It also, of course, easily oxidizes (rusts). Although iron melts at a much higher temperature than copper (1535° C), it can be worked when hot through forging and hammering, hence wrought iron, which can be made into knives. Cast iron, however, which also contains carbon, requires a furnace with bellows and a fuel source, such as coal, which puts out much greater heat and was only used in Italy from the Middle Ages on and in China from several centuries BC.

The real breakthrough, in terms of cutting technology, came with the development of steel, also dated to the Iron Age. Adding carbon to iron significantly lowers its melting point to about 1150°C, just about the limit of an ancient kiln. The more carbon is added, the harder the implement, but also the more brittle. Thus, most early knives are low-carbon steel, which can be easily worked and regularly sharpened. Steel can also be made much harder through controlled cooling or quenching, in which the heated metal is plunged into cold water, which in repetition tempers the metal. The advantage of steel implements was not only the ubiquity and low cost of iron but also its strength and durability as a cutting tool. To this day, many chefs prefer carbon steel, and although it can stain and pit, it can be easily sharpened. Modern high-carbon stainless steel is made with chromium, and although it may look nicer, it is more difficult to sharpen but does keep an edge well. A more recent invention is the use of ceramics for knives, which are harder and remain sharper much longer than metal, though they are brittle and can shatter if dropped.

as an implement here called a *raschiatore da bancho*. Today this is called a *mezzaluna* (half moon). It consists of a curved blade with a wooden handle at each end. To finely chop herbs or other ingredients, the blade is rocked back and forth on a board or in a wooden bowl.

Although there was a native blade-making industry in Italy, eventually Italians adopted the knife forms common throughout Europe, manufactured primarily in Germany and France. The basic form of the French chef's knife, with a curved cutting edge of eight to twelve inches, coming to a point, with a straight spine (back) and tapered from heel to tip, was also adopted in Mexico, introduced by the Spanish. Along with these there are also a variety of smaller utility knives, serrated bread knives, boning knives, paring knives, and the small, beak-shaped *tourné* knife used for turning vegetables and cutting food into elegant shapes, as well as peelers. The granton-edged knife is another recent innovation in which shallow depressions along the blade's side allow the knife to slice through meat without catching or tearing.

Scissors, shears, and **slicers** are all further variants of the basic cutting tool. Scissors are merely two modified knives riveted in such a way that they cut precisely without the need for a board. Heavy shears are used for bisecting joints. Slicers are a more recent invention, ranging from a flat plate of metal cut in the middle and sharpened, as in a cheese slicer, to more sophisticated mandolins, which have a fixed v-shape with razor-sharp blades set in a track over which food is passed. In Italy there are also sophisticated manual meat slicers, most prized for prosciutto (a dry, cured ham, the most famous of which comes from Parma), which is said to be ruined by the heat generated by an electric slicer.

Wires can also be used for cutting, especially cheeses through which a knife often sticks. Compound sets of wires also appear in contraptions like the egg slicer, as well as a unique Italian device, the *cittara*, which gets its name from the ancestor of the guitar. It is a frame of wood strung with evenly spaced metal wires (figure 2.2). A sheet of dough is laid on top and then a rolling pin is passed over it, resulting in neat strands of pasta.

Graters are relatives of cutting devices and are normally a piece of soft metal such as tin pierced roughly with an awl. Hard ingredients such as nutmeg or Parmigiano cheese are grated along its rough surface. Sugar, too, was once grated from a "loaf" or cone this way, as were breadcrumbs, commonly used to thicken sauces before the advent of fat-based sauces in the seventeenth century. In Asia graters have also been made of ceramic, which are useful for fibrous roots such as ginger, which would clog an ordinary perforated grater.

An adjunct to most cutting tools is the **cutting board**. Surprisingly, these vary widely from culture to culture. In China, the board is usually a round section of tree trunk cut from a hardwood such as locust or ginkgo or even laminated bamboo, which is remarkably strong and lasts for years. In Italy, cutting boards are made from practically any wood, either cut into planks or strips arranged vertically and bound together end up. The larger of these are huge, free-standing butcher blocks with table legs. The hardest woods such as oak and maple make the strongest cutting surface. Today boards are also

Figure 2.2. The metal strings of the *cittara* cut thin, even noodles.

make of plastic, which although they are said to resist bacteria better, also become ruined quickly. Glass and stone make terrible cutting surfaces because they not only dull knives but also food slips on them. Manufacturers persist in making them nonetheless.

Sharpening tools are another accompaniment to knife technology. Traditionally these have been a simple whetstone that grinds down the knife's dull edge. Today these come in advanced diamond-edged and electric models. A honing steel is also important for keeping the molecules on the knife's edge aligned and sharp, although technically it does not actually sharpen. The knife is merely passed quickly across the steel long shaft in an arc, on both sides, to maintain a sound cutting edge.

Grinding and Pounding Tools

As old, and probably predating cutting tools, are a whole series of pounding implements, the simplest of which, two **stones,** was almost certainly the culinary tool most often used in prehistoric times. These were used to bash fibrous plants and roots to make them more edible and eventually to grind cereals when agriculture was invented. The basic form of a **mortar and pestle** has scarcely changed in the intervening millennia (figure 2.3). A hard

Figure 2.3. A mortar is among the oldest and most versatile of kitchen tools.

bowl or cylindrical container holds the food, which is then pummeled with a long, narrow pestle. A grinding rock as found among Native Americans is a good example of its earliest forms. The container could also be made of wood—a hollowed vertical log with another log as the pestle. Later, stone was quarried and cut to make portable mortars of hard granite or marble, or indeed any stone that will not break.

An advance on this basic technology, the stone **quern**, was used to grind grain (figure 2.4). It consists of two circular grinding stones with a pouring hole in the middle and grooves cut along the inside surface of each. The top stone is rotated with a handle, and the ground meal pours out the sides. These could range from a small device weighing about fifty pounds to huge mills turned with oxen for larger tasks. It was such technology that made feeding huge armies in the empires of the ancient world possible. Similar mills were also used in ancient times to crush olives, though the stones were set farther apart to prevent breaking the pits (figure 2.5). The mashed olives would then be pressed for oil, in complex levered machines described in detail in early agricultural manuals like that of Cato the Elder of the second century BC. A major advance on this technology was the waterwheel-driven mill, proliferating after around AD 1000. A large wheel fitted with buckets was positioned under a water sluice, and as the water poured into the buckets it turned the wheel, which was connected to a shaft and gears

Figure 2.4. A hand quern.

Figure 2.5. An ancient olive oil press found in the Roman Gate, below the Damascus Gate in the Old City walls of Jerusalem. © Richard T. Nowitz/CORBIS.

that turned the millstone. Windmills use a similar mechanism. Until the invention of steam-powered mills with steel rollers, this was the basic way of processing grain in premodern times.

In the kitchen, however, and in many places around the world, the basic mortar and pestle continued to be the prime grinding device. In Mexico, for example, the *molcajete*, made of relatively soft volcanic stone, usually basalt, was used for pounding sauces, such as guacamole. The *molcajete* and its accompanying *tejolote* (pestle) are both terms taken from the Aztec language Nahuatl, in their original forms *mulcazitl* and *texolotl*. The latter is also called a *mano* today. Pre-Columbian examples were often made of ceramic, but the basic form remains the same: a large, rough-hewn bowl supported by three legs, with a small, blunt, handheld pestle. As with most stone mortars, the *molcajete* must be seasoned. Traditionally this is done by grinding dried corn or rice into the surface repeatedly to remove loose grit. When the grains crush cleanly, the bowl is seasoned. A larger *metate* was used for grinding nixtamalized corn. The *metate* is a curved stone surface on which a stone rolling pin is rocked back and forth vigorously to crush the softened corn kernels. It cannot be used with dry grains, and thus the available technology truly determined the way corn would be processed in pre-Columbian Mexico, into a masa for tortillas and tamales rather than meal for cornbread. Today, the *molino de maiz* has largely replaced the *metate*, but it, too, is a specialized implement for corn rather than wheat flour.

The mortar is equally important in Asian cuisines and is generally used for grinding spices or fresh herbs used in sauces. Particularly in South Asia, soups and stews and currylike dishes based on coconut milk usually begin with grinding herbs, onions, chilies, and other ingredients. In China, the **mortar and pestle** is equally important, particularly in preparing dried herbs and spices for medicinal soups. There was a natural crossover of technologies from pharmacy to cuisine because in all three traditions food was considered a type of medicine.

To this list of pounding devices we must add **mallets** and **meat tenderizers**. They break up the tough fibers of meat through pounding or, as in the case of the Italian *batticarne*, are used primarily for flattening meat into cutlets (figure 2.6). Unique in design, this implement consists of a heavy, smooth metal disk with a vertical handle in the middle. It pounds the meat without tearing, the way a textured mallet would.

A **food mill** is a device related to grinders and strainers, as it accomplishes both tasks (figure 2.7). The solid, or semiliquid, ingredients are placed in the basket at the bottom of which is a perforated disk. A crank then passes a sheet of metal tightly over the disk, mashing and pushing the smooth ingredients through while leaving the solids larger than the holes behind. It is used in Italy for making tomato sauce, as it removes seeds and skins. It can also be used for processing berries or for pureeing vegetables, and because it also strains, it is in many ways superior to the modern food processor. There are also a series of smaller mills specifically designed for hard, dried ingredients, such as the pepper mill, spice mill, and the hand-cranked coffee grinder.

Figure 2.6. The heavy, flat surface of an Italian *batticarne* flattens meat without tearing it.

Sieves and Strainers

The **sieve** has always been an important tool for both draining boiled foods such as noodles or solids from a broth and for passing foods through the fine mesh to obtain a smooth texture. They range from pierced metal baskets, which we would call a colander today, to conical "china hat" sieves made with a mesh screen. The original sieve, however, was a wooden hoop strung with tightly woven horsehair. Food would be passed through the weave with a spoon or flat, wooden wedge. Such sieves are still made today, though normally with metal screens. In medieval Italian cookery where fine, soft textures were desired, a sieve was indispensable, and remained so for making purees (food completely processed into a fine paste or smooth liquid) up to the present. Spaniards brought this aesthetic directly to Mexico, and sieves were always used to create the smoothest possible *molés* (sauces). Food processors have largely made such implements obsolete today. Related to these are also various **sifters** for flour as well as **bolting cloths** used to separate finely milled flour from the hull of grains. Without bolting technology the fine white bread used to demarcate the higher class in Italy would have been impossible.

Another variation on the strainer is the handheld, perforated **ladle** or **wire mesh basket** attached to a bamboo handle, which is common in Chinese cuisine and is often

Figure 2.7. Tomato sauce being passed through a food mill to remove skin and seeds.

called a spider. These are used for scooping cooked items from boiling water or bubbling fat. This allows the chef to add and remove ingredients from a wok full of oil without ever dumping the entire contents. In Europe, a "scummer" or **skimmer** serves the same function and is basically a shallow, perforated spoon on a long handle with which one might remove tortellini from a large, continuously boiling stockpot of water. Smaller strainers are also used for straining tea of loose leaves or for removing cooked morsels of food from smaller pots.

Spits and Roasting Equipment

Although far less common in kitchens today, **spits** were an essential part of kitchen technology around the world (figure 2.8). In their most basic form, they would have been a simple spear of green wood at first and later of forged iron. More complicated spits include tines to hold food in place, or pins that could be adjusted for each cut of meat to

Figure 2.8. A duck roasting slowly on a mechanical turnspit beside a roaring fire.

be skewered. Spits can be used horizontally for roasting a suckling pig (*cochinito* in Italian or *lechon* in Spanish), as well as vertically for roast goose or duck in China. They held not only large cuts of meat but also smaller morsels, vegetables, and even ingredients like eggs or ground-meat mixtures. In Mexico, these are *albóndigas*, in Italy *polpettone*. Smaller skewers could also be used over a grill, and all three cuisines have the equivalent of what we call kebabs. Smaller yet were various needles for trussing birds for roasting, or for **barding**, which was a method of sewing strips of fat into the surface of meats to keep it moist and self-basting as it turned on the spit. In China, skewers were often made of bamboo, holding a combination of foods cooked over a charcoal **brazier**, whose incarnation in twentieth-century Asian restaurants in the United States was the pu pu platter.

Braziers and Grills

Although in professional kitchens a **brasier** refers to a type of pot used for braising, the term itself comes from the French word for coals. The original "brazier" is nothing more than an iron container for holding hot coals over which food is cooked or kept warm. These date back thousands of years in the Old and New Worlds and remain a principal means of cooking outdoors, which in many crowded Chinese cities is preferred. They are important because with little space and a small amount of fuel one can prepare an entire meal. Street vendors also depend on them. The brazier is either fitted with a grate (like a hibachi) on which food is cooked with direct heat, or a pot can be placed on top for gentle cooking, as in the Genoese *bagna cauda* (hot bath)—a combination of oil and seasonings into which vegetables are dipped. A sterno tabletop heater and even a modern food warmer operate on the same principle. A simple brazier also served as the prototype of the cooking range, over which a pot would be placed for gentle cooking. The early cooking range consisted merely of indentations set into a stone or brick bank that held hot coals shoveled from the fire. In China, there are also a number of more complicated braziers that hold hot coals in a central chamber, which heats liquid contained in a surrounding vessel. These are mostly for making medicinal soups, which require long and gentle cooking, and they keep the contents warm on the table.

The grill is among the simplest of cooking utensils and began as a rack of iron bars with four legs placed directly above hot coals on the ground, often with a long handle. Grills today hold coals in a chamber and may be fitted with staged grilling surfaces, as well as a lid. The classic hemispherical "barbecue grill" and more sophisticated gas grills accomplish the same goals: searing food over direct heat while creating smoke from dripping fats, seasonings, or marinades. This technology is virtually universal throughout Italy, China, and Mexico, especially for cooking smaller cuts of meat, ribs, and sausages as well as vegetables. The classic Tuscan *bistecca alla Fiorentina* made from a huge ribeye of Chianina beef is always cooked *sulla griglia* (on the grill). In modern times, grilling has largely supplanted spit roasting as the preferred method of outdoor cookery, especially for the home.

Pots

The simplest of **pots** in prehistoric times were created from hollowed-out logs, stone-lined pits, or watertight baskets into which heated rocks were placed to cook food. But it took ceramics to make a true pot.

Clay is ubiquitous and cheap, so it was logically the first material used to make permanent cooking vessels, the oldest archaeological remains of which date back many millennia. Pots for cooking have always been rounded on the bottom, which is a necessary consequence both of the physics of the construction technique as well as the necessity of remaining intact when placed on a fire. A flat-bottomed vessel would crack at the angled junctures, but a round form evenly distributes the heat around the surface of the vessel, making it able to withstand direct heat. The simplest of pots are "pinched" from clay. The coiling technique, whereby long "snakes" of clay are coiled in a spiral upward from the base, forming the pot walls, allows for larger vessels. Normally the pot is left to dry slightly before another set of coils is added, and this prevents the walls from collapsing. The interior is then smoothed out with a stone or other implement, both compacting the clay surface and increasing its size. Early Mesoamerican, Mediterranean, and Asian cooking pots, sometimes looking remarkably similar, all used these basic techniques.

To fire such pots, either a pit or a rudimentary kiln was used—an enclosed chamber that retains heat, differing little from an oven. The temperatures achieved with this technology yield what is known as **earthenware**, which is ideally suited for cooking over an open flame. High-fired ceramics, in which the clay vitrifies, have too tight a molecular structure to expand and contract with the heat and thus have been unsuitable for cookware, until discoveries made in the twentieth century for high-fired clay bodies. In general, high-fired stoneware and porcelain are used primarily for serving and storage vessels, though they can be used to cook with indirect heat, as with a casserole placed in the oven. Europe obtained the necessary kiln technology for stoneware only in the Middle Ages and the ability to make porcelain only in the eighteenth century; China had these technologies about two millennia earlier.

Another crucial factor was the development of glazes, which make the interior surface, and sometimes the exterior as well, totally waterproof. Glazes are essentially powdered glass mixed with a flux that allows it to melt at a temperature that will keep the glaze on the pot surface and adhere after cooling without cracking. Before European contact, pots in the Americas were not glazed but burnished, which is a process of rubbing with an object that smooths and compacts the surface to a high sheen and makes the pot practically waterproof. Traditional Native American pottery of the Southwest is still made this way.

Whether it be the *olla* used in Mexico, the *pignatta* or terra-cotta *pentola* in Italy, or what is often called a sand pot (*Sha guo*) in Chinese cookery, the same basic principles apply. Food is cut up and placed in the pot with water, which then gently boils when the pot is set over glowing embers. The pot is sometimes kept steady with built-in legs or placed on a trivet or "spider" under which the coals are placed. Such a vessel is almost never placed over roaring flames. The advantage of this method is that the ingredients slowly stew, are not in danger of burning unless the liquid is overly reduced, and result in a richly flavored, slow-cooked soup or stew. Many vegetables, legumes, and tougher cuts of meat can only be successfully cooked this way, and many cooks to this day prefer the taste of food cooked in clay.

Cooking in metal pots is a very different process. As metal can withstand much greater temperatures without damage, it can be placed directly over a roaring flame, thus enabling the cook to bring liquids to a full boil and even to brown ingredients with fat before adding liquid. Cast bronze cooking pots were historically the most important in the ancient world, and ironically they often took the very same form as pottery vessels. We would call such vessels today cauldrons, though they can range in size from huge stockpots holding twenty gallons or more to small vessels holding only a few cups. Typically these would be suspended by a chain hanging directly over the flame or hung from a trammel, a kind of crane that swivels in and out of the hearth. In European cookery the cauldron was not only used to cook a stewlike mixture but also could contain boiling water in which many different foods were cooked all in their own separate casing or wrapped in cloth. Once cooking moved from the hearth to the range, the rounded-bottom cauldron form gave way to the flat-bottomed stockpot. The shape of the pot also had a direct influence on the development of cuisine. For example, some with straight edges were designed to cook large amounts of liquid slowly, whereas others with flared rims (like a *saucier*) expose a greater portion of the surface area and were best suited to rapid cooking and quick reduction. Such pans became prominent exactly as reduced sauces began to proliferate in the seventeenth century. Such pans are unusual in Asia, though, where a wok serves to reduce liquids, in the instances when this is called for.

Tin-lined copper has been another choice for cookware, proliferating especially in professional kitchens by the sixteenth century in Italy, if not earlier. Copper is an excellent conductor of heat but is highly reactive. If not properly maintained it can lead to verdigris poisoning. On its own it is usually only used for mixing bowls, such as those used for whipping egg whites, or for candy-making equipment. Normally copper is lined with tin, which although impervious to reaction with food, it does wear off with time and cannot be used with extremely high heat or it melts. Retinning replaces the interior coating when worn off, so that food is not in contact with the copper. The exterior of copper also oxidizes, leaving a green patina unless rigorously polished. The conductivity of copper is still used in high-quality cookware today, though normally encased in stainless steel. The use of copper pots predominantly in professional kitchens is partly due to their high cost and necessary maintenance, but they are one more factor that contributed to the precision in temperature and timing as well as efficiency in achieving predicable results that occur in the modern kitchen.

Aluminum has also been used in cooking ware since the mid-twentieth century and remains a popular and inexpensive option in all three regions discussed here. Its major downside is that it reacts with acidic foods and causes some to turn an unappealing gray. Anodized aluminum is nonetheless the lightest metal available and is often used in professional kitchens, especially for large stockpots. In China, one will even find inexpensive aluminum woks. Stainless steel is beginning to replace many older metals, mostly because it is extremely sturdy and remains shiny when cleaned regularly, though it is not necessarily the ideal metal for cooking.

Pans and Flat Cooking Tools

The most important cooking vessel in Italian cuisine is the **pan**, the hallmark of Italian cuisine. Pans range in size from a few inches in diameter to large skillets. The sides of a pan normally slope up gently so it can be used for sauces, cooking vegetables, and for flat cuts of meat, but there are also versions with straight sides, originally made from earthenware glazed only on the interior. These were known in Italy as a *cassola*, and in Spain, and thereafter in Mexico, as a *cazuela*. They are designed to withstand direct heat but can also be used in an oven, as with what in English is called a casserole. The major functional distinction is with metal pans with a long handle, which facilitates tossing, as well as cooking over direct and more powerful flames. As in most Mediterranean cooking, dishes often begin with a *sofrito*, a combination of vegetables such as celery, carrots, and garlic cooked in a pan, to which might be added tomatoes, ground meat, or stock, forming a sauce, which can be added to other ingredients or tossed with pasta. Such pans can also come with a lid, or instead of a long handle are fitted with two smaller handles, the equivalent of the French *rondeau* or *griswold*, and are somewhat between a pot and a pan. The largest of flat pans are used outdoors on a tripod over a wood or gas fire and resemble the Spanish *paella* pan.

A recent innovation in technology has been the Teflon-coated pan. Its completely nonstick cooking surface enables the use of less grease, in the interest of health, and makes most cleanup jobs effortless. However, heating at high temperature is said to release carcinogenic gases, and scratches in an old pan can end up as flakes of Teflon in the food. Furthermore, the caramelization of solids sticking to the pan, in French the *fond*, necessary for making many sauces, is practically impossible in Teflon. Nonetheless, they can be found in home kitchens in Italy, Mexico, and China and have influenced the development of rapid, relatively fat-free cooking over heavier-sauced dishes.

There are also a number of flat metal surfaces without sides designed to cook dry ingredients that need not be contained. They replicate what was probably one of the earliest cooking methods, over a hot, flat stone. The most important of these in Mexican cuisine in the *comal*, essentially a flat disk of cast iron sometimes with a loop handle, sometimes with a pan handle (figure 2.9). These are used primarily for cooking tortillas but also for charring chili peppers and other ingredients. They are seasoned like other cast iron implements and are said to improve with age and use, so that they are often handed down as heirlooms within families for generations. There are also oval *comals*, either for cooking several tortillas at once or for cooking quesadillas, or recently for fajitas, strips of meat and vegetables brought to the table sizzling in a hot, oval *comal*.

Steaming Tools

Although the basic principle of steaming is the same across the world, there are varied solutions to containing food over boiling water. In China the **bamboo steamer** is most common, a hoop of bamboo fitted with slats and covered, which is placed in a wok over boiling water. The steamers can be stacked on top of each other as well, making this an extremely efficient way of cooking. The Chinese believe that this form of cooking brings out the purest and most natural flavors of the ingredients. Usually the bamboo slats are oiled to prevent sticking, though food such as dumplings may be placed on cabbage leaves, or as with steamed buns on a round of paper. A whole dish of food may also be set in the steamer. In professional kitchens stackable metal steamers have replaced bamboo, though the resulting flavor according to some is inferior.

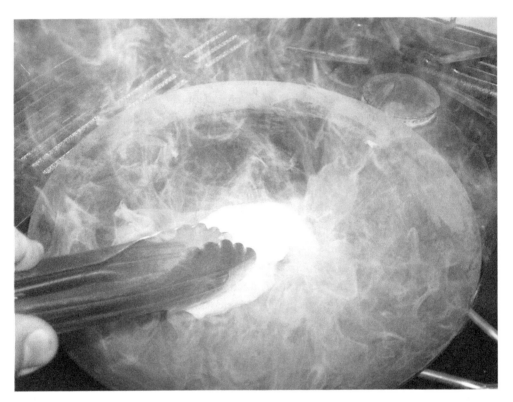

Figure 2.9. The fat placed on a red-hot *comal* creates a nonstick polymer.

Steaming is not commonly used in modern Italian cookery, though recently in the interest of health there are metal steaming inserts. A pot fitted with a perforated colander can be used as a steamer, as well as a collapsible metal steaming basket used for vegetables. More often Italian cooks will make a braise, which is a form of slow steaming in a stock, but the food is in direct contact with the liquid in a covered vessel.

In Mexico steaming of tamales is an ancient practice, at first with food set directly into a pot, wrapped in leaves or cornhusks and cooked slowly by force of steam. The pre-Columbian inhabitants of Mexico understood that this was the most efficient way to cook a large quantity of corn dough–based products, leaving them both light and toothsome. There is also a traditional *tamalero*, a large metal bucket holding as much as twelve gallons, today usually of galvanized steel, into which a steaming insert is set over boiling water. They are typically used at parties and family gatherings when a large number of tamales can be cooked at once.

Concave Cooking Implements

The signature cooking vessel of Chinese cuisine is the **wok**, a large, hemispherical dome traditionally made of hammered steel, which is placed over an intense fire, in which food is stir-fried, boiled, fried—in fact almost any cooking can take place in a wok. Today they often have long handles and even come with flat bottoms designed for use on Western burners, but traditionally they have only two small, ring-shaped handles protruding from either end. The technique of stir-frying is meant to quickly sear the ingredients while constantly scooping them up and mixing them around with a metal, spatulalike implement. The wok is designed especially for food cut into small pieces, which cook rapidly, and onto which sauce ingredients can be added toward the end of cooking. The wok is also used for deep frying, and even steamers are set in the wok over boiling water. Although archaeological remains of ancient woks are scarce because iron rusts and disintegrates, the wok is believed to date to the Han dynasty (206 BC–AD 220) or earlier, when steel-making technology was known in China. Interestingly, small, ceramic votive woks have been found in burial tombs for the dead to use in the afterlife, evidence that woks were commonly used in the ancient kitchen as well.

Like cast iron vessels, a wok must be seasoned before use (figure 2.10). This is done with oil or fat, rubbed into the surface of the heated wok repeatedly to create a polymerized and virtually nonstick surface. Today this can be done in an oven or even outdoors in a barbecue grill, because it creates a lot of smoke. Although professional cooks today use a wok set into a specially designed hob over an intense, searing heat, home cooks

Figure 2.10. The shiny surface of a newly seasoned wok.

have traditionally used small, hot fires, especially in places where firewood is scarce. Being a cooking technique suitable for very quick cooking, it is thus ideally adapted to the limited fuel resources of China, especially in an urban setting, and of course this in turn has determined the cutting techniques and ability to eat with chopsticks. That is, the entire approach to cooking in China is to a great extent determined by the scarcity and expense of wood (or nowadays gas), the development of the wok, and the need to cook and sauce ingredients quickly. It also meant that a knife would never be needed at the table because food was always cut into bite-sized pieces before service.

There are also a number of specialized pots and pans such as kettles for boiling water, oval pans for cooking fish whole, vessels used to cook with an enclosed chamber set into hot water, as with the bain marie. There are also vessels that steam grains above another pot of simmering ingredients, similar to the North African *couscousier* used to make couscous.

Ovens

It is useful to think of an oven as just another tool for transforming ingredients. Although ovens had been used since prehistoric times, they are a relatively expensive way to cook food, using a great amount of fuel, and thus were not common in most households until the modern era, even though there might be a communal oven for a village. In China they are practically never used in a home kitchen. Ovens can be freestanding outdoors, as in the Mexican *horno* or Italian bread/pizza oven, or they can be set into the side of a hearth, into which hot embers were placed from the fire. An oven usually has a domed chamber, sometimes with a flue, in which wood is burned and then raked out. The walls of the oven serve as insulators retaining heat and may be made of simple mud bricks or stone. The advantage of the oven is that some foods can be cooked rapidly during the early stages of cooking while other foods in covered pots can be left to simmer slowly. The Italian *cassola*, for example, can be placed in one part of the oven, while a *joint* of meat can roast in another, and a quick pizza in yet another. The flavor imparted by wood smoke is also unmistakable in Italian dishes *al forno* (cooked in the oven).

Oven technology changed dramatically in the modern era, with indoor ovens burning coal and later gas or electricity. Although baking traditions require an oven, neither Italian, Mexican, nor Chinese cookery rely on an oven for most daily recipes in the average household. This has changed recently as ovens have become more affordable, but most cooking is still done on the stovetop.

Baking Tins, Sheet Pans, and Molds

Tins, as the name suggests, were traditionally made of tin, a lightweight metal that can be rolled into sheets or cut and soldered into square shapes. Tins are essential for baking traditions in Italy, Mexico, and to a much lesser extent China. Think of brightly colored

Mexican wedding cookies, biscotti, and macaroons. All such confections depend on tin sheet pans and cookie sheets. Tins can also be rectangular with straight sides for baking bread or meat loaves, patés, and the like. They also come in various elaborate forms such as bundt pans, muffin tins, springform pans for cakes and tortes, pyramidal molds with rosette forms, and other extravagant shapes limited only by the imagination of the tinker. They can be used to mold gelatins and aspics (vegetables or meat set into cold, flavored gelatin), which when turned out retain the shape of the mold. Ceramic and glass forms work equally well. They may range in size from small ramekins used for flans to oval gratin dishes, to large casseroles used to hold lasagne or baked ziti, to enormous bombe forms used to make towering presentation dishes. To a great extent, Italy and Mexico retain some of the habits of ostentatious display as remnants of their aristocratic past, and at least for special occasions food molded into towering forms recalls bygone eras and opulence.

Spatulas and Cooking Spoons and Ladles

Spatulas are an indispensable utensil in the kitchen, far more important in fact in Chinese cooking than in the West. A long-handled metal spatula (*chan* or *guo chan*, meaning "wok spatula" in Mandarin), slightly concave and with small rims on the sides to hold food, is used not only for stir-frying but also for passing ingredients both dry and wet from bowls to the wok (figure 2.11). While the left hand steadies the wok by one small handle and occasionally makes a flipping motion, every other movement takes place with the spatula, which becomes a kind of extension of the arm. The Chinese ladle is its counterpart and may be used interchangeably depending on the consistency of the dish; soups obviously require a ladle. Restaurant cooks tend to prefer the ladle in general, especially for moving liquids and sauces into the wok while cooking. Ladles are used in all these cultures for serving soup.

In Italy spatulas come in many sizes, always with a short handle, some with holes in a solid square of metal. There they are really only used for flipping flat foods and transferring them from the pan to a plate. An **oven peel** is essentially a huge spatula, used for transferring bread or pizza in and out of an oven. Far more important than the spatula is the **cooking spoon**, traditionally wooden, ideally of olive wood. This is used for stirring sauces, moving cooking ingredients around in a pan—virtually any motion that requires the food to be stirred.

Whisks

For whipping cream, egg whites, or for incorporating batters, the **whisk** is an indispensable tool. Such preparations only became common in the past few hundred years, and so these are mostly modern implements. Whisks are usually made of thick wires bent into hoops and bound in a wooden handle; they may be large balloon whisks or tiny. There

Figure 2.11. A battery of tools used with cooking in a wok.

are newer plastic whisks as well, with various hoop patterns. In Mexico a related device dating back to the colonial period is used to froth chocolate: the *molinillo* (figure 2.12). It was invented as a way to replicate the froth so appreciated by the Aztecs on the top of a good pot of chocolate. The froth was achieved only by pouring steadily from a great height. This is a neater way to do it, as the *molinillo* is merely a stick of wood with several round disks at the end, and when rubbed back and forth between the hands creates a froth.

Rolling Pins and Related Devices

A **rolling pin** is basically just a thick dowel of wood used to roll out sheets of pastry or pasta. It can have handles that rotate using ball bearings and can also be made of marble, metal, or a synthetic material that prevents sticking to the dough. Rolling pins have always been used in Italy to make pasta, and there are even pins with vertical notches for cutting pasta and others with squared-off edges for cutting sheets of ravioli. Today there are also mechanical pasta machines that use two metal rollers into which the dough is fed in successively tighter settings, yielding a thin and even dough. Attachments for cutting

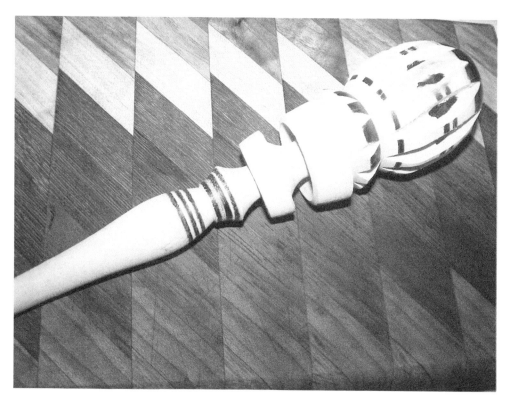

Figure 2.12. The *molinillo* used to froth up chocolate.

various widths of pasta, or for cutting ravioli shapes, are also available, and the machines can either be hand cranked or powered with a small electric engine. These are suitable only for flat pasta shapes, but there are also machines that extrude pasta with dies for macaroni and myriad other forms.

In China, rolling pins are comparatively smaller and are used usually for small knobs of dough, wrapped by hand into dumplings (figure 2.13). They are also darker and made of much harder wood than European rolling pins and naturally are not suited for rolling out large amounts of dough at once, which is unusual in Chinese cooking, except in professional kitchens and for rolling certain types of noodles. For some dumplings, however, the dough is not rolled but flattened with pressure with the side of a cleaver on an oiled board, which allows the use of a moister and more tender dough, usually made of rice starch and tapioca. The effect is comparable to rolling but does not require flouring the board.

Modern Mexicans also use rolling pins for pastry, but the traditional solution to making flattened dough, as with tortillas, is entirely different. Here a small, levered metal press is used (figure 2.14). A knob of dough is placed between the two disks, today lined with plastic, and the dough is squashed into a round shape.

Figure 2.13. A Chinese rolling pin is used to roll small knobs of dough rather than large sheets.

Extruders

An **extruder** is surprisingly not a modern invention, as various syringes have long been used in Italian and Mexican cooking. The principle is simple: a long cylinder, the extruder, is filled with a batter or other ingredient, which is forced through a narrow aperture by means of pressure exerted by a piston. Fritters are made when batter is pressed into hot oil; dumplings into water or soup. The ubiquitous Mexican churro is made either with a syringe or a pastry bag with a star tip. Among other common extrusion tools we might include the garlic press, the citrus press for squeezing lemons and limes, and even ricers for potatoes.

A **meat grinder** is also a kind of extrusion device, mincing food through die cutters with a crank and screw mechanism that is used for meatballs or filling sausage casings. In Italy, more complex extrusion systems were developed with cut brass dies and screws that force a more stiff pasta dough through the aperture. Today, these are mechanically driven and come in small models for household use or huge industrial-scale pasta makers (figure 2.15).

Figure 2.14. A tortilla press.

Modern Food-Processing Equipment

Revolutionary food-processing equipment was introduced in the twentieth century. The **blender**, used primarily for pureeing soups and sauces, has replaced the mortar in most recipes requiring liquefaction. For example, most Mexicans today use a blender for processing chilies for sauce. They also grind nuts, puree soft fruit, and so forth. The **food processor** is a more recent invention. With larger blades and a more-shallow bowl, it can puree and grind dry ingredients and, with various blade attachments, can slice, grate, or shred ingredients. It has not only rendered many laborious and time-consuming procedures of the past obsolete but also increasingly the archaic implements as well. Likewise for the electric mixer, which can handle large amounts of batter or dough effortlessly. These tools have made many formerly difficult procedures accessible to the home cook, but they have also influenced professional chefs who need no longer hire underlings to do the most tedious tasks. Though food processors do much of the most difficult work today, some purists insist that hand-processed ingredients are superior.

Figure 2.15. A late-nineteenth-century manual meat grinder with hand crank.

In twenty-first-century professional kitchens, a number of scientific instruments are also being used to process and cook food. **Centrifuges, lasers,** machines that freeze at subzero temperatures using liquid nitrogen or denature proteins, and countless other devices designed for the laboratory or food industries are now used in experimental cooking, known popularly as **molecular gastronomy.** The effect these have had is largely innovative, with ingredients transformed in ways unthinkable in the past. Compounded with the global availability of exotic ingredients and the tendency toward fusion cuisines, it is likely that this trend will spread from Spain, Britain, and the United States and eventually make its way to restaurants in our world cuisines. These impulses are tempered by the desire for local, fresh, straightforward, uncomplicated food, and these in dynamic interaction are the impetus for the evolution of cuisine. For every high-tech machine, there is a traditional technique being recovered.

Serving Technologies

A complete catalogue of serving vessels and cutlery would be endless and exhausting. Nonetheless, a number of important distinctions must be made among serving vessels around the world because to a great extent these determine not only the way food is consumed but also the texture and consistency of the food and ultimately how it is prepared. For example, a flat, fried cutlet of meat can only be served on a plate, whereas a stir-fry with bite-sized pieces of meat and ample sauce should be served in a handheld bowl on top of rice, even though it is incongruously eaten on a plate in the West. In every case, understanding how a food is served and consumed reveals why a particular cooking technique was chosen.

Bowls are the simplest and certainly most ubiquitous serving vessel on the planet. They can be made from virtually any material: wood, ceramic, metal, or plastic. But the subtle differences in bowl design are among the most important factors in understanding the internal logic of a cuisine and the physics of consumption. In Italy, for example, bowls generally have a broad foot because they are meant to remain stable on the table surface. They also have rims and may be so shallow as to resemble a concave plate, hence they are often referred to as a soup dish. This is the preferred vessel for serving *tortellini in brodo* (in broth) and other soups, offering a wide opening into which one can dip a soup spoon held in the hand horizontal to the table. As one carefully lifts the spoon to the mouth, the rim is designed to catch any drips. The culinary aesthetic likewise allows the diner to see the tortellini dispersed through the clear broth, perhaps garnished with sparse herbs. In some cases there may even be ingredients that require cutting, and this would only be possible in a shallow bowl with a flat base, which will not tip over as pressure is applied in cutting.

Fascinatingly, a comparable wonton soup in China is consumed in an entirely different fashion. Here the bowl has a narrow foot because the entire bowl is held in the hand, the foot resting on the fingers and stabilized by the thumb along the rim. This rule applies to bowls in general here. The bowl has no rim because it is held close to the mouth and the liquid and dumplings either scooped directly from bowl to mouth or the liquid sipped straight from the bowl. One would never consider a need to cut ingredients in the bowl, so a wide base is unnecessary and causes the soup to cool off much faster. There is also the aesthetic appeal of uncovering ingredients as one progresses through the bowl, a slice of pork here, a section of scallion there. The process of discovery is part of the pleasure, just as with the Italian bowl—it is seeing the arrangement in its entirety the minute it is set on the table.

Although there is no hard and fast rule about Mexican bowls, which have largely been influenced by Spanish fashions, bowls here are somewhere between the Italian and Chinese in form. That is, they normally do stay on the table and have a broad foot or flat base, but the opening is narrow, without a flat rim. The typical pre-Columbian earthenware bowl was normally a simple hemisphere, which can in fact be lifted directly to the lips held in both hands, in a way that a broad-rimmed soup plate never could be. How such vessels were used in the past must largely be guesswork, but the form nonetheless persists to this day. Many bowls also taper inward slightly at the rim, probably as a way to prevent the contents from spilling, though this may have been merely the result of forming the vessels by hand. When smoothing out the vessel walls from the interior, to enlarge the form, it is much more stable with an inward-turned rim than with a perfectly vertical wall.

Practically any bowl form can be found in all three regions, but these are the most sharply defined and salient differences. Large serving bowls must always have a broad, stable base. Smaller bowls for sauces and condiments likewise are shallow and enable the diner to dip in a spoon or piece of food. Either of these may also be furnished with a cover if the contents must stay warm, and with small handles or knobs if it is to be carried to the table. In many Asian cuisines, a smaller bowl is placed atop a larger one so food stays warm as it comes to the table.

Similar distinctions can be made in cup forms. The Chinese cup used for drinking tea is handleless and must be raised to the mouth, secured both by the foot and rim, with the hand positioned at a slight distance from the hot surface of the vessel. Firmly grasping the cup can also serve to warm the hands. Lifting the hot cup to the lips must be done carefully and with concentration. This enables the diner to reflect upon the tea as the aroma wafts from the bowl in proximity to the nose. Drinking tea is thus almost meditative.

Although Aztec and Maya chocolate cups were handleless and sometimes more like bowls, which probably facilitated the frothing achieved by pouring from a height, today the cup more closely resembles a mug, though the sides may be straight and the vessel cylindrical, or it can taper toward the base—a form common in colonial *majolica* (tin-glazed earthenware). A wide opening enables the diner to dip biscuits or churros into the drink. Thus, it is set on a saucer. In a way, as with many Mexican beverages such as *atole* (thickened with corn) or *champurrado* (with corn and chocolate), this is less a drink than a sustaining liquid meal. Thus, the cup is capacious and sturdy and interestingly differs significantly from European chocolate cups, which tend to be elegant, dainty, and more baroque. A wide variety of forms are nonetheless used nowadays in Mexico, but the **mug** appears to be the quintessential form.

Compare these forms with the ubiquitous **cup** of espresso in Italy. This is a tiny, shallow, cylindrical cup with an equally diminutive handle, designed to knock back a jolt of caffeine quickly while standing at a bar and then move on. Italians use a variety of cup forms for various beverages, as well as larger hemispherical coffee cups when milk is added or for cappuccino, but the espresso cup reveals something very characteristic about Italian culture and its pace. Coffee is a drink to be taken publicly, en route, in a busy city between other tasks, not something to linger over meditatively, and certainly not a nutritious meal. These differences have been described as extreme stereotypes, but they point not only to essential contrasts in the approach to food in these three countries but also why the cup form takes such radically different shapes.

Plates

The **plate** is the other universal serving implement, and anything flat can serve as a plate—from a large, sturdy leaf, to a wooden trencher (square or round plate), to more sophisticated ceramic and metal forms. Plates are round out of long-standing custom, a rudiment of once being turned on a pottery wheel, but they can also be oval, square, or any shape. The common plate was used in Italy since antiquity, and from the Late Middle Ages it also became a decorative surface for *majolica* painting. This tradition was also taken to Mexico with the Spanish, who in turn had learned it, along with other culinary traditions, from the Moors. But the idea of each diner receiving a plate at an individual place setting is an innovation of the Early Modern Era (c. 1500 to 1800).

Plates are vitally important in understanding the physics of eating with a knife and fork in Western culture. Food must either be flat or of a thick-enough consistency to adhere and not run off the edge. A broad rim and slightly depressed central surface prevent this when sauces are added; nonetheless, the plate enables the cook to serve larger pieces of food, which are cut up by the diner. In practice various prepared foods apportioned to each diner are kept distinct on the plate, sometimes with an accompanying starch or vegetable or a garnish. Thus, rather than mixing various ingredients to form a complex whole as in Asian cookery, plates encourage the separation of food modules into a centrally featured meat and accompanying "side dishes" that ideally complement it. There are many exceptions to this, a stew being the most obvious, but in general the Italian and Mexican mode of service is largely shaped by the use of plates. Furthermore, the modern custom of bringing a main course directly from the kitchen already arranged on a plate, what is called service à la Russe (Russian style), further fostered this as the predominant serving utensil. In households food is usually placed directly on the table and then portioned or passed around, more resembling the older service à la Française (French style), but the plate is still indispensable.

Plates also come in countless sizes, from small appetizer plates, bread plates and saucers, to large platters. They can be made of pewter or silver and can come equipped with a charger or larger plate underneath. In some respects, the ubiquity of the plate has actually hindered creativity in European cuisine. This is partly the result of the mass-produced aesthetic that requires that every diner receive the same setting on matching china, and this has limited the range of vessels that a restaurant or home would keep in stock. Inadvertently it has prevented experimentation with recipes that might be semi-liquid or that could contain various ingredients of contrasting textures and viscosities. On a plate the only way is upward, a trend that easily becomes exhausted. A vessel with compartments, or a hybrid between the bowl and plate, or indeed of any irregular shape, would open up infinite possibilities, which are only just beginning to be glimpsed in modern cuisine.

Cutlery

The standard table setting of **forks, knives,** and **spoons** is really a fairly recent innovation, dating back only to the sixteenth century in Italy. The use of the multitined fork as a regular eating implement was in fact invented in Italy. Its evolution from two-pronged kitchen forks with sharp tines is evident in that early forks, and forks in general in Europe, are somewhat sharp. They are held in the left hand with the prongs pointing downward and are used for spearing food, keeping it stable while the knife cuts the food in the right hand. The food is then transferred to the mouth with the left hand, the tines still pointing downward. This differs from American practice, in which the fork is switched with each cut to the right hand and food is shoveled into the mouth with the rather more flat tines pointing upward. Strangely, this is the result of colonial American culture evolving independently from Europe and before cutlery took its final form there. The difference is also evident in the manufacturer's hallmark, which in Europe is on the interior of the curve and thus not visible while eating. In the United States it is on the "bottom" of the fork as we conceive it.

Cutlery use in Mexico interestingly more closely reflects European practice, no doubt influenced directly by Spain and France. Nonetheless, a number of indigenous foods are still normally eaten with the hands, and remnants of earlier practices survive, especially in the use of a tortilla to wrap around foods or even to scoop them up. This predates the introduction of tableware entirely, and in deft hands makes the eating experience more direct and arguably of greater sensory pleasure. All the senses are engaged when one feels the tactile warmth of the tortilla in one's fingers and as the aromas waft upward when the bundle is brought close to the mouth. In a certain way, cutlery distances one from the immediacy of eating, placing a cold barrier of metal between the diner and the food.

The Chinese found an entirely different solution to eating elegantly without soiling one's fingers: by banishing the barbaric knife entirely from the table. Food is normally cut into bite-sized pieces, so it can merely be lifted with chopsticks. Chopsticks are usually made of bamboo or other wood or nowadays sometimes plastic and are blunted at the end, so they also serve as a kind of scoop for transferring rice from a bowl. That is, they are unlike wooden Japanese and Korean metal chopsticks, which are shorter and have more pointed ends. The two sticks are positioned in the right hand with the lower one stationary, resting on the crux of the thumb joint and side of the middle finger. The upper chopstick is parallel, held by the tips of the thumb and index finger, and it alone moves up and down to pick up morsels of food.

Although forks are superfluous when using chopsticks, there are spoons, mostly for drinking liquid soups. Chinese spoons are unique globally in being made of ceramic and having a flattened, boat-shaped bottom curving up into a concave handle. The shape is dictated entirely by the need to place them flat in a kiln for firing without the glaze sticking to the kiln shelf. Western spoons, as well as other eating utensils, can now be found everywhere, though. But the modern spoon is also a relatively recent implement, the result of advances in silversmithing technology and especially as they were hammered from a single flattened piece of metal, hence, flatware. Originally, up until at least the seventeenth century, spoons were quite different, with a teardrop-shaped bowl connected to a straight, cylindrical shaft and without the swoop of the modern spoon handle.

Last, knives have also undergone a subtle evolution. In the Middle Ages, each diner would bring his own knife, a sharp, pointed knife kept in a case, or later collapsible. This was used both for cutting food and sometimes for transferring food to the mouth. In the course of the early modern era, again with standardized tableware at each setting, the knife became blunt ended, presumably to pose less of a threat to the diner and companions, though steak knives remained pointed and sometimes are menacingly serrated.

Specialized eating implements also abound, mostly variations on the abovementioned ones: deep soup spoons, fish knives and butter knives, salad forks, and the more obscure long-handled marrow spoon, used for scooping out the interior of cooked bones. These mostly vary by size but sometimes have unique features, such as a sharpened edge, outward turning tine, or unique shape for eating specific dishes, such as the diminutive snail fork. These proliferated in the nineteenth century, but the tendency in recent decades has been to narrow the table service to the basic three or four utensils and even to combine them into all-purpose sporks.

Review Questions

1. Give examples of ways recipes are influenced by the vessels they are served in and the utensils with which they are eaten.
2. Explain why similar cooking methods arise in the three regions discussed here and how these relate to material resources such as cooking fuel.
3. How have developments in technology in the last century transformed the way the process of cooking is undertaken, for example, with timing and measurements?
4. In what ways has cooking technology influenced larger historical developments and social arrangements?
5. What kinds of sacrifices have been made in flavor and texture with the substitution of modern cooking implements? What advantages do they have nonetheless?
6. Compare cooking technologies in the professional kitchen with those used in private homes. How has the availability of time and labor influenced these distinctions?
7. Describe the way dining customs and implements determine the texture, consistency, and form of various recipes.
8. How are manners important in dining customs among these three culinary traditions? Why, for example, are certain implements not found on the table in some of the traditions?
9. How do the different serving vessels reflect different gastronomic values, in particular the speed at which one can consume the food?
10. How can serving vessels inadvertently hamper gastronomic appreciation of food or, on the contrary, aid it?

{3}
Grains and Starches

Learning Objectives

- Understand the central role of grains and starches in the three cuisines and how civilizations are built upon a grain staple.
- Understand why particular grains and starches predominate in each culture and how they spread to neighboring regions.
- Become familiar with ways grains and starches are used in each cuisine and why they differ based on material resources and grinding technologies.
- Learn how and why grains are also fermented into both raised and nutritionally superior products.
- Consider why certain grains have become relatively obsolete over the centuries.

Grains and Starches

Very broadly, a proper meal in Italy, China, and Mexico typically consists of three indispensable components. In fact, we might speculate that all food cultures construct meals out of three essential parts. First there is the starch or carbohydrate, historically providing the bulk of calories, at the center of the meal. The second part is the main dish, either a protein, vegetable, or beans, and last a condiment, which might be a sauce or flavorings in which the main dish was cooked. Should one part be missing it is not considered a real meal, but more of a snack, or at least an incomplete meal if it were merely starch and sauce or starch and a main dish without flavoring. But conceptually it can never make up any kind of meal without the starch base. Naturally, the meal can be elaborated to include several dishes and a multitude of side dishes and flavorings, but without the starch it cannot be a meal. Thus, we start our coverage of ingredients with starches, which in our three cuisines are principally grains but also tubers and other forms of carbohydrates. These are the foundation upon which these cuisines are based. In the Mediterranean the starch is **wheat**, in Mesoamerica **corn**, and in Asia **rice**, as well as the products made from these.

Moreover, the form in which these starches were prepared and served ultimately determined the physical layout of the meal, the kinds of recipes favored, and how people ate. Corn-based tamales and tortillas are meant to be eaten with the hands and are designed to hold the main ingredient and sauce inside. Bread, in contrast, is usually served beside the other foods, though it may sit beneath ingredients in a soup or be sliced and onto it other foods may be placed. Note, though, that bread is a separate, self-contained food, and this ultimately determined why in the West other foods are usually eaten apart from it, on a flat plate. Last, in Asia rice is served in a bowl as the base, literally and figuratively, for the whole meal. Without these starches as foundation and conveyance, the meal, and the entire cuisine, would not work. We might go further, to posit that the civilizations of Italy, China, and Mexico were only possible because of the domestication of wheat, rice, and corn, respectively.

This oversimplifies the story because all three starches are used extensively in the three cuisines. It would be remiss to neglect the importance of risotto (a rice dish) and polenta (corn porridge) in Italy; or wheat tortillas, bread, and Spanish rice in Mexico; or wheat and corn in China, for that matter. Nevertheless, the staple starches in each culture remain the foundation upon which every proper meal is based.

This chapter surveys the major starches in each cuisine, explaining how they are used in their native contexts and how each came to be prepared very differently, often by accident in cuisines elsewhere. That is, in the wake of global trade as it emerged in the sixteenth century, many ingredients were exported without the indigenous knowledge of how they should be prepared. For example, as we shall see, in Italy corn was not nixtamalized as in Mexico but ground into meal. This led to innovation as well as to the evolution of each cuisine, yet each remained true to its fundamental principles. This was not, therefore, a fusion of cuisines, but the borrowing and adaptation of basic ingredients following the logic inherent to each culinary system.

Wheat in Italy

Wheat was one of a handful of grains domesticated in the Fertile Crescent after 10000 BC. There were also several different species: *Triticum monococcum* or einkorn; *Triticum diococcum* or emmer; and many others, such as *T. durum* or semolina used to make pasta and *T. aestivum* used to make bread; and *T. spelta* or spelt, which was important in the prehistoric and ancient central European diet, though not in the Near East. The many wheat varieties used today for bread, characterized by a hexaploid genetic structure, are actually a cross between more primitive wheats and a wild grass, *Aegilops tauchii*. Only these have sufficient glutens, the protein strands that enable bread to rise. In other words, it was a happy genetic accident that enabled a type of wheat to be made into risen bread, the foundation of Italian cuisine.

Long before bread baking, however, grains were simply boiled whole, and they still are. Farro, a kind of emmer, is still cooked today. Emmer was the more important type of wheat in Classical times, along with barley. Emmer was also ground and made into a flat bread or brewed into beer. Grinding grains into meal and boiling them to form a kind of porridge or puls among the ancient Romans was the most typical early use of grains, especially for those without access to ovens. A more finely ground flour was made into flat breads, baked in ashes or on a flat griddle set over hot coals. These require no leavening and are still ubiquitous in the eastern Mediterranean. Even in Italy the *piadina* is a descendant, usually made only with olive oil, flour, and water. It is still pliable and can be folded over or rolled around cheese or meat, or equally flour can be made into a loose batter that is poured onto a hot surface, creating a kind of pancake, which in ancient times was called *placenta*, from the Greek meaning "flat cake."

The real miracle of bread, however, happens only with the entirely fortuitous confluence of flour, a wild fungus of the *Saccharomyces* species—or as we call them yeasts—and lactobacillus bacteria, which are also found everywhere in the wild. The earliest risen breads are easy to replicate at home. Simply leave flour and water out, replenish every day with more flour and water, and the wild yeasts will find and colonize the starter, making it strong enough to raise dough in a few weeks. Sometimes grapes or other fruit is also added; the powdery surface found on fruit is yeast. Until only a century ago, with the advent of monoculture yeasts, this is essentially how all bread was started, though sometimes brewer's yeast was used as well. There are many variations on this procedure, some using a wetter batter, others a "mother" or *poolish*, which is a wet mixture of flour, water, and yeast, while the Italian *biga* is a little drier. Nowadays they are made with commercial yeast, but in the past it would have been wild. The species present in the air differ greatly depending on the location and season. The sourness of certain sourdoughs can only be found in very specific microclimates, San Francisco's being but one of many.

Once the starter is ready, some of the raw dough is kept for subsequent batches to perpetuate the yeast. The light, aerated texture of bread is created by yeast as it ferments the carbohydrates and sugars in the flour, creating pockets of carbon dioxide gas held in place by the glutens. Kneading the dough elongates these protein strands, which supports larger pockets and achieves a greater rise. The gas also expands as the bread is heated in an oven, and the moisture remaining from swabbing out the ashes before baking creates a crisp crust. Salt is also usually added to bread dough to keep the yeasts from overfermenting, and it also adds flavor and improves texture. But some breads are made without salt, most commonly in Tuscany. Breads are also often slashed on the top before baking, which enables the expanding dough to rise farther upward rather than splitting at the weaker sides.

The first free-form breads probably resembled focaccia, which takes its name from the Latin *focus* (hearth). Unlike true flat breads these are made from risen dough but lack the internal structure to rise very high. Consequently they are cooked very quickly in a very hot oven. These are known by various names today; in Tuscany they are called *schiacciata*, often drizzled with olive oil. These are very different from the fluffy white bread baked on a sheet pan, doused in oil and herbs, and served as focaccia in the United States. Real focaccia is a thin round baked directly on the oven floor. It is also the ancestor of pizza, basically a flat bread with toppings. In the Calabria region of Italy there is also *pitta*, a round flat bread with a hole in the middle, a direct descendant of the ancient Roman *pictae*.

Pizza is now ubiquitous throughout Italy and around the world but is recognized as coming from Naples. In general pizza is quite simple in Italy, a round of flat dough—around Rome paper thin, thicker in the south, and very thick in the case of square Sicilian pizza baked in a tray. Normally the dough is tossed in the air using the upper knuckles of both hands until it stretches into a perfect round. It is then placed on a pizza peel—a flat wooden or metal square affixed to a long handle—and lightly swirled with tomato sauce using a ladle. A few slices of buffalo mozzarella or shredded cow's milk mozzarella are laid on top, and then a few basil leaves. This is the classic pizza Margherita. Other toppings are generally simple: a few anchovies, perhaps salami or prosciutto (pepperoni refers to a chili pepper in Italy), but almost never the mounds of ingredients found in the United States. Traditionally individual pizzas are eaten with a knife and fork, but today American style slices or square cuts meant to be eaten on the street can be found as well.

The range of risen bread forms in Italy is truly staggering. There are the thinnest *grissini* from Turin, a delicate breadstick, sturdy, football-shaped Pugliese breads from the south, and long *filone* resembling French bread. There is the "slipper"-shaped ciabatta, a relatively modern invention, with large holes and a firm texture, and light, finely crumbed, and sweet *panettone* studded with candied fruit and *pandoro* eaten during the Christmas season—something between a bread and a cake (figure 3.1). There is the huge round *pagnotta* cut with a hefty knife into thick slices, and the delicate rolls of the north called *michetta* made in a rose shape. Although more of a cake, *panforte*—a specialty of Siena—and *panpepato* are a thick, round "bread" made with dried fruits and orange peel, nuts, and spices and said to have been given to crusaders on their travels because they are virtually indestructible. Although bread is industrially mass produced and usually purchased in Italy, there has been a resurgence in traditional breads made in wood-fired ovens.

Interestingly and throughout history, the color of bread has been taken as an index of social status. Darker or whole-grain breads are the result of a simpler refining technique,

Figure 3.1. A fresh loaf of *panettone* cut to reveal the candied fruit.

and whiter breads are more finely sifted and bolted, and therefore more expensive (figure 3.2). Since at one time very fine white bread could only be afforded by the wealthy, it became a mark of distinction. Ironically, once modern milling techniques made white bread available to all, the simpler, more rustic dark peasant bread has been afforded a certain cachet, aided of course by the perceived health benefits of whole grains.

Bread is also used extensively in Italian cookery. It is simply sliced, toasted, and topped with a condiment such as tomatoes, pureed beans, or cheese as in bruschetta. It is used to form sandwiches, the now ubiquitous *panini*, which can be a simple filled roll with cheese or prosciutto or a more elaborate, press-grilled hot sandwich. Bread is also used to thicken soups, like the bean-based *ribollita* and *cacciucco*, a fish soup, or the simple *pancotto*—a bread soup. Breadcrumbs are used to top casseroles and to coat meat cutlets. There is even a bread salad called *panzanella* made with vegetables and olive oil. A meal without bread in some form is virtually unthinkable in Italy, and it is always automatically placed on the table in restaurants. Interestingly, according to custom, it is not placed on a separate bread plate as elsewhere.

The importance of bread in Italian culture cannot be overstressed, and this stems in part from the simple fact that bread is an indispensable element in the central Catholic

Figure 3.2. White flour is sifted several times in successively fine sifters.

ritual required by the Church for salvation: the holy mass or communion. The ritual commemorates the Last Supper in which Jesus asked his followers to remember him when they broke bread and drank wine. He said "this is my body" when referring to bread, and since the Lateran Council in 1215, this was taken to be a literal transubstantiation of bread into the flesh of Christ, which is in turn consumed by the communicants, infusing them with grace or forgiveness of sins. The host is normally a thin, consecrated wafer of wheat flour, in structure comparable to the Passover matzo, which would have been present at the Last Supper. Nonetheless, it is still a form of bread, and its consumption, weekly or more often, is a necessary sacrament. Even in the daily context of ordinary meals, bread is treated with respect and reverence. It is understood to be indispensable to life and is never wasted. It is the same respect afforded rice in Asia and corn in the Americas, and these are equally imbued with symbolic religious significance.

The other major way wheat is consumed in Italy is in the form of pasta. There is much debate among food historians regarding the origin of pasta, whether the flat sheets of dough called *tracta* and *laganae* used in Roman times crumbled into various dishes and cooked would qualify or whether pasta was introduced from the Arab world in the Middle Ages. In either case, it was well known by the time Italian explorer Marco Polo visited China, as he remarked that they have pasta just like at home. Fresh pasta is normally made with soft flour, which is low in protein and makes rolling easy, mixed with eggs, water, and salt. This can be cut into ribbons such as fettuccine in virtually any width, ranging from linguine to broader tagliatelle and pappardelle. The dough can also be rolled around a small iron rod to make hollow tubes called *ferretti* or *ferrettini* (*ferro* is the word for iron). Whole sheets can also be boiled, then baked with other ingredients, as in lasagne. Fresh pasta is also used to make filled dumplings such as ravioli, which means "little radish," presumably because these were originally bundled into little purses resembling radishes, though today they are square with serrated edges (figure 3.3). *Anolini* are little half-moon-shaped dumplings filled with braised beef from Emilia-Romagna. Tortellini are a pasta stuffed with cheese and prosciutto whose ends are pinched together so they resemble little belly buttons, served either in a broth or *asciutta* (dry), though with sauce. Fresh pasta dough can also be unrolled and worked by hand into *cavatelli*, which are ovals resembling little hot dog buns, little ear-shaped *orecchiette*, bell-shaped *campanelle*, hat-shaped *cappelletti*, and countless other shapes.

These fresh pastas are all distinguished from dry pasta shapes that are made with durum semolina flour, which will not crumble when dried and packaged. Dough made from it can also be extruded mechanically, and there are hundreds of shapes, including

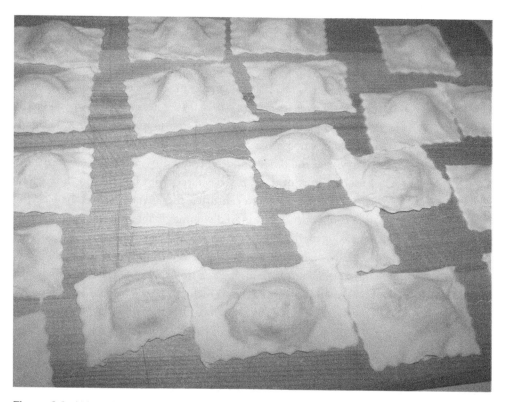

Figure 3.3. Although the name means "little radishes," the shape of ravioli has evolved into a pillow form.

GRAINS AND STARCHES

spaghetti; tubular macaroni; penne; rigatoni; ziti; and squiggly, fuselike fusilli; *bucatini* and *pici*, which are long, thick, hollow shapes; as well as farfalle (butterflies), commonly called bowties in English; *conchiglie* (shells); *rotelle* or *ruote* ("little wheels"); and radiatore ("radiators"). To add further confusion, similar pasta shapes may have completely different names from region to region, and some forms are entirely unique to one area, though this has largely changed as a few large companies now make most of the dried pasta eaten throughout Italy. The proliferation of shapes has simply been the result of ingenious manufacturing and marketing, and the invention of colorful names for each pasta has naturally followed suit. Italians do contend, though, in all seriousness, that certain shapes are designed for specific sauces. Concave shapes hold thick meat sauces better; long, smooth shapes are better suited for relatively homogenous and thinner sauces.

The important thing to remember, though, is that pasta is not a side dish in Italy but a first course, eaten before the main course. Thus, it is served in much smaller portions than is common in the United States and is rarely drenched in sauce. The flavor of the pasta is appreciated on its own, and the sauce merely adds flavor. Pasta can also appear as a main dish, especially in lasagne and other baked dishes (*al forno*), which can include various meats and sauce, or pasta even baked into a pie, as in the *timballo* of macaroni. Again, there is a logic to which sauce is used on each pasta. Fresh, filled pasta shapes are usually only napped with sage butter or a light cream sauce because they contain savory ingredients. Dried pasta shapes are usually served with heavier sauces; the more ridges or open space in the pasta, the thicker the sauce with more varied ingredients, which can be caught up in the folds. Thus, an *amatriciana* with pancetta and tomatoes would accompany macaroni; a ragù alla bolognese with meat sauce is best on ridged but unfilled pasta. In Italy the term *pasta asciutta* or *pastasciutta* refers not to dry pasta but to any pasta served in a sauce as opposed to being served in a soup. The latter are usually smaller shapes that can be picked up with a spoon, or filled tortellini.

Italian cakes and pastries are the other important uses of wheat. Italians were most likely the inventors of many forms of flaky pastries laced with butter, and the earliest recipes go back to the fifteenth and sixteenth centuries; these were used as the base for *torta* (open-faced pies) that could include any ingredient imaginable. The *torta pasqualina*, an Easter tart from Liguria, is a modern descendant containing beet greens, spinach, and ricotta, as well as whole, hard-boiled eggs symbolic of the Resurrection. The dough is usually made with oil rather than butter, though. *Pasta frolla*, a kind of shortbread, is also typically used for both sweet and savory tarts and includes butter, sugar, and egg yolk. It dates back to the Renaissance, if not earlier. So, too, does the *pasta sfoglia* comprising leafy sheets, though it is made very differently from French puff pastry and is more akin to Greek filo dough.

Pan di Spagna (sponge cake) was most likely introduced by the Spanish, as the name suggests, as was a kind of *pate a choux* laden with eggs. But the Italians were recognized as the masters of various kinds of fritters containing fruits or flowers. Fried dough also came in other forms, such as tube-shaped stuffed cannoli, or poufy *zeppole* from the south, a fried, light batter dusted with powdered sugar. A simple bread dough containing lard is also cut into lozenge shapes and fried in fat, called a *gnocco fritto* in Modena, not to be confused with boiled potato gnocchi.

There is also an enormous range of Italian cookies, flavored with anything from almonds to rosewater, anise, chocolate, and so forth. Amaretti are a well-known kind of macaroon made from egg whites and apricot kernels, not, as is often assumed, from almonds. Biscotti (or *cantuccini*) are another popular cookie, made by first cooking a loaf of dough, letting it cool, and then slicing it and recooking the slices until perfectly hard. *Biscotto* means "twice cooked," as does our English word *biscuit*. Savoiardi are delicate ladyfingers, used as the base for tiramisu. *Mostaccioli* are a cookie whose name seems to come from grape must (pressed juice, skin, and seeds of grapes before they ferment), with which they may have first been made. Later they became a spiced cookie perfumed with ambergris and musk; today, they are fairly plain or flavored with anise. *Sbrisolona* is a crumbly kind of tart with nuts and sometimes chocolate, from Mantua, resembling a kind of huge, crumbly chocolate chip cookie.

Wheat in China

Wheat is also an important starch in China, especially in the colder climates of the north where rice does not grow well or water is not abundant. It was introduced there from Southwest Asia in ancient times, when it began to displace millet as the dominant grain. Its adoption in the north and south is typical of Chinese cuisine, which is generally willing to accept any new ingredient but treats it in uniquely Chinese ways.

Although there are many types of risen bread, especially flat breads in western China, dough in the East is usually made into steamed buns rather than baked, no doubt because of the rarity of ovens (figure 3.4). The buns were thus an adaptation using a Chinese technology as well as distinctively Chinese flavorings. They are often stuffed with pork, vegetables, or sweet bean paste and are known as *baozi*. The simpler, plain, rounded forms popular in the north as street food are known as *mantou*. A story says that these earned their name because a general during the Three Kingdoms period (AD 220–280) leading a campaign against the barbarian Man tribe decided to make an offering of a head—in fact made of dough—which nonetheless enabled them to defeat their enemy. Today, the buns are smaller and may even be deep fried and served with a condensed milk dipping sauce.

More important uses for wheat include the range of Chinese noodles, which are almost certainly an adaptation of noodles made originally out of millet or rice starch. Noodle dough is formed into steamed dumplings, in many ways similar to their Italian counterparts, as well as noodles cut into long strands like spaghetti. Unique to this cuisine, however, is the pulled noodle (figure 3.5). Essentially a long snake of dough, often made with alkaline water containing lye, the dough is repeatedly stretched and folded over, creating longer and longer protein chains. It is then floured and stretched

Figure 3.4. Steamed buns for sale in Shanghai, China. © Michael Freeman/CORBIS.

between the arms over and over again, creating long skeins of thin noodles, progressively thinner with each round until they resemble wisps of hair. These may then be added to soup or flavored with soy and other ingredients. There are also fresh egg noodles, typically in soup or stir-fried, as in *lo mein* (*Lao mian* in Mandarin), combined with slivers of meat and vegetables. As the Chinese wheat noodle replicated forms that were already being made long before contact with the West, their development must be considered entirely independent of Western noodles, even though superficially some shapes seem similar. There are also noodles made of wheat starch alone or in combination with other starches, all of which are unknown in the West. Because they are long, *changshou mian* noodles are often eaten as a symbol of longevity for birthday celebrations or anniversaries or auspicious occasions like New Year's.

There are also a number of noodles made using utterly unique techniques and flavorings, such as *dao xiao mian* noodles from Shanxi, which are made by slicing them from a block of dough directly into the boiling water. The ancestor of instant noodles or ramen is also a Chinese invention, known as *yifu* noodles, which are deep fried and reconstituted in hot water or broth. Not to be forgotten are the *dandan* noodle dishes from Sichuan peddled in the streets by men with a pole (*dan*) over their shoulder with a stove on one side and noodles on the other.

Wheat starch is also used to form the dough base for a number of steamed dumplings, often in combination with another starch such as tapioca. Starch is obtained from the wheat

Figure 3.5. Man making traditional Chinese noodles in his small shop, Guanzhang, Henan Province, China. © John Stanmeyer/VII/CORBIS.

by soaking and separating and was known also in the West as *amida*, but it was only used as a thickener, not a noodle base. Only Asians, used to making noodles from rice starch, figured out how to use it so. Although they may resemble Italian ravioli, wheat starch dumplings are formed quite differently. In one case, the starch is mixed with hot water and kneaded well. It is then rolled into a long, thin snake and divided into equal nubbins. Each lump of dough is then squashed into a thin round with the side of an oiled cleaver and peeled carefully from the board. It is then stuffed, in the case of *ha gao* (*Xia jiao* in Mandarin), with finely chopped raw shrimp (and might also include bamboo shoots). This is enclosed with an elegant pleat on the top and steamed in a cabbage leaf–lined basket.

This is only one shape among the innumerable dim sum recipes using wheat flour. The generic horn-shaped (or crescent-moon-shaped) dumpling is called *jiaozi*, which is especially popular in the north (figure 3.6). It is made with a thick round of rolled wheat flour dough. These can be boiled, steamed, or even fried as a potsticker (*guotie*). They can be filled with vegetables, meat, or fish and are usually dipped in a soy-based sauce including vinegar, perhaps ginger, rice wine, and sesame oil. According to legend, they were created by a famous physician, Zhang Zhongjing (AD 150–219), to treat frostbitten ears. Another story claims that they resemble the golden ingots used as money in the Song dynasty. They are eaten in the New Year for prosperity.

The thinner, square wonton and spring roll wrapper is also made of wheat flour and more closely resembles a Western noodle, as it is a stiff dough rolled out and cut,

GRAINS AND STARCHES

Figure 3.6. Chinese dumplings such as *jiaozi* are often pleated on top and steamed.

and used fresh. In the case of the former they are filled and added to simmering broth with other ingredients. With the latter, including what is in the United States called an egg roll (*Chun juan*), the square is rolled with the pointed end facing upward and the ends folded in, enclosing the contents in a tube shape. The filling might include cabbage, scallions, shrimp or pork, mushrooms, or virtually anything. When rolled, they are deep fried and served hot with a dipping sauce.

Wheat gluten, although attributed to vegetarian Buddhist monks in the Middle Ages, is today a modern industrial product, consisting essentially of the proteins extracted and textured. It serves as a meat substitute known as *mian jin*. The gluten is cut and flavored to resemble chicken, fish, and other meats, sometimes with striking realism.

Wheat flour is also used in a variety of flat breads typical of western China and in pancakelike wrappers, the most familiar of which is used for Peking duck or *mu shu* (*Muxi rou*) pork, which is made from a hot water–based dough rolled out rather than poured like a batter-based pancake. Wheat is also typically used in combination with soybeans to make soy sauce.

There are also a variety of Chinese pastries and confections based on wheat. For example, the mooncake, filled with red bean paste and a salted egg, or in some regions with lotus seed paste or nut paste, is formed of a wheat flour crust stamped with a symbol for longevity (figures 3.7 and 3.8). The crust can be either sweet and chewy as in Canton, or more like a flaky puff pastry or shortbread, though based on oil or lard. There are also a number of Chinese cookies, such as the sandy-textured almond cookie, which is normally made with a Chinese almond, a variety of apricot kernel. Fortune cookies are an invention of the Chinese American community, though their prototype may actually be Japanese.

Just as in Italy, and as we shall see in Mexico, the Chinese are also very fond of fried dough, doughnut-like confections and fritters. One of the more popular is called *youtiao* (*Yauh ja gwai* in Cantonese, meaning "oil-fried ghost"). It is a kind of cruller of two twisted strips of risen dough, served for breakfast or as a street snack. Legend has it that the dish commemorates a husband-and-wife pair of conspirators during the Song dynasty (c. 1142), who are eaten in symbolic effigy in the fried dough.

Wheat in Mexico

Wheat was introduced to Mexico by Spanish conquerors in the sixteenth century and prevailed in the colonial era. It was always considered by the upper classes to be more

Figure 3.7. A cut mooncake, revealing the insides.

Figure 3.8. The writing imprinted on this mooncake says it comes from Jing Du (the capital city).

refined and elegant than corn-based tortillas and tamales eaten by the indigenous population. Even after centuries of intermarriage and mixing and attempts to retain a racially distinct identity as Spaniards, among *criollos* (creoles) and *mestizos* (those with Indian and European blood), wheat products retained a social cachet. To this day bread is just as popular as corn-based dishes, and Mexicans regularly purchase baked goods at their neighborhood *panadería* and *pastelería*. Mexican breads are mostly direct descendants of forms first introduced by the Spanish in the colonial era. They tend to be sweet, incorporating eggs, and the lighter and fluffier the better. Little rolls (*bolillos*) are a common accompaniment to meals or eaten for breakfast. There are also various kinds of *pan dulce* (sweet bread), which are often sold covered with a sweet pink or yellow icing. But Mexicans are also fond of the same kinds of bread enjoyed in Europe, from a French-inspired brioche to the shiny *brilloso* and ring-shaped *rosca*, as well as a fluffy, mass-manufactured white bread called *Pan Bimbo*, the equivalent of Wonder bread.

Not surprisingly, many recipes using bread are based on medieval Spanish dishes, the *capirotada* being the best example. It is a kind of sweet bread pudding with raisins and almonds, or savory with cheese and spices. The original would also have included meats. The Spanish also introduced many other sweet wheat flour–based pastries;

doughnut-like fritters (*buñuelos*) are fairly ubiquitous. So too are *churros*, which are derived from early modern syringe-extruded choux-paste fritters and are now a standard breakfast food, dipped into hot drinking chocolate. *Pastel* is also a word used through Latin America to refer either to a cake or a kind of pie. When smaller and folded over, the name is *empanada*. However, in the Caribbean, that word is used for a corn-based kind of steamed pie.

Perhaps the most original culinary invention of Mexico is the flour tortilla, typical of the north. In form and usage it may differ little from flat breads of the Middle East and India as a tool to pick up foods, but it is also wrapped around food the way a corn tortilla would be. Thus were born various inventions on both sides of the northern border, such as the burrito, a wrapped flour tortilla containing meat, beans, vegetables, cheese, or any combination of these, and the chimichanga, the fried version. The huge overstuffed burrito is a U.S. invention, though.

Rice in China

In southern China rice and food are synonymous. *Chi fan* (literally "eat rice") means simply "to eat." When greeted one asks not "How are you?" but "Have you eaten rice?"

As with wheat, rice is practically sacred and is understood as the foundation of the entire cuisine and the supporter of life. In the south especially rice provides the bulk of calories and is present at every meal. It is considered one of the seven basic necessities in every household, along with firewood, tea, oil, soy sauce, salt, and vinegar. There are very good reasons why rice became the staple in China, not only because of available technology for farming in wet rice paddies, which create a kind of miniature ecosystem including fish and water fowl, but also because rice can support more people per square acre than any other grain staple. The presence of rice as a dependable crop is both a cause for rapid population growth and a suitable means for supporting it.

The Chinese almost always steam their rice without flavorings. Dehulled and polished, rice is white after cooking. This whiteness was used as a mark of distinction, as the poor ate brown or unpolished rice. This so-called polished rice has been problematic, though, as much of the nutrients are stripped in the modern processing, which has led to the vitamin deficiency disease beri beri. Ironically, the cheaper, less-processed rice is more nutritious.

Rice is a grass in the genus *Oryza*, with many wild species scattered around the globe, though the cultivated forms are native to Asia. Normally Chinese cuisine uses a medium-grained rice (*Oryza indica*, as distinguished from *japonica*, which is even shorter and stickier) that still sticks together slightly so it can be eaten with chopsticks, and it forms the base in a bowl onto which other foods are placed for eating. This rice should be distinguished from sticky rice or glutinous rice (*O. sativa*, var. *glutinosa*), which is a shorter and sweeter grain popular in Thailand and Laos, but it is also used in many Chinese dishes, particularly sweets. There are also varieties of red and black rice used mostly in sweets, whose color comes from the intact hull, which in most rice is brown. Red rice is also fermented in Fujian on the coast into a product called *hong zao*, which tastes sweet and sour and is used in other dishes.

Rice is "steamed," so to speak, by first washing and draining it several times, then putting it in a pot with water about an inch above the level of the rice. Actually, the water level should rise the length of one segment of the index finger. The amount of rice used and the size of the vessel are unimportant, but a large pot works best to prevent boiling over. This is brought to a boil, then lowered to a simmer for about fifteen minutes and left to rest before serving. Technically only the last few minutes should be considered steaming, once the water has been almost completely absorbed. In some places there is still practiced what might be the earliest form of rice cookery. Rice and water are poured into a segment of bamboo, sealed, and then roasted in the fire. The bamboo is sliced open and served. Similarly, packages of rice can be wrapped in banana leaves or another large leaf with various other ingredients, and then steamed.

Rice is also made into a breakfast porridge called *juk* (*zhou* in Mandarin) or sometimes called *congee*. The raw rice is basically just boiled for several hours until it disintegrates. To this is then added meat or fish, preserved vegetables, or spicy, chili-based condiments. It is smooth and comforting, considered very nutritious and easy to digest. It is also the first food given to infants.

As with bread in the Mediterranean, rice is never wasted. Children are warned that if they leave rice in their bowl they will grow up to marry a pockmarked girl, or if they accidentally spill some on the table, the god of thunder will punish them. There is no waste in the kitchen either: leftover water from boiled rice is saved and used as a cooling drink. Leftover rice, especially after it has dried out a bit, can be stir-fried. The rice is placed into a hot wok and stirred with oil until golden. Then one adds scallions, peas, or other vegetables, some cut-up egg omelet, perhaps some diced roast pork or ham, soy sauce, and rice wine. Strangely, the end result is distantly similar to other rice dishes around the world using a melange of vegetables, meat, and beans, though the technique is entirely different, and so, too, is the flavor.

Rice is also made into a number of other foods and ingredients. It is ground into rice flour and rice starch, which are used to make noodles and rice paper. Rice paper is actually edible and is used for the bottom of sticky candies. Rice is also fermented into wine, consumed as a beverage and as a cooking wine called Shaoxing (or just *huangjiu*—yellow wine), and made into vinegar, some dark and deeply flavored. It should be remembered, though, that although rice is eaten as the primary staple by roughly half to two-thirds of the Chinese, wheat still predominates in the north, and many regions also depend on millet. Sweet potatoes are eaten in the southeast and corn in the southwest.

Basic rice noodles are sold in dry form but can also be formed from a fresh dough. In this case, rice starch is processed with hot water similar to the process for wheat starch noodles. They are often bought in large, fresh sheets (*he fun*), sliced when added to soups or stir-fries, and have an appealingly chewy texture. Sticky rice flour is also used to make dishes such as *nian gao*, a flat, steamed cake sandwiched around a layer of bean paste, eaten on New Year's because the word for this sticky cake sounds like "grow taller every year." *Tang yuan* are another sweet for the Winter Solstice and Lantern Festival. They are made from glutinous rice flour and are basically a round dumpling filled with sesame paste or peanuts, red bean paste, jujubes (a datelike fruit), or even a pork meatball. They are served in the water in which they were cooked, resulting in a kind of soup, often sweetened. The name sounds like *tuan yuan* in Mandarin, which means to unite the family, and this is precisely why it is served during reunions and special holidays.

Another traditional rice recipe uses sticky rice wrapped in bamboo fronds and steamed. Inside it may contain bits of pork or dates, lotus seeds, red bean paste, or a salted egg yolk. In form and function it is quite analogous to the Mexican tamale, especially as wrapping the contents was usually done communally by women and the skills passed down from generation to generation this way. These four-sided or conical, carefully wrapped parcels, called *zongzi*, are universally eaten during the Dragon Boat Festival. Legend has it that the dish was created for the poet Qu Yuan (340–273 BC), whose warnings of imminent invasion went unheeded by the ruler, who was jealous of the poet's popularity. In desperation, the poet drowned himself, and his admiring followers created these parcels and threw them in the river so fish would not eat his body but eat the *zongzi* instead.

Rice in Mexico

Mexico essentially inherited the rice traditions of medieval Spain, which in turn derive from the Persian pilaf introduced by the Moors. Although Catalonia and Valencia on the east coast of Spain have their own indigenous tradition of paella, a thick, saffron-laden, short-grain rice dish with chorizo sausages, chicken, shellfish, and peas, this dish can be found in Mexico. More typically Mexicans prepare what is called Spanish rice; that is, a long-grain rice that is first soaked in hot water and drained, then fried in lard or oil until golden, to which is added tomato puree, onions, carrots, and peas, and sometimes annatto oil to color it yellow. This is cooked undisturbed with the cover on, again, much like a pilaf or Indian pilau. In Mexico it is often served as a separate course, after the soup and before the main course, but in popular North American versions it appears heaped next to refried beans as a ubiquitous side starch accompanying every single "main" course.

Rice in Italy

Although rice was known in the Middle Ages in Italy and was a highly valued import, it was mostly used in quasi-medicinal recipes ultimately inherited from Persian cuisine. Many medieval cookbooks include some form of rice pudding, cooked with milk and sugar, or as in the case of Martino of Como's fifteenth-century recipe, with almond milk for Lent. His recipes also call for rice flour, which was used as a thickener in, for example, the ubiquitous blancmange, a dish usually made with shredded chicken, almond milk, sugar, and rice starch.

It was not until the late fifteenth century that rice was extensively cultivated in northern Italy, specifically a fat, rounded variety that today is distinguished as several different types: vialone nano, carnaroli, or arborio. These are used to make the classic risotto, and the method is completely unique to Italy. Unlike the pilafs of Persian cuisine often considered to be an ancestor, the rice in a risotto is not fried to a golden brown first. The advantage of this technique in Persian and Indian cuisines is to keep the long-grained rice separate and dry—the pan is covered and the rice gently steamed. In the risotto, in contrast, onions and chopped vegetables are lightly fried first and the rice is added afterward, followed by ladlefuls of broth, continually added with the pan uncovered. The intention is to allow the rice to become smooth and creamy while retaining a bit of bite in the center, and Italian short-grained varieties are ideally suited to doing this. In the classic risotto Milanese, mushrooms are the main flavoring, though any number of ingredients can also be used. Often finished with a nap of butter and a grating of Parmigiano, the risotto is, like pasta, considered a first course.

Leftover risotto in used in a variety of ways—cooked into a tart, or better yet rolled into balls stuffed with cheese, dipped in egg and breadcrumbs, and then deep fried. These *suppli al telefono* (telephone cords) are so called because when bitten into the cheese creates long, stringy cords, though presumably in the age of cell phones this reference will become obscure. They are a common site on bar counters in Italy.

Corn

Corn, as it is known in the United States from the generic British term for grain, is a shortened version of the original "Indian corn" or "Turkey corn." More properly it should be called maize (*Zea mays*), like the rest of the world calls it. Corn is the result of a strange genetic accident to the plant teosinte, in which its reproductive organs moved to the side of the stalk rather than the top of the plant. The advantage of having its kernels enclosed in a protected ear meant that it was relatively free from predation by birds. Presumably early Mesoamericans recognized this, carefully selectively bred corn for increased size and yield, and after hundreds of years created what we know as corn, a plant completely unable to replicate itself in the wild.

Corn or maize was the staple crop of indigenous Mexican peoples and remains the most important carbohydrate source there today. As with grain staples elsewhere, it was treated with reverence, never wasted, and even handled carefully to prevent spilling. Although today, in the wake of NAFTA treaties, U.S. hybridized corn has begun to flood the market, Mexicans still appreciate the particular qualities of their specific maize breeds, long adapted to local climates and soils and adapted to specific culinary uses. Traditionally rural families kept a small garden plot (*milpa*) where they would grow their own corn and beans and would generation after generation use the land in ways handed down from elders. Modern agronomists divide corn into several different basic types: dent and flint, flour corn, sweet corn eaten fresh, and popcorn. These also come in myriad different colors (figure 3.9). This diversity may prove to be essential to the survival of the species, which now depends entirely on human propagation, especially as a narrowing range of hybridized or genetically engineered corn types have come to dominate the market.

Maize is most often processed by nixtamalization, which involves soaking the kernels in lime (calcium hydroxide) and a strong pH base (alkali) and then rinsing and rubbing off the skins. This is then ground by hand or machine into a dough (*masa*), and then either steamed in a corn-husk casing as a tamale filled with various ingredients or with a slightly differently processed dough, pressed flat into a round tortilla and cooked on a griddle (*comal*) (figure 3.10). Today, a dry instant *masa harina* mix is often used as well, though it is more difficult to use and tastes relatively uninteresting compared to fresh dough.

The tortilla is as ubiquitous in Mexican cuisine as bread is to European and rice to Asian cuisine, and in many ways its uses directly descend from pre-Columbian cuisine without influence from Spain. The tortilla is used as a basic, all-purpose bread accompanying other foods, or it can be folded in half as a taco or rolled into an enchilada, in

Figure 3.9. A diminutive blue corn used to make *masa*.

Figure 3.10. A woman cooks tortillas on a *comal* in Zinacantan, near San Cristobal de las Casas, Chiapas, Mexico. In the background, another woman rolls corn on a *metate*. © Melvyn Longhurst/CORBIS.

which the tortilla is fried and then moistened with a chili-based sauce. These dishes are the lasting legacy of pre-Columbian cuisine, and conquistadors in their accounts of Montezuma and his great feasts in Aztec times describe the same basic tortillas. The tortilla also serves as a kind of plate with ingredients on top as a tostada. Tortillas can be deep fried when rolled or cut up into tortilla chips and fried (*totopos*, from the Nahuatl *totopotza*, meaning "toasted"). Fried wedges or strips of leftover tortillas can also be simmered in a spicy green or red salsa, topped with shredded chicken or other ingredients such as cheese or a fried egg, and served for breakfast as *chilaquiles*, a word that descends directly from Nahuatl. These make a fine breakfast. Tortillas can also be cut up and added to soups, layered as a kind of casserole, or used as an ingredient themselves, as in sauces. There are also myriad *antojitos* (little snacks) made from corn masa shaped into little rounds covered in cheese, canoe forms, or bite-sized morsels.

Corn is also eaten fresh in Mexico, and one typically sees roasted corn on the cob (*elotes*) on a stick sold in the streets (figure 3.11). These are often slathered in mayonnaise coated with cotija cheese and sprinkled with chili powder, salt, and lime juice. The combination makes a lot of sense gastronomically, as the flavors all play nicely against each other. Or the corn can be cut off the cob (*esquites*) and served in a cup with similar flavorings. When considering this particular flavor profile, think of how the salty balances

CHAPTER 3

Figure 3.11. Grilled corn at a public market, Mexico. © Keith Dannemiller/CORBIS.

with the spicy, the sweetness of corn with the sour of the lime and the creaminess of cheese and mayonnaise or sour cream. This is a basic flavor principle that runs through much of the cuisine.

As a kind of by-product of corn, *huitlacoche* (*Ustilago maydis*) is a fungus that attacks the individual corn kernel, causing it to swell and darken. This is highly appreciated and used much as mushrooms are, in stews and wrapped in tacos. Its flavor is deep and savory.

Corn in China

Chinese cuisine may be one of the last places one can picture corn as a prominent feature, except perhaps as miniature "baby corn" tossed into a stir-fry. Nonetheless, corn is one

GRAINS AND STARCHES

of the leading crops in the warmer southwestern China, and it has thrived there for the past several hundred years since its introduction, probably by the Portuguese in Macao. Some scholars have even attributed the great population boom of the Qing dynasty in part to the arrival of corn. Corn is generally consumed cracked and boiled as a grain, in steamed cakes found in the northwest, or as a plain gruel replacing rice in a kind of *juk* (rice porridge). Fresh corn on the cob is also common. There have even been noodles made of corn starch, again a good illustration of how ingredients are normally adapted to local food culture without knowledge of their original context. In Shandong Lu–style cooking, corn is also cut off the cob and stir-fried.

Corn in Italy

As with so many ingredients taken out of their original context, corn was adopted in northern Italy in the course of the sixteenth century and misidentified as *granoturco*—coming from Turkey. It was used precisely as other lesser grains had been since antiquity: dried, ground, and cooked as a thick porridge (polenta). In places where corn became a staple, dangerous niacin deficiency diseases such as pellagra (meaning "rough skin") became common. Unlike the Native American practice of nixtamalization, coarse milling and boiling leaves much of the vitamin B content unavailable. As a side dish, accompanied by other foods, polenta can be remarkably delicious. Traditionally, the coarse meal is sprinkled by hand into a large copper pot filled with boiling water or broth. It is then slowly and continuously stirred, which nonetheless leaves a delicious, crusty layer stuck to the bottom of the pot. It can be supplemented with butter, and perhaps grated cheese, and then the whole pot is turned out onto a wooden board, or in truly rustic settings right on the table. Diners help themselves and top the polenta with sausages, stew, or roast birds. When leftover polenta cools, it stiffens and can be sliced and fried or grilled, a truly magnificent dish.

Potatoes

Potatoes are one of the most important starchy foods in Europe and North America today, but this was not always the case. They come from South America and were introduced to Europe after the sweet potato, probably around the middle of the sixteenth century, where they met with little enthusiasm. There are several possible reasons for this: difficulty growing them in the hot climate of Spain and Italy where they were first planted by botanists; the similarity to poisonous relatives in the nightshade family, which were familiar to these same botanists; and perhaps most importantly the lack of a culinary niche for potatoes. There was virtually nothing like them commonly eaten by Europeans. Food writers compared them, unfavorably, to truffles, considered excrements of the earth. It is no wonder that it took several centuries before potatoes were adopted, and then mostly in northern Europe. But they were eventually planted in Italy too, especially in the colder mountainous north, and as with so many other foreign foods, they were used exactly as other starches would have been.

The best-known potato dish of Italy is thus *gnocchi di patate*. Gnocchi merely means any solid but light dumpling, traditionally made of breadcrumbs or semolina flour in the shape of a knuckle (*nocco*), according to some etymologists. In this case, a peeled and cooked potato is mashed with a small amount of wheat flour, the resulting dough is rolled into a snake, cut into segments, and these are rolled across the back of a fork's tines or a cheese grater, creating ridged indentations in which sauce collects. These are boiled gently, and then served with a sauce or baked in a casserole with butter and cheese and served as a first course. Potatoes have also come to figure in Italian cuisine in other, simpler ways. They are simply roasted, fried, and also tossed into stews with other ingredients. They are also fried in a *frico*, a kind of pancake with melted Montasio cheese, onions, and stock.

Potatoes, ironically, were introduced to Mexico from South America by the Spanish. And typically they are consumed in much the same way as in Spain, in a potato omelet, called confusingly a tortilla, which means "little egg tart." They can also be used with fried chorizo sausages as a filling for tacos and other wrapped foods. Of course, today French fries and potato chips are common in Mexico but naturally should not be considered part of the native cuisine.

China is another place where potatoes were adopted only reluctantly, probably as late as the eighteenth and nineteenth centuries, in particular in the warmer, wetter mountainous regions along the Tibet border and around Sichuan, introduced there mostly by French missionaries. They are used much as any other vegetable would be, boiled or added to a simmering stew or even julienned and stir-fried in Sichuan and Hunan cooking. In most of the rest of China, though, they are relatively unimportant.

Sweet Potatoes

Botanically, **sweet potatoes** are completely unrelated to white potatoes and are the tubers of a flower related to morning glories (*Ipomoea batatas*). As the Latin nomenclature suggests, it is from these that we derive the name for potato, as they were the first encountered by the Spanish in the Caribbean and Mexico. In the United States we also confuse these typically with the yam (*Dioscorea*), which should only refer to an African tuber but is colloquially used for orange-fleshed sweet potatoes as well, probably because African slaves were the ones cooking and serving them. In Mexico the sweet potato is native and has been used since ancient times. *Camotes* in Spanish comes directly from the word *camotli* in Nahuatl. They can be roasted simply or peeled and cut up, then added to stews or soups. They feature most prominently in the Mexican Caribbean, especially Veracruz, where they can even be mixed into dough for *chalupas* (a crispy, deep-fried container onto which other ingredients are placed). The most famous use is in sweets such as *camotes de Santa Clara*, a soft candy from Puebla.

Spanish conquistadors frequently mentioned the sweet potato, enjoyed its flavor, which reminded them of jam, and were eager to introduce it to Europe. It was, in fact, the first American tuber to arrive in Spain, and it soon gained a reputation for being an aphrodisiac. The sweet potatoes never really found a permanent place in Italian cuisine, though. Surprisingly, sweet potatoes are very important in the Chinese diet. This has been the case since the late sixteenth century when they were introduced from the Philippines (where the Spanish had brought sweet potatoes) to stave off famine. Some scholars attribute the dramatic increase in the population of China thereafter to sweet potatoes, among other plants from the Americas. They are grown most extensively in the south and east, especially in the poor, sandy soil of the coastline. Nonetheless, sweet potatoes retain the reputation of being a famine food and are generally disliked. Naturally with such a reputation, they rarely figure in the more elaborate forms of Chinese cuisine but are usually simply boiled, or slices are dried and cooked with other foods. However, they do have a reputation for being good for the digestive system, and they are sometimes served in nostalgic Cultural Revolution–themed restaurants.

Other Starches

There are a number of other minor starches, some of which are very important around the world, particularly in the tropics, but therefore receive little attention in our three areas of focus. These include **taro** root (*Colocasia*), which comes from Southeast Asia and spread through Polynesia, though it is also grown in southern China. It was also known to the ancient Romans and is still grown in the eastern Mediterranean. There is also an American species of taro (*Xanthosoma*), commonly called *malanga* in the Caribbean. A similar American root, of enormous economic importance globally, is **manioc**, also known as yucca and cassava. There are both sweet and bitter varieties, the latter of which must be processed with laborious grating and soaking to remove the toxins. It is relatively unknown in the Western world, except in the form of tapioca. It has been used in southern Mexico for thousands of years, though, and was actually one of the first plants encountered by Columbus. Again, as a tropical plant, its use in the temperate Mediterranean and China has been negligible. The same is true of **sago**, a starch taken from the pith of palm trees. Nonetheless, the starch of these plants, as well as arrowroot, is also used as a thickener and may be used to make noodle wrappings, or dried in chips and fried, as in the case of shrimp chips.

Last, among the starches there are also a number of other grains or pseudograins, some of which remained isolated in their region of origin, others that have spread worldwide, though they remain of relatively minor importance. **Millet** is one, which, although among the most important of foods in ancient times, is today known mostly as birdseed or as an invasive weed (figure 3.12). There are a wide variety of millet species around the world as well. Foxtail millet (*Setaria*) was one of the first grains used to make polenta in Italy, and there are still some traditional cakes and cookies that incorporate it, such as *migliaccio*, originally made with millet and blood with sugar and nowadays usually

Figure 3.12. A once important grain, millet is now mostly used for birdseed.

much more tame and made with corn. In China, millet was most likely the first domesticated plant, more important than rice at first. The Celestial Emperor Huangdi (or Yellow Emperor, c. 2697–2597 BC) included it among his essential plants and is said to have initiated the art of cookery by teaching people to steam millet. In fact, the oldest noodles ever discovered by archaeologists were made of millet. Here millet and panic (a similar tiny, round grain) are made into porridge and various snacks and are even used for brewing alcoholic beverages. There are also species of American millet, which may have been domesticated even before corn in Mexico. In many locales, they still form an important part of the diet.

Sorghum, an African native, is another of the minor cereals that was widely cultivated, especially in China, before the introduction of corn. In fact, the plants look similar when growing, except that the seeds of sorghum are on top rather than enclosed in an ear. Its stalks are also a source of syrup and sugar and even used as a building material in the northwest. The seeds can be boiled in a porridge, or after processing to remove the outer coats (pearling), cooked like rice. Sorghum is also used extensively to make distilled alcohol.

Buckwheat is a minor starch but can be a staple in the colder north of China. It is actually a relative of rhubarb and sorrel, rather than a proper grain, but its tiny, dried, triangular fruits are cooked like grain and can even be made into noodles, which are especially popular in Japan as soba, but are also known in China. Buckwheat in Italy is known as *grano saraceno*, in a supposition of its origins, thought to be among the Saracens (Muslims), but it is actually from Asia, most likely Siberia and Manchuria. Although best known as kasha in Russia, or in the form of *galettes* (pancakes made of buckwheat) in western France, Italians especially in the alpine north (Valtellina) use buckwheat extensively as well. There buckwheat noodles called *pizzoccheri* are combined with potatoes, cabbage, and bitto or fontina cheese in a rustic dish. The grain can also be coarsely ground and cooked as a dark and fragrant polenta.

Amaranth is another grain formerly very important in the New World but largely displaced in the modern era. It is believed that because amaranth held an important place in the Aztec religion that Spanish conquerors consciously sought to end its cultivation in an effort to promote Christianity. This is unfortunate, not only because the grains are extraordinarily nutritious but also because even the leaves can be eaten as a vegetable. Normally the grains are boiled as a porridge, but they can also be popped like popcorn or fermented into a beverage.

Last, mention should be made of a few other grains that have always been of limited importance in our three regions: **oats** and **rye** in particular. These are more common in Northern Europe and the United States but have limited use elsewhere. There is, however, a *pane nero* (black bread) made with rye in the Aosta valley of northern Italy. These grains and all the aforementioned grains are gaining attention recently as health foods or as alternatives for those with celiac allergies to wheat gluten. Rye is fairly unknown in China, though oats are sometimes grown in mountainous regions. Job's tears (*Coix lacryma-jobi*) is a kind of grass whose leaves resemble corn. It bears round seeds that are used in necklaces and for food and medicine; it is mentioned in Huangdi's ancient medical classic. It is sometimes confusingly labeled Chinese pearl barley, though it is in no way related to barley. Liquor is also sometimes distilled from it.

Chestnuts are another important starch, perhaps best discussed in this context. They have been an important food, usually gathered from wild trees in Italy, since prehistoric times. They remained an important food up to modern times, especially in mountainous regions. Chestnuts can be simply roasted or boiled, added to soups, or even ground into a flour. A *castagnaccio* is a kind of chestnut flour cake, often laden with spices, raisins, and nuts. There is also a chestnut native to China (*Castanea mollissima*), which was consumed in prehistoric times and is found in Han tombs and ancient literature. Chestnuts were typically roasted or boiled, but as in the West, they were made into flour as well as pickled and conserved in syrup. Chestnuts are also used in many Chinese dishes, especially as a stuffing for fowl.

Study Questions

1. Give examples of ingredients that travel without the technology that accompanies them. In what ways has this contributed to the evolution of cuisine? Were there any downsides?

2. Explain the consequences of the historic shift to a starch-based diet. How did this affect social and political development in general?
3. Why have such disparate cultures invented such similar recipes such as noodles and fried dough? Is this mere coincidence or evidence of contact?
4. Explain how food can be used as a marker of social status and why sometimes this valuation reverses completely. Does it have something to do with a change in perception about food processing?
5. Speculate why certain labor-intensive cooking processes are usually done communally by women and why this contributes to their use in festive occasions, as well as their tendency to persevere.
6. Why are some foods adopted globally so easily but not others?
7. Describe the religious importance of grains and why, as a culinary staple, these were invested with such sanctity.

Recipes

Porridge

POLENTA/ITALY

The simplest way to cook ground grains is as a porridge, and not surprisingly diverse cultures arrive at similar ways of cooking completely different grains, and they figure as comfort food in similar ways. This porridge from Italy would originally have been made with barley or millet, which was replaced with coarsely ground cornmeal in the course of the early modern period, so much so that it became a staple in the north. To cook this, get a capacious pot filled two-thirds with water, lightly salted, and bring to a gently rolling boil.

Sprinkle in the polenta a handful at a time while stirring all the time. This prevents lumps. Note that regular cornmeal is not a good substitute; neither are hominy grits, which have been nixtamalized, though the end product is similar. Keep adding more handfuls until you have a smooth, soupy mixture. This needs to be stirred continuously for about 15 to 20 minutes until the meal is cooked through. Remove from the heat and stir in butter and Parmigiano if you like. This is poured directly onto a board traditionally, along with meat, though you can certainly serve it from a bowl. If you have leftovers, spread it evenly on a baking pan and put into the refrigerator. The next day, cut the solidified polenta into squares and brown these in butter, or brush with olive oil and grill them or fry them in a pan.

CONGEE (RICE PORRIDGE)/CHINA

This dish is known by several names; *juk* (*zhou*) is most common. It is the quintessential comfort food in China, usually eaten for breakfast, and deemed suitable for convalescents and infants. The basic recipe involves boiling rice in many times the amount of water until it falls apart into a gruel or thick soup. It is eaten as is or garnished with soy sauce or chili sauce, salted duck eggs or pickles, and sometimes chunks of meat or fish and salted peanuts. Unlike Western breakfast porridge, it is sometimes sweetened, but more often salted. Practically anything can go into this dish, but especially nice are slices of parboiled lotus root, goji berries, ginkgo nuts, wood ear fungus, or lily bulbs. Each has its own reputed medicinal virtues.

PANATA (BREAD PORRIDGE)/ITALY

In Italy, the medicinal equivalent of the Chinese *congee* is made not from grain but from leftover bread, though there is an ancient therapeutic barley-based gruel called *ptisan*. Essentially, *panata* is bread cooked in a clear chicken broth until it falls apart. Considered a temperate food in the system of humoral physiology, it was believed to offer great nourishment without taxing the digestive system and without the possibility of throwing off the humoral balance. Thus, it contains no extreme flavors at all; it is neither sweet, sour, bitter, nor salty. There are no extreme colors, which might likewise alter a person's complexion; it is always white. It can also be enriched with milk or egg yolks, which are likewise humorally neutral, nutritious, and easy to digest—a food fit for infants and the infirm. From a gastronomic point of view, it is very comforting, especially if livened up with a little butter and some lemon juice.

ATOLE (CORN PORRIDGE)/MEXICO

Although this can be considered as much a drink as a porridge, its function in Mexican gastronomy is also as a breakfast or comfort food and is derived directly from pre-Columbian *atolli*. It is simply nixtamalized corn-based *masa* cooked down in water. It is best when quite thick. Many people use instant *masa* flour instead of fresh dough. It is sweetened with *piloncillo* (a little cone of brown, unrefined sugar) and usually flavored with a cinnamon stick and perhaps vanilla. When chocolate is added, it is called *champurrado*. Fruit can be added at the end. It is also eaten pretty much any time of day.

Pasta

BASIC PASTA DOUGH

Fresh pasta dough is just about the simplest thing to do with flour. You can use extrafine 00 flour from Italy, but regular all-purpose flour works well. Do not use bread flour, or it will be very difficult to roll out. In general, European flours have lower gluten content and are easier to roll. Set a large wooden board out and pour onto it a mound of flour. The proportions do not matter here, and, depending on the weather, you will need more or less water, so it is best to just get the feel of the dough without exact measurement. Make a well in the top of the mound and break an egg into it. Add a good pinch or two of salt. Then with a fork gradually pull in the flour to the egg, mixing and adding water as you go to bring in more and more flour. You can do it in a bowl, too, if this sounds daunting. Continue until all the flour is used up, then knead until you have a smooth dough. Wrap this and set aside for about an hour.

Dust your board with flour and roll out a fist-sized ball into a thin sheet. Professional pasta makers use a very long, thin dowel and are able to roll out a huge sheet at once, but doing it in batches works well, too. Then with a ruler or other straight edge, cut the sheet of dough into thin strips for tagliatelle, wider ones for pappardelle, or whatever width you like. Toss them with a little more flour and set aside. Alternately, roll the dough up and cut with scissors, or use a *chittara*. You lay the dough on top and roll over the tightly wound metal strings, which fall into the box below. You can also use a pasta machine with a crank, feeding the knob of dough into the rollers at successively tighter settings until you get the desired thickness. Then feed the sheets through the cutter on the other side and you will have professional looking pasta. There are also electric pasta machines, though that seems to take all the soul out of the process.

When ready to eat, put the pasta into well-salted boiling water, without oil, or the pasta will not stick to the sauce. It should take just a few minutes to cook. With tongs or a skimmer transfer the pasta directly into a pan of sauce and toss until well coated, perhaps adding a little of the pasta cooking water if necessary to create a smooth cohesion.

This same dough can be formed by hand into little ear-shaped orecchiette, twisted gemelli, or any other shape you can dream up. Note that dried pasta uses a completely different kind of flour, durum semolina, which you can use to make pasta by hand, but most shapes are extruded and require sophisticated machines.

CHINESE PULLED NOODLES

Although they may seem similar, the technique for making fresh noodles, or actually the many various techniques, could not be more different in China and Italy. Chinese pulled noodles are called *la mian* or, as is often seen in the United States, lo mein. It takes professionals many years to perfect the procedure and be able to pull the finest of threads, the dragon's beard (*longxu mian*), but it is possible to make decent ones with a little practice. The proper chewy texture is achieved by using a highly alkaline well water (*kansui*), though today a premixed powder is used composed of sodium carbonate and potassium carbonate. A pinch of sodium bicarbonate (baking soda) works almost as well, especially if baked at a low temperature first. Start with regular all-purpose flour. Add a pinch of salt and baking powder. Mix with enough water and a tiny amount of oil to make a pliable dough, and set aside for an hour or so. Then knead the dough for about 20 minutes, slapping it down on a wooden board, folding it over and over again until it is springy and will not tear when stretched. Then roll it into a long cylinder. Have nearby a pile of flour, as you will be continually dusting this as you go along. With one end of the dough in each hand, start to stretch the dough by flinging it up and down, as you would a jump rope. Slap it down on the board periodically. This is not for dramatic effect—it actually helps the whole process. When it is the length of your outstretched arms, fold in half and continue stretching, flinging, and slapping it down, dusting with flour each time. Continue folding, as many times as you can, until you have the desired thickness. The ends stuck in your hands you can just cut off and use again. Then boil the noodles, which will take only a few minutes, and use in a recipe. They can go into a soup or be stir-fried with vegetables.

PIZZOCCHERI DELLA VALTELLINA/ITALY

This recipe hails from the northern central region of the Valtellina at the foot of the Alps and is made from buckwheat flour (*grano saraceno*), which is rarely used elsewhere in Italy. The pasta is made from two parts fine buckwheat flour to one part wheat flour. If you have whole buckwheat groats, they can easily be ground finely in a blender. Mix the flour with water and salt until a firm dough develops. Roll out and cut into short noodle lengths.

In a pot of boiling water cook two peeled and cubed potatoes and a small head of cabbage shredded until nearly tender. Add in the pasta and cook a few minutes further. Drain this and assemble on a plate layered with cubes of cheese, such as Casera from this region or a similar cheese such as Montasio or Emmenthaler. Each layer should be drizzled with a little melted butter flavored with sage and garlic, a sprinkle of Parmigiano, and some cheese, which will melt from the heat of the mixture.

Filled Dumplings and Dishes

TORTELLINI IN BRODO/ITALY

The best way to serve tortellini is in a simple, clear but intense chicken broth. This is understood in Italian cuisine as well as Chinese, where the wonton and *jiaozi* are direct analogues of tortellini. Legend says that the form of tortellini was invented by an innkeeper gazing through a keyhole at the navel of Venus, which they do resemble. This is merely one of dozens of different filled pasta shapes known as *tortelli*, of which this is a diminutive form native to Emilia-Romagna and claimed by both Bologna and Modena, where they are called *cappelletti* ("little hats") and are filled with cheese.

Start by making a chicken broth. A whole chicken, preferably an old stewing hen, is simmered in a pot with chopped carrots, onions, celery, parsley, and perhaps a rind of Parmigiano that is used up. You can also add a beef bone for extra flavor. Skim the top of impurities, and cook gently for about 3 to 4 hours. Strain out the solids. Pick the meat off the bones and discard everything else. Strain a second time through a double layer of cheesecloth to clarify. Salt to taste.

The pasta filling can be made with the chicken, but the legally defined tortellino includes cooked pork loin, prosciutto, mortadella, and Parmigiano all finely chopped together and bound with eggs and a touch of nutmeg. You can also use veal or any delicately flavored meat.

Roll out sheets of dough made with flour, water, and an egg, as thin as possible. Cut out circles with a small glass. Top each with a tiny pinch of filling, and fold over. Then connect the two points and press together firmly. Flour these and set aside. Bring your broth to the boil and add your tortellini one at a time, making sure they do not stick to the bottom or to each other. They will take about 2 minutes to cook through, so be ready to serve.

Chinese Dumplings

Although round and square-shaped wonton wrappers are easy to find anywhere, they are so simple to make, it is worth the effort to have them fresh. Simply mix flour and water with a pinch of salt until you have a firm dough that is not too sticky. Set this aside to rest a while. Then roll it into a long cylinder and cut off knobs of equal size as you need them, perhaps five or six at a time, and keep the dough covered with a towel while you work. Take the individual knob and flatten it out with the small Chinese rolling pin. Hold the dough on the edge with your left hand and roll with the pin in your right. Your left hand will be turning the dough around a little after each turn until you have a circle of dough. This is the traditional way to do it, but of course you can just roll it with two hands on the pin. The major difference in technique is that each wrapper is rolled separately, rather than rolling out a large sheet and cutting it. This is more efficient in a small space and creates less waste in dough scraps.

You can make a pile of wrappers and then fill them, but it is easier to have a bowl of filling and make each dumpling as you go along. The filling can be simply finely chopped raw shrimp seasoned with soy, ginger, garlic, sesame oil, rice wine, and a little cornstarch. These are *jiaozi*. Put a knob of filling in the center, lightly moisten one edge of the dough with some beaten egg or just water, and then seal, pressing out the air, into a half-moon shape. Then stand the dumpling on its side and pleat the edge by just folding over a little and pressing until stuck and continuing all the way down the edge. The more pleats the better; ten is considered well made. This can then be steamed on a little piece of cabbage leaf or browned in a pan with oil, and then steamed in the same pan by just pouring in some water. These are pot stickers. There are also countless other ways of folding the dough over ingredients: some are closed, some opened at the top and steamed like *siu mai* made with pork. Most are eaten as dim sum, in small servings rather than a large plateful of one kind.

Some dumplings, like the ubiquitous *har gau*, are made with a very different kind of dough, using hot water, wheat starch, and tapioca starch. Instead of rolling, knobs of this dough are flattened with the side of an oiled cleaver, usually a dull one kept just for this purpose. The circles are filled and pleated as above, but the dough is translucent and chewy.

APICIAN PATINA (APICIUS 4.2.14)/ITALY

Apician Patina make thusly: bits of cooked udder, flesh of fish, flesh of chicken, fig peckers or breast of cooked thrush, and whatever would be the best. Chop all this diligently, except the fig peckers. Mix up fresh raw eggs with oil. Crush pepper and lovage; add liquamen, wine, raisin wine, and put it in a pot so that it heats and bind with starch. First however, put in all the chopped flesh and let it boil. And when it is cooked, remove it with the juices into a pie dish with a ladle, in double layers separated by whole peppercorns and pine nuts, with each layer spread out like a laganum, however many lagana you use put on a full ladle. One single laganum pierce with a straw and put on top. Sprinkle with pepper. Before this, however, you should have bound the meat mixture with the cracked eggs put in the pan with the rest.

The principal interest of this recipe derives from the jumbled order of procedures, the result, it seems, of a cook dictating the directions, realizing with each sentence that he has left out an important step. The proper order should be: chop the meat, add the seasonings and boil, then thicken with starch and eggs, then create the "double" layers with the *lagana* and pine nuts and meat. There has been a great deal of contention about what exactly these *lagana* might be, primarily because the word itself, derived from Greek, is the ancestor of lasagna. These thin sheets of dough were normally baked and crumbled into stews as a thickener. There is no evidence that they were boiled, and thus they should not be considered lasagna as we know it. However, these crackerlike sheets of dough (simple to make from fresh pasta dough rolled out in round sheets), when layered and cooked this way, do soften and the overall effect is very much lasagnalike. In the end, the entire dish solidifies due to the eggs and can be sliced for service. Whole peppercorns would make the dish difficult to consume, and it is likely that the author meant to write pepper and whole pine nuts. As for the other ingredients, the dish can certainly be made without the sow's udder. In lieu of fig peckers and thrushes, another small bird such as quail can be used. If lovage cannot be found, celery leaves are nearly identical. For the passum (raisin wine), any sweet dessert wine in which raisins have been soaked will yield a good approximation. For the liquamen, an Asian fish sauce, such as *nuoc mam* or *nam pla*, may be used.

Consider how this dish would have been eaten before the advent of forks, and while diners were reclining on couches. Diligence with the fingers would certainly be required, and it is probably because of these Roman dining habits that the dish is made solid, with sheets of dough and egg.

LASAGNE AL FORNO/ITALY

Lasagne, if not in the form we now know it, certainly has roots in ancient times. The root of the word is *laganon* in Greek and *laganae* in Latin. The dish was then made with thin strips of dough that were baked and then crumbled into a dish as a thickener. There is no evidence that they were boiled as modern pasta would be, but in the cookbook of Apicius (see above) there is certainly an ancestor of lasagne as we know it. Descendants of dishes much like this original do survive. Sheets of thin crackers (*carta di musica*) or *pane carasau*, made with a thin disk of dough that puffs up in the oven and then is split, are a common food in Sardinia. They are sometimes cooked in dishes like the *pane frattau*, in which the sheets are moistened in broth and then covered with sauce and cheese and sometimes a poached egg. It is a distant relative of the ancient lasagne.

The modern form of the dish differs in many respects from its immediate ancestor. A fifteenth-century Latin cookbook, the *Liber de Coquina*, explains that to make *lasanas* you take a fermented dough, roll it out as thinly as possible, cut it into squares, and boil it in broth. This is then sprinkled with grated cheese. As an alternative, it also suggests layering the sheets of pasta with pounded spices. Lasagne can still be made this way, with a fermented or fresh dough.

It is only one step from this to the modern form layered with ricotta and laden with tomato sauce or a rich meat ragù. The most recent addition, almost certainly introduced via the French in the nineteenth century, is the upper layer of bechamel sauce. This is essentially a roux of flour and butter into which is whisked scalded milk. A classic *lasagne al forno* (baked lasagna) must include this layer, but it strikes the historically minded as such an intrinsically foreign intrusion that omitting it and merely using a layer of mozzarella on top will elicit no cry of protest from these quarters.

To start, you can use either fresh pasta or dried durum wheat pasta. Neither is to be preferred, though the former is decidedly more delicate. To make a fresh dough, pour out about three cups of flour directly onto a wooden board in the shape of a mountain. Make a little volcano opening in the top and break in an egg and a good pinch of salt. Add a little water. Gradually work in the liquid with a circular stirring motion. You will probably need to add more water until you get a cohesive dough. Measurements in this case are completely superfluous. A few practice runs and you will get the feel for making fresh pasta dough. Let the dough rest in a covered bowl while you start the sauce.

Next, start a big pot of well-salted water to boil.

To make the sauce: fresh tomatoes are preferred, but by all means when they are out of season, use good canned San Marzano (or as they are often called generically in the United States, Roma) tomatoes. For fresh, ripe tomatoes, cut a notch in the bottom of each, dunk into boiling water for about 5 seconds and then immediately into a bowl of ice water. Once cool, you should be able to slip off the skins. Cut in half and remove the seeds, if you like.

Next, at the highest heat in a deep pan, add some olive oil, then a *battuto*, which is a mix of finely chopped onion, celery, and carrot. Cook until golden. Then add the tomatoes. If you are using canned, reserve the liquid. Cook these on high until the tomatoes disintegrate. Add some red wine, the reserved tomato liquid if you have it, and fresh herbs such as basil, oregano, and parsley. Let this simmer on low heat. Add salt and pepper to taste. If you like, at this point you can add ground beef, crumbled directly into the sauce and cooked. Divide your dough into four parts, and roll these out on your board as thin as you can. Cut these into wide strips, with a fancy pastry cutter if you like, although it does not improve the flavor much. Plunge these into the boiling water for about 2 minutes, one or two at a time. Remove them carefully with tongs, and lay on a cloth towel until cool. Continue with all the pasta sheets.

Take a casserole and put some sauce on the bottom and then a layer of noodles. Spread on top of this a thin layer of ricotta (or ricotta mixed with eggs, which ensures firmness in the final dish) and then another layer of noodles. Add sauce and some grated mozzarella, and continue with the layers until everything is used up. Top this with bechamel sauce or a final layer of mozzarella. The bechamel is simply equal weights of flour and butter gently cooked together until golden, into which is stirred milk that has been scalded. Cover the casserole, and bake in the oven at 350 degrees until bubbly, about 35 minutes.

If you decide to use dried pasta, simply boil these according to the manufacturer's instructions and proceed with the recipe (figure 3.13).

Steamed, Grain-based Foods

TAMALES/MEXICO

This recipe is a reconstruction of how tamales might have been made before contact with Europeans. Today slaked lime (calcium hydroxide or "cal") would be used, which could have been mined in the past, but lye made from ashes would have been more likely.

Take a pound of dried white dent or field corn (*cacahuazintle*), and boil gently for 15 minutes in an earthenware *olla* (pot) with 1½ tablespoons of lye. Let the grain soak for several hours, overnight or until the hulls begin to become detached. Wash the swollen grains repeatedly in a colander to remove the lime, rubbing with your hands

Figure 3.13. Two small lasagne baked in ramekins.

to remove the hulls completely. You now have nixtamal. Next, crush the softened kernels on a *metate*, repeatedly rolling back and forth until you have a fine *masa* dough. Next, cook one cup of beans (previously soaked overnight and drained) in water with ground chilies and *epazote*. When they begin to become tender, add salt. When fully cooked and the liquid has been absorbed, mash them in a *molcajete* (mortar) and set aside. Next, take soaked cornhusks, spread some of the dough along the interior, and dab some of the bean paste along a line down the middle. Carefully roll the husk inward so the bean is completely covered by the dough and the dough is covered by the sides of the husk. Then fold over the flaps of the husk on top and bottom, and tie securely with a strip of cornhusk. Repeat until all the dough and beans are used up. Place the tamales in the *olla* and pour in about a cup of water. Gently steam for 2 hours or until they are tender. Serve.

BAOZI AND MANTOU/CHINA

These are names for the classic steamed bun referred to as Mantou in northern China when unfilled, although this name can be used for all varieties in the south. It is made with a light and fluffy yeast-risen dough. These differ widely from place to place and can be filled with meat, vegetables, bean paste, lotus seed paste, or black sesame paste and are usually eaten for breakfast or are bought as a takeaway item. This version is called dou-sha bao and is filled with red bean paste. You can buy red bean paste, but it is very easy to make. The beans you want are called adzuki beans in the United States in the *Vigna* genus, but dark red kidney beans will work too. Simply cook them down until they fall apart, add sugar to taste, and cook gently until a nearly solid paste.

The dough can be made like any other bread dough: yeast started with a pinch of sugar and warm water. Flour is added, and a sticky dough is formed, which is allowed to rise for about an hour. Divide the dough into balls about the size of a peach. Make a depression in each ball, fill with a heaping tablespoon of bean paste, and seal the bun. Place them on squares of parchment paper and allow them to rise again for another hour or so. Then steam them in the basket of a bamboo steamer. Eat hot.

Mantou are similar and most popular in the north, where they are typically unfilled. They are sometimes also deep fried and dipped in condensed milk. Similar steamed buns are found throughout Asia, including the *momo*, and similar forms as far away as Persia and Turkey.

XIAOLONGBAO (STEAMED PORK BUNS FROM SHANGHAI)/CHINA

Although often translated as a "dumpling" outside China, these are distinct from dumplings, though they look similar and also appear among dim sum offerings. *Xiaolong* means to be steamed in a small basket and *bao* means a bun. They are distinctive in that they are closed by twisting the top of the dough around the filling, creating a pointed spiral. There are many different fillings, but usually they contain gelatinous aspic and pork. The aspic melts when steamed, hence these are often called soup dumplings in the

West, the soup being inside. The dough is not quite a fluffy bread dough or a thin noodle dough, but somewhere in between.

The dough is made very differently from pasta or bread dough. First you need high-gluten flour, which is mixed with water and kneaded to form a dough, as well as a small amount of regular flour, which is mixed with boiling water, also made into a dough. The two are kneaded together and left to rest for a few hours. Next make the soup/aspic, which becomes gelatinous because of the pork skin. Mix the skin with some pork meat trimmings, preferably with bones, some chicken wings or backs, ginger, and scallions. Boil these together for a few hours until the skin is soft and the liquid reduced. Strain the broth, discarding all the solids but the skin, which should be finely pounded with the liquid or whizzed in a blender or food processor until smooth. Set this in a shallow bowl in the refrigerator to solidify. Alternately, you can mix good chicken broth with some gelatin to let it solidify into an aspic.

When you are ready to assemble, fry some ground pork and flavor with soy sauce, Shiaoxing wine (a brown, oxidized, rice-based wine), ginger, and scallions. Cook until dry, and add a little starch so the mixture holds together. Let this cool thoroughly in the freezer. Then form your dough into long coils, and divide into small knobs about the size of a hazelnut. With a small, oiled Chinese rolling pin (they are about 8 inches to a foot long and thin) roll out a knob into a circle about 4 inches in diameter. Place a small amount of the gelatin in the wrapper and a small circle of pork filling. Carefully twist the top, sixteen times, until both are encased. Be sure to move quickly so the gelatin does not melt. As you make them, put in the freezer to keep cold. When all the little buns are formed, steam them either in little baskets, which is traditional, or in a conventional steamer on a leaf of cabbage so they do not stick. When the dough has become translucent, serve immediately, and if all goes well, diners can pop the entire bun into their mouths and enjoy a gush of hot soup as the bun is chewed. These are also often served with a simple dipping sauce of black vinegar, soy, ginger, sesame oil, and scallions.

Baked Bread

GRISSINI TORINESI/ITALY

Although *grissini torinesi* (breadsticks) are mass manufactured, individually packaged, and served in restaurants everywhere in Italy, few people would think of making these at home. Nonetheless, they are well worth the effort. You are basically making an absolutely simple bread dough. Use one cup of warm water, a pinch of sugar, a packet of yeast, a few tablespoons of olive oil, and about 3 cups of flour. Add in a teaspoon of salt, then knead for about 10 minutes and leave to rest in an oiled bowl for about an hour to rise, covered. Then turn out onto a wooden board, flatten the dough slightly, and cut off thin pieces from the end with a pastry cutter or knife. Stretch them individually until long and very thin, and place on a baking sheet. Bake at 350 degrees until lightly golden. Let cool on a rack. In the United States, we think of breadsticks adorned with sesame, poppy, or something else, or worse they are fat, greasy, and soft. True *grissini* are very light and delicate and merely pique the appetite before a meal rather than weigh it down. If anything, you will see as an antipasto a thin slice of *prosciutto crudo* around the breadstick. The only other ingredient, which is an unqualified success, is a little very finely chopped fresh rosemary in the dough. Commercial manufacturers also use malt syrup, which adds a little more depth of flavor to the dough.

PIZZA MARGHERITA/ITALY

There are as many types of pizza as there are pizzerias. Neapolitans take theirs more seriously than anyone, even designating restaurants that prepare it according to traditional methods with legal *di origine controllata* (DOC) status. Such pizza is cooked in a wood-burning brick oven, has a thin and blistered crust, and is covered sparsely with ingredients: a wisp of tomato sauce, real *mozzarella di bufala*. It bears little relation to its thicker, more opulent cousins on this side of the Atlantic. Afficionados will tell you pizza can only be made with exacting methods. The truth is, pizza is one of the simplest foods to make, and authentic or not, what you make at home will be infinitely superior to most of what you can buy elsewhere, and with most commercial chains and frozen pizza there is no comparison (figure 3.14). If you start with the basic recipe, feel free to experiment and top with whatever suits your fancy.

For two thin-crust pizzas or one big thick-crust pizza, begin with these proportions. Take one packet of dried yeast or one cake of fresh. The superquick varieties are not ideal unless you are in a super hurry. In a bowl, add the yeast, a pinch of sugar, and a cup of body-temperature water. Let sit until frothy. Then add to this roughly 3 cups of flour and a teaspoon of salt. The dough should be slightly firm and not too sticky. Knead on a floured board for about 10 minutes, and put in a covered bowl to rise. Lightly rub with olive oil so it does not stick.

Then, for thin pizza, divide into two and begin to toss each ball back and forth between your hands to stretch it. When ready to cook, heat your oven as hot as it will go, which for home ovens is about 550 degrees. A baking stone is invaluable. So, too, is a small ramekin (a shallow, cylindrical ceramic dish) of water in the oven to keep it humid. Place the dough on a lightly floured baker's peel (a wooden or metal wedge attached to a long handle). Some people use cornmeal, but it makes an odd, corn-encrusted bottom one would not find in Italy. With a ladle, gently swirl the sauce over the surface of the pizza so you have an exposed spiral of dough. Be sparing with sauce. This will prevent the cheese from sliding off. Add a good handful of grated mozzarella. In this case, regular low-moisture American mozzarella is preferable, as the oven will probably not get hot enough to cook buffalo mozzarella without it exuding liquid into the dough. Add a few leaves of basil. Then slide into the oven onto your baking stone, just catching the edge of the pizza on the farthest end of the stone, then sliding the peel gently out from underneath. Bake until bubbly and the crust is blistered. This is the classic pizza Margherita, named for Margherita of Savoy, who apparently was served

Figure 3.14. A simple, classic pizza cut into slices.

it on a visit to Naples in 1889, and its colors are said to commemorate the Italian flag—red, white, and green.

At this point feel free to make the crust as thick or thin as you like, and add whatever toppings appeal to you. Another trick, should you find the peel a little daunting at first, is to place a sheet of parchment paper on the peel before you lay down the dough. This way it will slide in and out of the oven easily.

BOLILLOS (ROLLS)/MEXICO

These are plain, oval-shaped rolls ubiquitous in Mexico and often bought at a bakery, but they are also simple to make. Oddly enough, they are descended from the French baguette, supposedly introduced in the era of Emperor Maximilian in the 1860s, though there were certainly forms of white bread previously introduced by the Spanish. To make them: take a large bowl and place in it a cup of water at 110 degrees—or just judge, it should not be as hot as coffee, but not lukewarm either. Add a teaspoon of sugar and a packet of regular dried yeast or a cake of fresh, broken up in the water. Wait about 10 minutes until it is frothy and smells yeasty. Then add about ¼ cup of melted and cooled lard, though this is not strictly necessary. Lard rendered at home tastes much better than the hydrogenated blocks you see at the store. Then add about 3 cups of flour to start. You will probably need another half cup or more, but this is entirely dependent on the weather and the moisture content in your flour. Make a firm but still slightly sticky dough, and knead for about 10 minutes. Grease the bowl and put the dough in, covered with plastic wrap, for about an hour until risen. Then shape your rolls into ovals, slightly pointed at the ends. Let them rise again about 30 minutes, on a parchment-lined baking sheet. Heat your oven to 400 degrees. Slash the rolls twice vertically with a very sharp knife or spin if you like in vertical cuts with a scissor. This is necessary for the final lift in the oven. You can also brush the tops with egg wash if you want a sheen. Spritz the inside of your oven with some water, and put in the rolls. They should bake about 30 minutes, and they should be golden and sound hollow when tapped. Let them cool, then consume.

The *bolillo* or its close relative, the more rounded *telera*, is most often used to make what is called in Mexico a *torta*, not a flat bread or cake, but a kind of sandwich. These are filled with scrambled eggs and *carnitas*, which are cooked, and also shredded pork shoulder or shredded beef, or ham, sausages—just about anything that can be placed inside a roll. These can be further garnished with lettuce, pickled chili peppers, avocado, and so forth. They are also commonly sold on the streets or from food trucks, as they are portable.

BRUSCHETTA AND CROSTINI/ITALY

Although the following sounds like the simplest of recipes, Italians take it very seriously and lavish great care to the toppings that go on bruschetta and crostini, which are essentially quite elegant slices of toast. The two names are used interchangeably. They have been popular for at least five hundred years. The sixteenth-century cookbook writer Domenico Romoli even took as his nickname *Panunto*, which was the original name for the dish, which just means anointed bread. You will want to start with a plain, large round of rustic white bread. Slice it in half and, laying each half face-down, cut into thin slices. A more narrow bread can also be used, simply sliced on the bias, but the broader base of a slice of round bread holds the ingredients better. Then toast these either on a wood fire on a grill or on a flat-top range, until golden. In Italy, a perforated stovetop grill (*brustolina*) is sometimes used, but a toaster also works. Anoint these scantily with butter, oil, or even lard. Sprinkle with a little salt and perhaps some oregano. These toasts are fine as is but will be even better topped with a thick puree of cooked beans, a slice of tomato, and some torn fresh basil, or half a clove of garlic rubbed lightly over the surface (after they are toasted or the garlic will become acrid). Chopped liver is also quite common. Prosciutto or cheese is also delicious. All of these together, as one often sees in the United States, is rather too much, though.

PANDORO (GOLDEN CAKE)/ITALY

A simpler version of the better-known Italian *panettone* studded with candied fruit, this basic sweet dough can also be used to make Mexican *pan dulce*, though that is normally made in fanciful shapes and covered in sweet colored icing, sprinkles, and other decorations. The Mexican version is certainly related to French sweet breads such as brioche, which in turn are related to much older forms of bread common both in Italy and Spain. This version attempts to capture the earliest roots of this entire family, and it is a version still made in Italy, especially around Verona, which is definitely more of a bread than a cake.

You can use dried yeast with this recipe, but for the full flavor to develop, it is best to use a sourdough starter. This can be made simply by leaving out a bowl with some flour and water, adding a little more flour and water every day. After about ten days the bowl should be nearly full, and it will smell sour and yeasty, having been colonized both by ambient yeast and lactobacillus bacteria. You can either keep this starter going if you intend to bake often or put it in the refrigerator, or as is common in Italy, just use it all up to bake at first but reserve a portion of dough with every batch. But again, you will need to bake very regularly with this last option.

Pour 2 cups of starter into a large bowl. Add an equal amount of water. Crumble in some saffron threads and let sit about 20 minutes until everything is yellow. Add 2 eggs to this mixture, a stick of melted and cooled butter, and a cup of sugar. Add enough bread flour to make a firm but still sticky dough. Knead this for about 10 minutes and set aside in a warm place, covered with a towel, for about 2 hours. Then knock down the dough and knead a few more times. Nowadays this is usually baked in a star-shaped mold, but traditionally it was made in a pot-shaped earthenware vessel. You can actually use a clean clay pot, greased and dusted with flour so the bread will not stick. It is actually the same way the ancient Egyptians baked bread. Only fill the pot about half to two-thirds the way up, as the dough will rise significantly. Put the dough in and let rise, which will take about 10 hours. It will be quicker in hotter weather and slower in the winter. Bake this in a 350-degree oven for about 45 minutes to an hour. Turn out and let cool thoroughly. To serve, dust with powdered sugar to resemble snow-covered Alps. You can also hollow out a section in the middle and fill with pastry cream, but it is really perfect without any adornment.

Note that if you are using instant yeast, the rising time will be significantly quicker, only about an hour or two. For Mexican *pan dulce*, it is better to use instant yeast, as you really do not want a slightly sour flavor. Also, saffron is often replaced with yellow food coloring, or you could use *achiote* paste.

CORNETTI (PASTRIES)/ITALY

Absolutely ubiquitous in Italy and eaten for breakfast with *caffè latte* or cappuccino, *cornetti* resemble croissants but are drier and are usually glazed with a clear sugar mixture (figure 3.15). Perfectly dreadful versions can be bought anywhere, but baking them fresh—they are something truly remarkable. Start with a basic bread recipe made with a packet of yeast, a cup of water, about 3 cups of flour, a ¼ cup of sugar, and a pinch of salt. Mix this into a dough. Knead and let rise, covered, for about an hour or more. Next, flour a wooden board, and roll the dough out into a single large sheet. Then take

Figure 3.15. *Cornetti* just removed from the oven.

a stick of unsalted European butter (which has a higher fat content than butter in the United States) and roll it out into a thin sheet between two pieces of plastic wrap. Lay this on top of the dough and fold over twice. Place this in the refrigerator for about 20 minutes. Roll the dough out and fold it over again. Repeat this process another three or four times, or even more if you have the patience. Then roll it out one last time, and cut into strips about 4 inches wide. Cut the strips diagonally so you have two triangles. Roll up the triangles, starting at the wider base. Let these rise for about an hour until poufy. Preheat the oven to 350 degrees. Put in the *cornetti* and bake for about 15 minutes. Then brush the tops with sugar syrup (made with equal parts sugar and water, boiled) and return to the oven for another 10 to 15 minutes or until golden and crisp. If you are feeling decadent, slice these in half horizontally and smear on Nutella, which is a chocolate hazelnut spread.

Grilled and Fried Grains

TORTILLAS/MEXICO

Tortillas are the staple food in Mexico, and although most Mexicans buy them fresh every day, they can be made at home (figure 3.16). To start you need about 2 pounds of

Figure 3.16. A coarse tortilla nixtamalized and ground into *masa* by hand.

dried field corn grown specifically for *masa*. The corn is boiled in a big pot, with ¼ cup of cal (calcium hydroxide), also called lime, which is a white powder that comes in a little packet. Let it soak overnight. The alkaline water will cause the kernels to swell, loosening the seed coat. Run these under water thoroughly to wash away the cal, and then rub the kernels between your fingers until the husks come off. What you have now is *posole* (hominy). You can cook with it as is, in a soup or stew. The kernels dried and ground, incidentally, are grits. You can either grind these moist, softened kernels with a *metate*, or if you have a mechanical grinder for moist dough, use that. A regular grain mill will not work.

Alternately, you can, as most people do, buy instant *masa* flour, mix it with water and lard, and proceed from there, but it does not have quite the same flavor. Making the tortillas is simple. With a tortilla press and two sheets of thick plastic (you can cut apart a heavy-duty freezer bag), roll a small circle of the ground *masa* and place it in the center slightly away from the hinge side of the press, between the two sheets of plastic. Then swing over the arm and press down firmly. There are also old-fashioned wooden presses, but the metal ones are most easily found. Take this small round tortilla and put it on a hot *comal* (griddle) until slightly toasted on both sides. You can also place them directly over an open flame for a few seconds on each side, until lightly charred, for extra flavor. Stack them in a dish covered with towels or in a tortilla warmer.

Tortillas are served with practically every meal, eaten as is, or used to pick up other foods in a soft taco. They can be cut up and fried for *totopos*, or shredded and fried for *chilaquiles* (a kind of breakfast tortilla dish), or wrapped around other foods. A classic treatment is the enchilada. Simply fry the tortilla in a pan of oil for just a few seconds per side, and then dip into another pan filled with chili sauce. (This is usually made with dried chilies, toasted, soaked, and blended, with water, garlic, maybe tomatoes and herbs, then fried.) Next, fill the enchilada with roasted meat, or shredded chicken, or beans, and wrap up. Line a casserole with these, and if you like, cover with cheese and a little extra sauce. Bake them until bubbly.

In the streets, Mexican vendors sell what are called *tacos de canasta*, the canasta being the big basket that holds and steams them. The tacos can be filled with anything, but after about an hour of "sweating" they become very soft and delicate.

The very same dough can be used for *sopes*, which are smaller, thicker rounds placed under other foods.

FRIED CRULLERS (YOUTIAO)/CHINA

These are basically just pairs of fried dough, eaten for breakfast with rice porridge or hot soy milk. Recipes often call for using ammonia and alum in the dough, which is said to give it a lighter texture and crispy exterior, and you can certainly experiment with these if you like. But a basic sweet yeast dough with some alkaline baking soda gives good results. Mix a cup of warm water with a packet of yeast and a teaspoon of sugar to activate the yeast. Add this to 5 cups of flour, ¼ cup sugar, ½ tsp salt, ¼ tsp baking soda, and ¼ cup oil. Add enough water to make a sticky dough. Knead this and then leave to rise, covered, for about an hour. Then roll out the dough and cut it into long strips. Ideally they should be a foot in length or more, but unless you have a professional fryer, cut them into strips that will fit in your largest pan. Heat a few inches of peanut oil in your pan. Then take two strips, lay one on top of the other, and place a chopstick along the center and press the two together. Fry these until golden. *Youtiao* in Mandarin means "oil strip." In Cantonese, they are called "oil ghosts," recalling the story of a traitorous couple.

ZEPPOLE (FRITTERS)/ITALY

Also called *sfinge*, these fritters are usually served on St. Joseph's Day (March 19). The dish itself goes back to ancient Roman times when *frictilia* (fritters) were eaten during a festival called *Liberalia*, which falls about the same time as the Catholic festival. Today *zeppole* vary greatly from place to place, sometimes round, sometimes football shaped, and even the dough base is different, sometimes using a *choux* paste (made with flour and water boiled together, to which eggs are added). There are even savory versions with anchovies. To make the dough, combine a stick of butter (or lard, if you like), a cup of water, and a pinch of sugar in a pot. Once the butter is melted, remove the pot from the heat and stir in a cup of flour. Return the pot to the heat and stir constantly until the mixture comes together as a ball. Remove from the heat again, and add four eggs, one at a time, still stirring constantly, until incorporated. Heat a deep pot with several inches of olive oil. Then using two large spoons, turn football-shaped globs of dough and drop directly into the hot oil. Turn over as necessary until golden on both sides (figure 3.17). Drain on paper towels. Sprinkle with powdered sugar.

CHURROS

Churros have a very distinguished pedigree, having been brought to Mexico from Spain, where they appeared in many early printed cookbooks such as Francisco Martinez Montiño's *Arte de Cocina* (Art of Cooking), where they are called *buñuelos de viento*. The author explains that this dough can be used to make syringe-extruded fritters. Similar fried dough forms might even be traced back to medieval Persian cooking, which also influenced India in the opposite direction. Some contend that their origin is in China and that they are a modified version of the aforementioned youtiao, brought back by the Portuguese in the sixteenth century. They are made from what is called a pate a choux paste that is squeezed from a syringe into hot fat, in which they puff up and get crispy. Most typically they are eaten for breakfast with hot chocolate. Make the batter by mixing a cup of water with 5 tablespoons butter and 2 tablespoons brown sugar in a

Figure 3.17. *Zeppole* frying in oil.

pot and bring to a boil. Turn off the heat and stir in a cup of flour. Then mix in 2 eggs and ¼ teaspoon vanilla extract. Put a small amount of batter into a star-tipped syringe and squeeze into a pot of hot fat. Each churro should be as long as will fit into your pot without them sticking together. Remove with a slotted spoon or tongs and let drain for a minute or two, and then sprinkle with sugar and cinnamon. They can also be made with a pastry bag with a star tip. Because they are so light and airy, in many early cookbooks they are called whore's farts or nun's farts. The name churro does not seem to be used much until the latter nineteenth century, though. Here is an early Mexican recipe from *El Cocinero Mexico* (vol. III, México, 1831, p. 562. Reprinted 2001 in *Colección Recetarios Antiguos*, Concaculta).

BUÑUELOS DE JERINGA

> Se echa la leche correpondiente a una libra de harina, que se mezclará con ella y con doce huevos, claras y yemas; se pone a la lumbre hasta que haya tomado el punto conveniente, y entonces se le añade tantita agua de anís y un pedacito de azúcar, amasándose bien hasta que despegue del cazo. Se van poniendo en la jeringa echándoles manteca para que despeguen.

Take milk equal to a pound of flour, which you mix with it and with twelve eggs, whites and yolks; put it on the fire until it comes to the right stage, and then add a little anise extract and a bit of sugar, mixing it well until it comes away from the pot. Go and put it in the syringe, greased with lard so it does not stick.

FRIED RICE (YANGZHOU CHAO FAN)/CHINA

This is, not surprisingly, one of the most familiar Chinese dishes to have spread to the United States and around the world. First make the rice. Wash the rice well in several changes of water. Then put it in a pot with water one index finger's digit above the level of the rice. Bring to a boil, and then lower the heat and let simmer gently for 20 minutes. Turn off the heat and let sit for another 10 minutes. Uncover and let cool completely. You can also put it in the refrigerator until the next day. Carefully fluff the rice with a fork and spread it out on a towel. You want to make sure the grains are all unbroken, separate, and dry. Then heat a few tablespoons of oil in a wok or large frying pan. Add the rice, at first leaving it to brown on one side, then flipping it with your spatula. To this add diced carrots, slivers of ginger, and some finely chopped onions. Continue to stir until these begin to brown. Next, beat an egg in a bowl with a little salt. Move the rice and vegetables to one side of the wok and pour in the egg, swirling it around and breaking up the larger pieces. Then add some green peas and some slices of Chinese barbecued pork, the one with the red edges. Keep stirring until everything is nicely golden. Add some soy sauce to taste, a little chicken stock, and continue to stir until all the flavors are blended. Perhaps garnish with some sliced scallions and serve.

RISOTTO ALLA MILANESE/ITALY

This is normally a first-course dish eaten on its own, though it can be served alongside *ossobuco* (veal shank) in Milan. You need a short-grained rice, such as arborio, vialone nano, or carnaroli for this. They are all slightly different, some larger, some smaller; some give you a creamier dish. Traditionally it would be made with beef broth or knobs of beef marrow and lard, but more commonly today it is made with chicken broth and butter. Heat a small pot of good chicken broth with a pinch of saffron. In a large pan gently cook a chopped onion in olive oil. Pour in the rice and stir around until completely coated, and let cook a few minutes. Then pour in some dry white wine and wait until it is absorbed. Then add a half ladleful of broth at a time, stirring constantly on medium-low heat. Add the broth only when most of the liquid has been absorbed, and continue until the rice is still barely al dente ("to the teeth," that is, chewy). It should be creamy on the outside and still slightly "toothsome." Before serving add a knob of butter and a handful of grated Parmigiano and mix in. Serve in a wide, low-rimmed soup dish.

SPANISH RICE (ARROZ ROJO)/MEXICO

Rice was introduced to Spain in the Middle Ages by the Moors and then in the sixteenth century to Mexico. It is not called Spanish rice in Mexico, though, only in the United States. The technique is quite different from risotto, and the final dish should be light and dry rather than creamy. In some respects it is actually closer to the Moorish recipe, a kind of pilaf, with the exception that it includes tomato. Heat oil in a pan first, then add the rice. Here a long-grain rice is preferable. Stir constantly on medium heat until the grains begin to brown and become opaque. Then add chopped onions and stir so onions are cooked through. Add some chopped tomatoes and green bell peppers. Then, all at once, add hot chicken broth or plain water, turn the heat down to low, and cover the pot. Let the rice cook for about 20 minutes. Check to see whether it needs more liquid, and if so, add it. Let the rice rest a few minutes, covered, before serving.

Sweets Made from Grain

SBRISOLONA/ITALY

This is rather like a large, crumbly cookie. It is a specialty of the city of Mantua, where many varieties are sold in shops. It was originally a dessert born of poverty, using mostly cornmeal and lard, and can still be made that way with excellent results. Nowadays, it is always made with butter and sometimes chocolate as well. This simple version goes perfectly with a glass of dessert wine.

On a large wooden board mix together by hand 3½ cups flour, 1½ cup fine cornmeal, 3 cups butter, and 2 cups sugar. Add 3 egg yolks, a little grated orange peel, and a pinch each of cinnamon and nutmeg. There is no particular order to this. Just incorporate the whole with your hands. You can also add a teaspoon of baking powder if you like. You want this to be a very crumbly mixture, not a solid dough, so just keep breaking up the flour and butter if necessary. Very gently press this mixture into two average-sized pie pans. Decorate with unpeeled almonds. Bake at 350 degrees for about 25 minutes or longer, until hard.

MEXICAN WEDDING COOKIES (PASTELITOS DE BODA)

Take a cup of blanched, peeled almonds and toast them lightly in a hot pan, and then crush them in a mortar. You do not want a fine paste, just well-crushed nuts. You will need about half a cup ground, tossed with a little flour to keep them dry. Then cream 2 sticks of butter with ½ cup sugar. Add in a teaspoon of vanilla extract, 2 cups flour gradually added, and a touch each of cinnamon and salt. Roll into little balls or form into crescent shapes and bake in a 350-degree oven for about 15 minutes. Let cool, and dust them with powdered sugar that has been sifted through a sieve.

CHINESE PEANUT COOKIES

These are surprisingly similar to the Mexican cookies above but are made in a very different way. Pound or process finely a cup of salted roasted peanuts. Then take 1½ cup flour, ¾ cup powdered sugar, and ½ tsp baking powder and mix together well. Drizzle in oil and rub in with your fingers until crumbly. You will need about ¼ cup or less in all. Then mix in the finely crushed peanuts. Roll these into balls or press into a small cylindrical mold or shot glass. You can decorate the tops by pressing in a little circle if you like. Brush the top with beaten egg yolk, and bake at 350 degrees for about 20 minutes.

SESAME BALLS/CHINA

These have such an intriguing chewy texture, with a soft bean paste center and crunchy outside, that they are very addictive and festive. They are served on New Year's and for birthdays. Begin by making the bean paste. This is simply red beans soaked overnight, then boiled until soft, then drained and mashed with sugar to taste. Let cool. The store-bought bean pastes sold in a can are usually tooth-achingly sweet. Then mix a few cups of glutinous rice flour with cold water that has been sweetened with sugar, and mix quickly until a sticky dough forms. With hands dusted with the rice flour, roll into small balls, make a depression in each, fill with the bean paste, and seal. Sprinkle these with sesame seeds. Fry them in a deep pot of oil until crisp and golden brown, about 10 minutes.

MOONCAKES/CHINA

The same bean filling as above works beautifully in these, though there are any number of other fillings as well, most popularly lotus seed paste (which you can buy canned) with a salted duck egg yolk in the center. Most people buy these from bakeries, specifically for the Mid-Autumn Festival, but they are pretty easy to make. The only piece of equipment you will need for a traditional cake is a mold with the Chinese letters for longevity and harmony set in a floral pattern. But you can make them without this, too. A wooden butter mold works as well. There are also many different types of crust across China, some a short crust, some tender and chewy, others flaky. This is one that works well, though the ones you will normally find include alkaline lye water to give them a particular chewiness, but a pinch of baking soda will have a similar effect. Simply take cool lard, add sugar, and rub in flour with a little cold water until it comes together as a dough. Refrigerate until ready to use. Mold the bean or lotus seed paste into a hockey puck–shaped disk with a salted egg yolk in the center. Lay it on a round of rolled-out dough and enclose, sealing the bottom. If you have a mold, now would be the time to impress the letters on the upper side, or you can use a ramekin that is well floured to get a perfect circle. Turn out, brush with egg yolk, and bake them in a 350-degree oven until crispy and deep brown.

Starches

POTATO GNOCCHI/ITALY

The original ancestor of this dish was made with either breadcrumbs or semolina flour, and versions like this still exist, as do modern variations with ricotta or spinach, which are heavenly. But this simple version is also very satisfying. Boil some mealy potatoes (russets are fine) in their skins. You can also bake them or even poke them with a few holes and microwave them for about 10 minutes. The drier the potato, the less flour you'll need to add and the lighter they are in the end. Scrape the potato from the skin, and pass through a ricer (a cylindrical press that passes the cooked potato through small holes) or mash very finely. Add a pat of butter or two and a little salt. Add enough flour to the potato to make a firm dough that will hold its shape when you roll it up into little nubbins. The traditional way to do this is to roll a long snake of dough, cut off little nubbins, and then roll them over the back of a fork, along the tines, so you get ridges. There are also little wooden paddles with ridges designed to do just this. When the gnocchi are formed, boil them in simmering water for just a few minutes. Taste them until they are cooked through—they should be a little chewy, but light. You can serve these immediately with your favorite sauce, but even more interesting is to put them in a pan with melted butter, perhaps a sage leaf or two, and a lot of grated Parmigiano. Toss over the heat for a few minutes, and then put into a hot oven. You can even pour over a little cream if you want an unctuous sauce. Perhaps add little flecks of pancetta (unsmoked, rolled Italian bacon) and some fresh green peas.

CAMOTES FRITOS (FRIED SWEET POTATOES)/MEXICO

Sweet potatoes with orange flesh are much more common in Mexico than white potatoes and are native as well. *Camotes fritos* are just sweet potatoes cut into slices with the peel on and deep fried. You can also pan fry them a few at a time in just an inch or two of oil. They can be thick or thin, cut in rounds or like French fries, simply seasoned with salt or with spices, or even sugar and cinnamon. It is up to you.

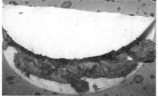

⁕{ 4 }⁕
Vegetables

Learning Objectives

- Understand the various uses of vegetables in the construction of a complete meal in each world cuisine.
- Appreciate the nutritional role vegetables play in the human diet.
- Consider why vegetables have been relatively deemphasized in the developing world, how their status has changed, and the consequences of this shift.
- Recognize the diminished diversity of plant species eaten in modern times and its cause.
- Understand how certain vegetables provide flavor and texture contrasts in what would otherwise be a monotonous diet.

THIS CHAPTER covers the most important plant foods that have been historically incorporated into the three world cuisines, as well as those that have gained prominence more recently. Culinary usage governs the organization here rather than botanical classification. Thus, tomatoes are found with the vegetables rather than fruits in the following chapter. Likewise, some foods, such as olives, have no other place than with their botanical relatives, even though they are not thought of as fruits. Within each category the ingredients are grouped together usually by taxonomic family, so that all the cabbage relatives are together and the onion family is in one vegetable group rather than with seasonings. Sometimes a group is based exclusively on culinary usage—all root vegetables are together even though they might be unrelated. Herbs, as their role is not as a food but a seasoning, are discussed in chapter 8. As with other chapters, the items discussed are not meant to be exhaustive of every species on earth or even all those used in Mexico, Italy, and China, rather merely those that are most widely used.

Vegetables play a central role in the diet of each world cuisine, historically unlike in the modern Western diet. That is, they are not an afterthought or side dish, but for most people have taken center stage. In Italy, this central place for vegetables has come to be known as the Mediterranean diet, hailed for its nutritional balance and dependence on grains and vegetables and a small amount of meat, fats, and sweets. China's traditional diet has often been described in similar terms, and so, too, has Mexico's before modern times. It should be noted that these patterns were partly the result of poverty, ill-considered government policy, and plain social snobbery. Hard manual labor (exercise) was also largely responsible for the longevity attributed to these peoples. Whether the perception that they enjoyed robust health can be borne out by historical facts is another matter, but it remains true nonetheless that vegetables have played and continue to play a central role in each of these cuisines, far more so than in the United States and northern Europe. It should also be remembered that there were sharp distinctions among dietary patterns between rich and poor. The latter ate little meat only because they could not afford it.

Regarding health, vegetables have certainly been the source of many vitamins and minerals now lacking in the diet of modern Western culture and that are largely supplied through vitamin fortification or supplements. For example, niacin (B_3) has been added to flour and bread since the 1930s in the West, though in traditional societies it was amply provided by leafy vegetables, asparagus, and legumes.

Cabbages (Crucifers)

Members of the **cabbage** family are among the great unsung heroes of world cuisines, serving tables both exalted and lowly. The Platonic cabbage form in Western minds is the hard rotund pale-green cabbage, but its ancestors were scragglier, longer-leafed plants, more resembling kale, a variety of the same species. The protean nature of this clan is frankly remarkable, including the diminutive Brussels sprout, which buds on a Christmas tree–like stalk, the blanched white floret of cauliflower, the treelike inflorescence of broccoli, and the tumescent stem of the kohlrabi. There are newer crosses, such as green or romanesco cauliflower with its hive of conical buds, tender broccolini, or the *cavolo nero* used in the "reboiled" Italian soup *ribollita*. These are all the same species of *Brassica oleracea*, bred by humans over centuries into different forms. In China there are leafier Chinese kale, and a slew of relatives from the *Brassica rapa* (turnip) side of the family: large and small bok choy (*bai cai*), yellow flowering cabbage (*cai xin*), what is sometimes called Chinese broccoli (*jie lan*), and the small-leafed tatsoi. Fortunately, in culinary terms, these are mostly interchangeable. The turnip side of the family is a little more bitter, having never been as rigorously attended to by breeders intent on developing milder flavor. In fact, many appreciate rapini or broccoli rabe precisely for its bitter flavor (figure 4.1).

Tradition dictates that the tougher of these greens are cooked slowly and simply. In Italy that means simmering in a frying pan with olive oil until nicely wilted. The Italians, perhaps surprisingly, have been devoted cabbage eaters for millennia. There is a whole section of the ancient agricultural manual by Cato the Elder (d. 149 BC) devoted to the culinary and medicinal uses for cabbage, though this section was probably not written by him. In China the lighter cabbages may be stir-fried, chopped, and used to fill dumplings and spring rolls, whereas the tougher and more bitter types are slowly simmered in a soup. There are also various types of stuffed cabbage, the large leaves lending themselves perfectly to enclosing other ingredients. In China, these might be filled with mushrooms and onions or ground pork, steamed, and then served with a piquant sauce. In Italy, they are usually stewed in tomato sauce. Cabbage plays a minor role in historic

Figure 4.1. The bitter flavor of broccoli rabe, which are flowering turnip greens, complements rich dishes very well.

VEGETABLES

Mexican cooking, as it is an Old World crop, but it often appears today shredded where one might expect to find lettuce, in a taco or as a fresh salad ingredient.

Another basic treatment for cabbage, especially the thicker, sturdier varieties, is lactic fermentation. Sauerkraut is made and used extensively in the Alpine regions of Italy, but other cabbages are also pickled whole elsewhere. For example, cauliflower is usually included in the vinegar-based (*sotto aceti*) pickled Italian vegetable mix called *giardiniera*, and Brussels sprouts also pickle well. In Asia pickled cabbage, while firmly associated with Korea's fiery *kim chi*, is no less important in China, called *pao cai*. Pickling in all these cases not only gives a pleasant sour crunch to the vegetables but also has for centuries been an essential way to keep a supply of vegetables through the winter. All such naturally fermented cabbages are merely left to soak in brine, submerged in a stoneware or glass crock. They ferment with ambient lactobacilli after a few weeks, though they may also be further aged. As with many fermented products, health benefits accrue partly from increased intestinal flora but also because many nutrients are more easily absorbed by the body when foods are pickled.

Root Vegetables

Root vegetables are often closely related to cabbages, or are the same species in the case of turnips. They can also be a source of oilseed, for *Brassica rapa* (rapeseed, known in North America as canola). The role of tubers in sustaining peasant populations cannot be overstated. This is primarily, once again, because they can be stored through the winter, providing necessary vitamins and nutrients when nothing else will grow. There are beets, whose leaves can be eaten in spring and summer and roots in fall and winter; carrots and parsnips; celery root; plus a few lesser-known roots such as skirrets (*Sium sisarum*) and *scorzonera* ("black root"), and closely related salsify or oyster plant, whose name derives from its supposedly oysterlike flavor. In Europe as well as Asia, most of these are either simply boiled, mashed into a soup, or roasted whole, which brings out their natural sweetness. Carrots are among the few root vegetables that can also be eaten raw whole, sliced, or grated into a salad. Interestingly, the orange carrot is a recent development, the original root normally being white or red.

To this group should be added the water chestnut (*Eleocharis dulcis*), which when fresh is a sweet and crunchy addition to stir-fries (figure 4.2). When fresh they merely need to be peeled, washed, and sliced and have a flavor very distinct from canned water chestnuts. Similar in flavor and larger, but totally unrelated, is jicama, a Mexican root of a bean plant (*Pachyrhizus erosus*), which is also eaten raw, with lime juice and chili powder. In China, jiamcas are often pickled or used as a substitute for water chestnuts. As one of the few root crops of world importance hailing from eastern North America, the so-called Jerusalem artichoke, whose name is a corruption of the Italian *girasole* ("turn toward the sun"), is actually a variety of sunflower. In Europe they are often called *Topinambours*, mistakenly, after a Brazilian tribe. The newer American marketing name

Figure 4.2. Fresh water chestnuts are crisp and sweet, unlike the canned version.

of sunchoke is a little closer to the mark. These, too, can be eaten peeled raw or cooked in a puree or soup. They are also, incidentally, regarded as the most flatulence-producing of all vegetables. Similar to these is another root sometimes called Chinese artichoke (*Stachys affinis* and similar *spp.*), which resembles a little beige grub, or in the more poetic Chinese imagination "jade beads." In Europe, they are known by the name of the location where they were first cultivated in France in the nineteenth century, Crosnes. They are related to the herbs betony and lamb's ears. In China they are often pickled, but in Europe they are more often simply sautéed, like a Jerusalem artichoke.

Another root, common in Chinese cuisine, is the lotus (*Nelumbo nucifera*), a kind of water lily. The nuts of these, known in the West mostly as a dried, decorative plant, are eaten, but more important are the roots, with their familiar interior pattern of holes. Sliced and cooked, these lend a stunning visual element to dishes, and they can even be fried as chips or put into soups. The importance of the plant itself stems from its association with Buddha, who is often depicted seated on a pink lotus blossom. The lotus represents purity, wisdom, and, depending on the color, Buddhist doctrines such as the noble eight-fold path.

Burdock (*Arctium lappa*) is another European root, though ironically most used in Japanese cuisine where it is known as *gobo*, and is typically cooked or pickled. It is also used in China to some extent, and in Italy in soups, though in the Western tradition it was more commonly used as medicine. The leaves are also edible and are cooked as most other leafy greens, slowly braised or chopped and simmered in broth.

Last, there is a large troupe of radishes (*Raphanus sativus*), including the familiar red varieties, and large white radishes such as the daikon. These are usually used raw or pickled, or in the case of horseradish (*Armoracia rusticana*) and wasabi grated and used as a condiment. Like mustard, all of these share a common culinary trait—a quick but fleeting jolt of heat that ascends into the nasal passages and head, unlike the burning sensation that lingers on the tongue, as is found in chili peppers. Isothiocyanates are the phytochemical responsible and are only activated when the raw vegetable is chewed, which is why these are rarely cooked. These are currently gaining wide scientific attention in fighting cancer.

The Onion Family

The **onion** family (**alliums**) are extraordinarily versatile, being used both as a flavoring for every manner of food as well as a vegetable in their own right. One might even go so far to say that they are one of the very few plants used in most savory dishes in our three world cuisines. Conceptually and aesthetically, they form one of the major building blocks of all flavor profiles. Along with salt, sour, sweet, and bitter, the acrid flavor of alliums should be considered a fundamental component of flavor, along with the pungent heat of chili, and perhaps smoke as well as the aromatic punch of fresh herbs. Of course, these are normally considered aromas rather than flavors perceived by the tongue. But flavor should be considered the combination of taste, aroma, and mouthfeel. In any case, the alliums are central to cooking in these three traditions.

Onions and garlic are the most familiar. We could scarcely even imagine Italian, Chinese, or Mexican food without them, but we should also include delicate shallots, scallions, chives, the so-called Chinese chive, the Welsh onion (actually Chinese), and the proper leek, which is purportedly Welsh. Some of these are merely sliced and eaten raw, diced and incorporated into salsa in Mexico, or snipped as a garnish in the case of chives. In some types it is the bulb, or stalk, or even flower head only that is eaten. Most alliums are cooked, often as the first step to preparing another dish, which tames the eye-burning sulphurous compounds and brings out their sweetness. Thus, they may be fried along with celery and carrots in a *sofrito* as the start of a Northern Italian sauce or stew. They are stir-fried quickly with other ingredients in Chinese dishes, or even pickled. Garlic is also a ubiquitous flavoring in other foods such as salami and the Chinese cured sausage *lap chung* (*la chang* in Mandarin). The New World does have many of its own native allium species, the ramp or wild leek among them, but in culinary usage the Old World alliums today dominate most places on earth.

Cipolline are an unrelated bulb (actually a grape hyacinth) used in Italy either slowly braised in a sweet and sour sauce (*agrodolce*) or pickled. There are also some edible types of lily bulbs. In China, where these are called *baihe*, they are sold dried and after soaking are added to soups and stir-fries and are believed to be good against heat during the summer.

Greens

Greens are a generic term referring to any leafy green plant that is eaten either raw in a salad or cooked. In Italy there are hundreds of salad greens both gathered in the wild and cultivated. Some are mild such as the lettuces, others bitter such as arugula, or pungent like watercress, or even sour like sorrel. The ideal is to match the various ingredients so they create a balance of flavors. The same is true for the dressing placed on them, which is typically merely salt, vinegar, olive oil, and perhaps a grind or two of pepper. The name *salad* itself (*insalata*) refers only to the salt, and as one seventeenth-century writer, Salvatore Massonio, quipped in his encyclopedic book on salad, it should be called *herba salolacetaria*, to account for the oil, vinegar, and vegetables, too. Salads can be eaten either before a meal as an antipasto or after as a refreshing closure, which is common practice today, although not in the past.

Lettuce often forms the base of the salad, though the supercrisp, bland, round iceberg is not preferred outside the United States and northern Europe. In Italy there are more flavorful varieties like romaine (*Romana*) and *lollo bianco* or *lollo rosso* with curly leaves. The name lettuce (*Lactuca sativa*) comes from *lac*, meaning "milk" in Latin, and the plant stem does exude a milky latex when cut. Lettuce is usually mixed with other salad greens such as those in the chicory family: escarole, white Belgian endive—a spearhead–shaped compact leaf cluster that is grown in the dark and traditionally with dirt mounded up around it—and radicchio, a bitter, red-leafed plant that comes in little heads from Chioggia or longer torpedo shapes, famously from Treviso. These can also be braised or grilled. It should also be noted that lettuce is commonly cooked too, especially in China. There is also in China a variety of long-stemmed lettuce, which is cooked.

Purslane (*Portulaca oleracea*) is another salad green of tiny, fleshy, slightly mucilaginous leaves with a pungent taste. In Mexico it is called *verdolagas*, eaten as a garnish in tacos and other dishes. Dandelion, though considered a weed in the United States, is also a delicious salad vegetable in Italy, eaten when young and tender. Orache (*Atriplex hortensis*) is a purplish-green plant with triangular serrated leaves, which are a cleansing salad ingredient or can be cooked. A related genus is *Chenopodium*, which includes goosefoot and lamb's quarters and Good King Henry, as well as the Mexican herb *epazote*, and the South American pseudo-cereal quinoa. There is also a plant called *huazontles* in Mexico, little green seeds lacing a vegetable sprig, also in the same family. Its name from Nahuatl means "grainlike hair." These are cooked in a kind of fritter (*torta*). The vegetable is tied in little bunches and blanched, a lump of cheese is placed in the center, and it is dipped in flour and egg batter and then fried. It is served with a roasted green tomatillo salsa. *Romeritos* (*Suaeda torreyana*) are probably named for their resemblance to rosemary but are nonetheless eaten as a vegetable in Mexico. They are also combined with dried shrimp and potatoes and cooked in little patties (*revoltijo de romerito*), served with molé sauce, for Christmas.

Another green common in Italy and elsewhere in Europe, though fairly unknown elsewhere, is *salicornia* or glasswort, so called because its ashes, high in sodium carbonate, were used in glassmaking. Its stems are long and crunchy, and it is sometimes found in the United States, where it is called sea beans. It grows along shores and in salt marshes. It can be eaten raw or lightly cooked and has a pleasant green, vegetal flavor. There is also a closely related species in China called *yan jiao cao shu*, eaten in much the same way. This should not be confused with samphire (*Chrithmum maritimum*), a similar coastal plant, usually eaten pickled. Sometimes this latter is called rock samphire, whereas *salicornia* is called marsh samphire.

The mallow is another European plant, often collected wild, that serves either as a raw salad ingredient or a cooked vegetable. It has a pleasant, mucilaginous texture and was enjoyed as a simple rustic food by the ancient poet Horace. In China, there is a close relative, *Malva verticillata* (*dong han cai* in Mandarin), also eaten as a vegetable and valued for its cleansing properties. These should not be confused with the marshmallow (*Althea officinalis*), whose root was in fact used in confections and in medicine as an emollient, before the modern sugary candy took its place.

The best known of the delicate leaves, equally good raw or cooked, is spinach, which comes originally from Persia and spread from there eastward to China in the Early Middle Ages and thereafter to Moorish Spain and the rest of Europe. Spinach was considered a very elegant ingredient in Italy, where it was cooked in fritters or baked into savory tarts. In fact, any dish stuffed with spinach is now called *alla fiorentina*, meaning in the style of Florence. Fish, poultry, pasta, and any number of foods can be spinach laden, including thin pancakes rolled up with spinach and ricotta (*crespelle alla fiorentina*). The famous dumplings from Trentino called *strangolapreti* (priest chokers) are made with breadcrumbs soaked in milk, blanched spinach, eggs, and flour. Pasta, too, can be made with spinach to color it green. In China, spinach is equally popular, cooked in soups, stir-fried with other vegetables, or used as stuffing in much the same way it is in Italy, or wrapped around other ingredients. Also, it is normally blanched first and squeezed dry before cooking, and the wok is one of the few kitchen vessels that can handle a large bunch of spinach that can be easy turned and cooked down.

There is also a plant in China called water spinach or *tongxin cai* (*Ipomoea aquatica*), which is related to the sweet potato but does not form tubers. Despite the name, it can also be grown on dry land. A red amaranth in China is known as Chinese spinach or *jin choi* (*xian cai* in Mandarin), whose young leaves can be eaten raw but more often are cooked in soups or mixed dishes. In the Caribbean, it is known as callaloo. Mexico, too, has various spinachlike greens: *quintoniles* in the amaranth family, which are eaten in a stew (*guisados*) with chipotle chilies and small fish. There is also *chaya* (*Cnidoscolus chayamansa*), native to the Yucatan and sometimes called tree spinach. It is used in Maya tamales and to make a drink. It has a five-sided maple leaf shape but is poisonous when raw. All these are used as cooked greens on their own, or as stuffing.

In lieu of spinach a number of other tougher greens are used cooked, including chard, which is actually a variety of beet green; kale; or collard greens, which are a type of cabbage; mustard greens; and turnip greens (figure 4.3). All are on the bitter, assertive side and are thus usually cooked longer than spinach, in soups or braises, with results that are in the end much more flavorful than spinach. Bitterness in a vegetable in many cultures is considered a purgative that cleans out the body during spring after much inactivity and heavier meals of the winter season.

Beans

There are **bean** species (*fabaceae*) practically everywhere on earth, and as a reliable source of protein that can be stored in dried or fermented form, they have been of immense importance to our three culinary traditions since antiquity (figure 4.4). In fact, the most important beans in the world today come from Asia, in the case of soy, and Mexico, in the case of the *Phaseolus* species. Italy adopted cultivated beans such as the fava, chickpea,

Figure 4.3. Beet greens grown for their large leaves are known as Swiss chard in the United States.

Figure 4.4. The variety of bean species from around the world is staggering.

and lentil from the Middle East, but they have been of no less importance there since prehistoric times. If we were to describe each of these cuisines in terms of its basic fundamental structure, we would have say that beans are second only to the staple grain in providing calories and nutrients, with vegetables following behind.

Mexico is perhaps the place most readily associated with beans as an indispensable part of the diet. American diners are familiar with *frijoles refritos*, mistakenly translated as refried beans, when in fact they are only fried once. The name really means something like "fried-up beans," though they are also mashed and flavored with onions and lard. These, however, are only one of the hundreds of ways beans are served in Mexico. *Phaseolus* species here also come in an amazing range of colors and sizes: black, white, brown, mottled, and every stripe in between. Beans are cooked simply in an earthenware pot (*olla*) or cooked in soups, mashed into fritters, used to stuff tacos and tamales, or even made into sauces and dips. In Mexico, they also enjoy Old World beans such as chickpeas, lentils, and fava beans. Here beans are not only an essential part of their culinary identity but also most commonly provide protein in place of meat. There are a number of other native American legumes worth mentioning: the tepary, well suited to desert conditions, as is mesquite, whose pods Native Americans ground whole to make a flour, as well as the sieva or butterbean, which is a variety of lima bean, though independently domesticated in Mexico.

In Europe *Phaseolus* species were enthusiastically adopted in the sixteenth century, to a great extent replacing Old World beans in soups, fritters, and casseroles. Despite the fact that Italy treasures several varieties as their own, such as the lamon and zolfino, borlotti and cannellini, they, too, are originally from America. Origin aside, nowhere do they pride themselves as being bean eaters (*mangiafagioli*) as in Tuscany. There beans are cooked simply stewed; *in fiasco* (in a bulbous Chianti bottle set in hot coals); or served with pasta, as in pasta fagioli or in the classic soup *ribollita* made with bread, olive oil, and *cavolo nero*; or even simply mashed and served on toasted bread. Beans are also eaten green and immature as whole pods, with several varieties, such as Romano beans, especially prized. The importance of beans in Italian culture depends on their low cost and suitability for fast days, in particular Lent, during which meat and meat products were forbidden. Wealthier people could afford fish at Lent and, during the rest of the year, meat and poultry. Beans, therefore, became strongly associated with poverty, and ironically it is for this very reason that they have been reappropriated today as a healthy, rustic peasant food, an essential part of the so-called Mediterranean diet.

In China, the native species of prime importance is soy. Although soybeans can be eaten fresh and immature, as with edamame from Japan, normally soybeans are dried or processed, which actually removes the nutritional properties of fresh soy. They can be soaked and pulverized into soy milk, which is then "curdled" the way cow's milk would be, traditionally with sea salt, later with gypsum or magnesium chloride. The solids are precipitated, drained, and then pressed into blocks of tofu. Tofu can also be further fermented into what is only half-jokingly called stinky tofu—a dark and richly aromatic ingredient. Of even greater importance are the condiments made from soy. Originally in China this was a kind of fermented bean paste called *jiang*. It has been discovered in ancient tombs, and there is a solid record of its use going back centuries. It is also the ancestor of what is known commonly as miso paste. Also a derivative of *jiang*, the liquid drained off is soy sauce. There is no other ingredient so firmly associated with Chinese cuisine, and rightfully so. It is the most common flavoring and source of salt in Chinese cuisine, used by some estimates in as many as 70 percent of all dishes.

The Chinese also used a number of other beans, especially those in the *Vigna* genus. Among these are the red beans, familiarly known as adzuki beans, cooked down into a sweet paste and used to stuff buns and mooncakes. A relative are the so-called yardlong beans, which are eaten fresh like any green bean. Another member of this genus is the mung bean, which is the most common source of bean sprouts, a ubiquitous ingredient in stir-fries. These beans are also made into thin, transparent noodles that are sometimes called cellophane noodles, which also go into stir-fried dishes and soups. The Chinese are also avid eaters of fava beans, whose starchy texture goes well in long-simmered dishes; they are also fried as a popular crunchy snack.

Last, we should include the humble pea, native to southwestern Asia but important across Europe and Asia, as well as the Americas since the sixteenth century. Dried field peas were normally eaten like other beans, long cooked in soups, but garden peas are the ones eaten fresh or lightly cooked, mixed into rice dishes like risotto in Italy or stir-fried rice in China. The snap pea and snow pea are more recent cultivars that can be eaten whole with the pod and have become ubiquitous in Chinese cuisine.

Squashes and Gourds

Another large and important family of ingredients are in the *Cucurbitaca* family, most of which come from the Americas. Cucumbers are an exception, but most other **squashes** are from the Americas, including, surprisingly, zucchini, which are so firmly associated with Italy, from where they were introduced into the United States. All the other varieties of pumpkin, winter squashes such as acorn, hubbard, crookneck, spaghetti, and even kabocha, are also from the Americas originally. The word *squash* comes directly from a Narragansett Indian word. Their distant African relatives in the *Lagenaria* genus should properly be called **gourds**, though they have been eaten for millennia in much the same way squashes have been since the sixteenth century. Interestingly, in Spanish the word *calabaza* was applied to these, which originally meant calabash, or gourd, but was also used for the American squashes. Even more confusingly, there are ancient archaeological remains of gourds found in Mexico, which are supposed to have floated across the Atlantic, if not brought by people. Whatever the origins, the culinary use of these different species follow the same patterns.

Squashes can be cooked down into a purée (and used to stuff ravioli or flavor risotto in Italy) or more gently cooked in a stew or casserole. They were a traditional accompaniment in the garden and on the plate with corn and beans in Mexico. The beans would twine up the corn stalks, and the squash vines would trail along the ground, keeping moisture in and preventing weeds from popping up. They could also be sliced into strips and dried for later use in cooked dishes.

Of equal importance are the seeds of pumpkins and squashes, baked or fried and eaten as a snack, called *pepitas*. The seeds were also ground and used as a thickener and flavoring in sauces. They can also be pressed for a dark, rich, nutty oil used widely in Mexico and in Europe today. Even the squash blossoms are edible—they can be battered and fried as fritters or stuffed with other ingredients and sautéed, a practice common now in Italy, as it has been for ages in Mexico.

There are a number of other species also used in similar ways—the American chayote (*Sechium edule*) is one that can be sautéed or stuffed and baked. The plant resembles squashes, though they are unrelated. There is even a spiny variety, which is either peeled first or baked and the flesh scooped out. The so-called bitter gourd or bitter melon (*Momordica charantia*) also looks like a knobby cucumber. It is used extensively in Chinese cooking and in medicine as a purgative and classified as a "cold" medicine useful to counteract the summer's heat. For food, it is usually boiled or steamed to rid it of some of the bitterness. The taste, though an acquired one, is alluring. It can also be seeded, finely sliced, and stir-fried or lightly braised for true aficionados. As for regular cucumbers, they are members of a genus (*Cucumis*) that are normally eaten as fruits—that is, melons—but cucumbers are usually eaten either raw in a salad or pickled. In China they are also cooked as a vegetable. Here there are also several other Cucurbits, generally called *gua* in Chinese, that are treated similarly, such as the wax gourd or winter melon because it stores very well (*Benincasa hispida*), and the variety of this species known as hairy melon, which is eaten mostly in soup. The former are often made into a soup as well with fish or chicken, known as winter melon soup.

Okra is a fingerlike fruit of a plant native to West Africa whose name in Bantu, *ngombo*, gives us the word for the Creole dish *gumbo*. When sliced, its mucilaginous texture gives thickness and a slippery mouthfeel to stewed dishes. When left whole and cooked, these properties are mitigated. Although okra is relatively uncommon in Italian cooking, and only marginal in Mexican, it does play a role in China and Southeast Asia in general. Here it is added to soups, stews, and stir-fries much as any other vegetable would be, though its texture is especially prized as a conveyor of other flavors. In recent times a popular, industrially produced snack food is the okra pod dried and fried so it is light and crunchy. This plant should not be confused with what is called Chinese okra, though, which is actually a luffa (or, commonly, loofa), the same plant used as a bath sponge when dried. When eaten immature and green, it looks like a ridged cucumber, and it is cut up and cooked like a summer squash.

Fungi

It would take an entire book to catalogue the species of edible mushrooms used in our three cuisines. But there are some broad categories that may be applied to those cultivated and in most common use. In Italy, the porcini is absolutely revered as the most flavorful of **fungi**. It can be used fresh in dishes like risotto or dried and reconstituted in soups. It may grace a dish of *taglierini* (flat-cut noodles), or be stuffed with breadcrumbs and cheese and baked. As important are truffles, not only the renowned white truffles of Alba but throughout the Apennines and especially in Tuscany (San Miniato) and Umbria, spectacularly aromatic truffles are used in a wide variety of dishes. Unlike their black Perigord cousins, which are often merely shaved over a dish, white truffles are also gently cooked or used to flavor wild fowl such as pheasants. They pair perfectly with eggs, fragrant cheese like pecorino, and other antipasti. They can also be roasted whole in hot embers, a common practice in the past but prohibitively expensive today. Growing wild, there are also chanterelles (*finferlo*), often served as a side dish on their own—perhaps sautéed with garlic and parsley and sprinkled with lemon juice—called *trifolati*. There are also morels (*spugnola*) and other species like the honey agaric (*chiodino*), which is poisonous if not cooked, and the Imperial mushroom (*Amanita caesaria*), a close relative of the highly poisonous death cap. Italians are keen wild mushroom hunters, once the season begins in August or September, but there are also many cultivated varieties such as the ordinary field mushroom (*prataiolo*).

There are many native species of mushrooms in Mexico, used long before contact with Europe for both culinary and hallucinogenic purposes. Psilocybin was used in sacred rituals and considered flesh of the gods (*teonanacatl*), and was consequently almost eradicated by Catholic officials. The corn smut (*huitlacoche*) mentioned in the previous chapter is one culinary fungus, and there are many other wild varieties, generally called *setas*, though specific types also have their own names. Today, there are also cultivated familiar species such as oyster mushrooms and portabellas. Mushrooms are basically considered a meat substitute and thus replace it anywhere where one might ordinarily find meat, such as in tamales, tacos, stews, and with marvelous effect in quesadillas. Mushrooms are even made into a *molé de hongos* (mushroom sauce) and are paired with other ingredients that spring up in the rainy season, like *epazote* (an aromatic herb, *Dysphania ambrosioides*), greens, and fresh corn.

In China, the variety of mushrooms used is staggering, including more than a hundred species used in food and medicine. They range from lacy wood ear (*mu er*) fungus and cloud ear (*yun er*), used classically in sweet and sour soup, to delicate straw mushrooms, to larger dried robust shiitake known in China as *hua gu* (*chao gu*) or flower mushroom (figure 4.5). Their importance, as with all vegetables, stems in part from Buddhist vegetarianism. In cooking they flavor soups, are stir-fried with other vegetables, or are included in steamed dishes.

Stalks, Buds, and Shoots

Last, a number of vegetables fit into no neat grouping but are still very important in their respective regions. A few of them have found global appeal, but most are identified mostly or even exclusively with their land of origin.

Artichokes are a member of the thistle family, and along with cardoons—the edible **stalk** of a close relative—are a regular part of Mediterranean food culture. The advantage of artichokes is that they offer several separate harvests, including the small, immature **buds** and larger, tougher ones, which are usually boiled. Surprisingly, Italian artichokes are also thinly sliced and eaten raw in salads. There is also a classic fried artichoke dish from the Jewish community in Rome, which involves cleaning out the hairy inner core, flattening the artichoke with pressure from the stem end, and frying it, so it spreads out like a crunchy flower bud. Cardoons are eaten raw, as when dipped in the Ligurian *bagna cauda* (a hot mixture of olive oil, butter, and anchovies), or they can be trimmed, battered, and fried. The taste is subtle but addictive.

Celery is a similar tough, fibrous stalk that is used primarily in salads raw or diced finely along with onions and carrots as a base for sauces and other dishes. Both the stalks and leaves are also used in Chinese soups and stir-fries, as is a similar plant sometimes

Figure 4.5. Wood ear mushrooms are an indispensable ingredient in hot and sour soup.

called Chinese celery or water dropwart (*Oenanthe javanica*), which looks like the familiar parsley, but several members of this genus are highly poisonous. Mention here should also be made of a plant indigenous to northern Asia, whose primary importance was medicinal as a purgative and febrifuge: rhubarb. Only the red stalks are used, as the leaves contain dangerous concentrations of oxalic acid. In the West it is used primarily like a fruit because of its sourness, in pies and jams, usually paired with strawberry.

Asparagus is an important Mediterranean plant, which has been introduced in recent centuries to China and Mexico, though it is its wild relatives whose roots are used in Chinese medicine. Normally asparagus is cooked as quickly as possible, blanched or sautéed to retain its fresh green flavor, and there was even an ancient Roman proverb, "*citius quam asparagus coquntur*" (faster than asparagus cooks), which suggests that its particular culinary charms were appreciated millennia ago. Asparagus spears can also be mounded with soil while growing to prevent the formation of chlorophyll, leaving them white, a specialty of Bassano del Grappa in the Veneto. Recently there have even been purple varieties cultivated, which actually lose their color in cooking, like many purple vegetables. Once interchangeable with asparagus in terms of cuisine, hop **shoots** were widely eaten as a kind of springtime tonic vegetable. In the same context, vine tendrils were also eaten and are a pleasant sour, crunchy addition to salads.

A uniquely Asian vegetable, popular in many forms, are the shoots of several bamboo species (*Phyllostachys spp.*, *Bambusa spp.*, *Dendrocalamus spp.*). Although hard and woody when fully grown, the thick shoots when they first emerge from the soil are tender and sweet after being blanched to remove bitterness. They are then finely sliced and added to stir-fries and soups, but they can also be pickled or even candied. Apart from being an ornamental plant, bamboo has also been used for centuries as a building material and for household and kitchen tools, a use which is catching on now in the West, especially for cutting boards, steamers, spoons, and even durable flooring material. Perhaps most important, chopsticks are traditionally made of bamboo—an easily renewable resource.

Seaweeds and Their Relatives

Another plant—or technically a large alga—used extensively, though not exclusively, in Asian cuisines is the family of **seaweeds**. Although normally associated with Japan, the use of seaweed in China is of great antiquity. The most commonly used forms are types of kelp, known by the Japanese names *kombu* and *wakame*, and purple laver, though there are hundreds of edible species. Fresh or dried seaweed is used mostly in soups or as a kind of medicinal tea. Agar-agar is a seaweed derivative, used as a thickener in desserts and other dishes, which is similar to gelatin in effect. It is sold as a powder or dried strips and has the advantage not only of being vegetarian but also of setting at room temperature, unlike animal-based gelatin. Strands of soaked agar are even used as salad, though its most common use today is in science as a culture medium. It has also very recently drawn attention among practitioners of modern scientific cooking, though it has long been an ingredient in industrial food manufacture.

There is also a kind of microorganism, once considered a blue-green alga, now classified as a cyanobacterium, called hair vegetable (*Nostoc commune* or *fa cai* in Mandarin), which is used in Chinese soups and also as a thickener. Because it often survives underground and only appears after a rainfall, it is also called fallen star or star jelly. It has no visible roots, thus in China it was compared to the "heavenly immortals" who wander the earth at will. Its name in Chinese also sounds similar to the phrase "acquiring wealth," so it is eaten on New Year's for prosperity, though it is increasingly hard to find nowadays. Another cyanobacterium was an important part of the Aztec diet: *teocuitlatl* ("stone's excrement"), commonly called spirulina today. It was described by the colonial Spanish botanist Hernández as a kind of pond slime, dried in the sun and formed into cakes, which he described as tasting like cheese. Ironically, although neglected as a food today, it is an important dietary supplement because it has a very high percentage of protein and all the amino acids, more so than any other plant foods.

Cactus

Another important vegetable in Mexico is the **cactus**. Cacti (*Opuntia*) yield an edible fruit known as tunas or prickly pears but also a vegetable from the nopal cactus. The spines are removed from the pads of the cactus, which are sliced and marinated in an intriguingly tasty and slimy salad, and are also used to fill tacos (figure 4.6). The fruits are made into jellies, candy, and sour drinks. The name in Nahuatl is *tenochtli*, which gives the name for the Aztec capital Tenochtitlan, which in myth was chosen when an eagle bearing a snake landed on a cactus plant. This symbol is still on the Mexican flag. Surprisingly, these cacti were introduced into southern Italy and Sicily, where they are called *figu d'India* (Indian fig).

Capsicums

Also of crucial importance to Mexican cuisine in the past and today is the chili pepper (**Capsicum**). These peppers might easily be classified as a flavoring, but their use is just as often as a vegetable, so they are included here. They have been of global importance since the sixteenth century, being one of the few foods adopted almost immediately in places as far-flung as Africa, Turkey, and Hungary (paprika), and in Southeast Asia and China, as well as Spain, where the pods are smoked, and in southern Italy where crushed red pepper is a ubiquitous spice. So, too, are various forms of "bell" peppers eaten sautéed, stuffed, marinated, and so forth.

Figure 4.6. Sliced and marinated cactus paddles are known as *nopalitos*.

VEGETABLES

In Mexico, there are hundreds of chili varieties, eaten fresh either green and immature or ripe, as well as many that are dried. The word in Mexico is spelled *chile*, in the United States *chili*, and sometimes elsewhere *chilli*. Some are fairly mild; others incendiary. Some are roasted and eaten as a vegetable, stuffed, or as a taco filling; others are pounded into vibrant sauces. The hottest are fresh red or green habañeros, shaped like and similar to the little scotch bonnets familiar in the Caribbean, which are usually used roasted, in sauces, and cooked dishes. Jalapeños are on the other end of the heat scale and are thus often pickled or chopped into raw sauces. The smaller and hotter serrano can also be pickled but is also roasted or charred and served as a condiment. The poblano is a larger, green, bell-shaped pepper, usually stuffed because it is large, though when it is dried it becomes an ancho. Large, fresh chilies are roasted on a *comal* (griddle) or charred directly over flames. This enables the skin to be rubbed off, the seeds removed, and then the flesh stuffed (as in chiles rellenos). A dried chili pepper can be ground whole, or briefly toasted, reconstituted in water, seeded, and then pounded or processed in a blender as the base for a sauce, which is more typical. In this category the most familiar types are long, brown, prunelike *pasillas*, the rounder mulato, also brown and used in molé, leathery smoked chipotles, as well as small, thin, red chilies *de arbol*, which are similar to Thai or bird chilies. Although the nomenclature can be bewilderingly different from region to region, there are also small, round cascabels; large, red *guajillos*; smoked, red *moras*; and more. Chilies should be considered one of the basic building blocks of Mexican cuisine flavor profiles.

As mentioned, the chili pepper was immediately embraced in China as an inexpensive substitute for black pepper, and a plant that could be grown domestically. Peppers have become signature flavors in Sichuan and Hunan cooking, and though also dried, are usually either simply tossed into a cooked dish, perhaps halved and seeded, or ground into powder. Unique to this cuisine, they are used to flavor chili oil, which is a widely used condiment in cooked dishes and on the table. Chilies are also pickled and made into sauces and fermented paste.

Chilies are as important in Italy, though as a spice they are merely coarsely ground and added to piquant sauces or pasta, such as the tomato-based *arrabbiata* (angry/rabid) sauce. Fresh peppers are char roasted as in Mexico, but then marinated in oil and garlic and eaten as an appetizer or stuffed with meat and baked whole. They can also be sliced and fried with onions as an accompaniment to sausages or as a pasta sauce.

Solanaceae

Solanaceae is a widely diverse family of plants with members on both sides of the Atlantic. Most of its European members, known there as nightshade and belladonna, serve only as medicine or poison, but the eggplant is one major exception. Probably native to India, it spread both to China and to Italy with the Moors in the Middle Ages. The name in U.S. English originally referred to a small, white, egg-shaped variety. Elsewhere it is called an aubergine, which comes from Arabic, as does the Spanish *berenjena*.

The Italian word *melanzane* is fancifully thought to derive from the word *meli-insani* ("unhealthy," if not insane, "fruit"). In Italy eggplant is normally fried, either on its own or with breading. Laden with tomato sauce and cheese it then becomes the classic *melanzane alla Parmigiana*. Eggplant are also added to slow-cooked stews such as *caponata* and similar ratatouille-like dishes. Slices of eggplant can also be rolled around ingredients like cheese or baked in casseroles. In Asia, which has its own distinctive varieties, some are long and black like Japanese eggplants or tiny, green, and bitter ones common in Southeast Asia. These, too, are long stewed or sliced finely and stir-fried with typical flavorings such as soy and ginger. Eggplant also naturally makes a great substitute for meat in vegetarian dishes and as such is often steamed in place of pork, with black beans and chili flakes.

Mexico has never taken a great liking for eggplant, though it was introduced there by the Spanish and does figure in some Mexican dishes today. The New World has its own *Solanaceae* species as well: the potato; tobacco (both have trumpetlike flowers typical of the species); the pepino, a fruit of South America; and the tomato.

As a good segue to the fruits, the tomato is native to South and Central America. In Mexico, confusingly, the word *tomato*, coming directly from Nahuatl, refers both to what is bizarrely called in botanical Latin *Lycopersicon esculentum* (the edible wolf peach) as well as the small green tomatillo (*Physalis ixocarpa*), called *tomate verde*, although it is not a green tomato. Confusingly, the tomato is referred to in Mexico as *jitomate*, though in Spain it is simply *tomate*. The tomatillo has an unmistakable papery husk, which is removed before cooking (figure 4.7). Tomatillos can be roasted and then chopped in a fresh, green salsa with garlic and chilies or stewed gently until soft for a sour green sauce. For mysterious reasons, they are not generally used outside of Mexico, probably because they have specific growing requirements.

The original wild tomato is a tiny, cherrylike fruit full of seeds. In Mexico tomatoes of all shapes and sizes are eaten raw sliced or chopped in a *pico de gallo* sauce (almost a fine salad really) with cilantro and chilies, or pounded in a *molcahete* (stone mortar) in a fresh salsa. They are also cooked into sauces, either simmered briefly in water and then sometimes skinned before cooking with other ingredients, or roasted, crushed, and strained, or merely blended into a sauce skin and all. This is how a salsa ranchera is made: the raw tomatoes, garlic, and chilies being blended and then fried in oil, then added to other dishes such as *huevos rancheros*. They can also be halved and grated, which removes the skin and seeds, though this technique is more common in the Mediterranean.

Interestingly, tomatoes were introduced to Europe in the sixteenth century and were planted by botanists, but they only appear in cookbooks at the end of the seventeenth century, in a Neapolitan cookbook by Antonio Latini, in the form of a "salsa" very similar to that in Mexico. The first proper cooked Italian tomato sauces appear in the eighteenth-century cookbooks of Vincenzo Corrado, also from Naples. Obviously, tomatoes subsequently became a signature ingredient in Italian cookery, especially in the south. In northern Italy tomatoes tend to be cooked in long-simmered sauces using meat, as in

Figure 4.7. Unrelated to tomatoes, the tomatillo is used in Mexican green sauces.

the classic Bolognese ragù. Tomato sauce comprised primarily of tomatoes is more common in the south and graces pasta, pizza, cooked vegetables, meat or fish, soups—just about anything. They are also, of course, served raw in salads, such as the classic insalata Caprese of tomato slices, buffalo mozzarella, and basil leaves. Tomatoes are also stuffed with breadcrumbs and cheese and baked, or they can be sun dried and marinated in oil or lightly grilled and drizzled with oil as an appetizer. It is hard to even imagine Italian cuisine without the Mexican tomato. Although the most famous of Italian varieties is the San Marzano tomato, often called in the United States a Roma tomato, whose fleshy interior is ideal for sauces, there are dozens of other varieties as well. Italian families will often preserve their summer tomatoes in jars or dry them as a *passata* (concentrated paste). Canned tomatoes are also perfectly acceptable out of season.

As ubiquitous in Italian food as tomatoes are, they seem utterly foreign to Chinese food, but actually this is not the case, although the tomato is called a "foreign eggplant" in Chinese. Tomatoes are most common in Cantonese cooking, where they are cut into wedges and cooked into soup or stir-fries, but rarely as a sauce. One principal exception would be ketchup, an adaptation of a Southeast Asian condiment now also made with tomatoes, which has found a legitimate place in Chinese cookery, especially in sweet and sour dishes. Another surprising use of tomatoes in China, as perhaps the most popular of comfort foods, is chopped with scrambled eggs.

Study Questions

1. Consider how technologies meant originally only to aid in preservation may have also contributed to human health.
2. Why have some vegetables been thought of as fit for elite diners while others are considered rustic and lowly? Why do some people esteem them nonetheless?
3. What role do plant breeders play in furnishing cuisine with new varieties, and how is this important for the evolution of cuisine?
4. How do the basic flavor profiles of these three cuisines follow the same structure in use of salt, sweet, sour, bitter, aromatic, even though widely different ingredients are used?
5. Why are some plants assigned medicinal properties, and in what ways has modern science confirmed similar values?
6. Describe the role of "mouthfeel" and various slippery textures in cooking, but also as a conveyor of other flavors.
7. How are vegetables used as a substitute for meat, and why?
8. What vegetables, though non-native, have been so thoroughly adopted elsewhere that they have become signature ingredients? How and why has this happened?

Recipes

Salads

INSALATA MISTA (MIXED SALAD)/ITALY

The salad is ubiquitous in Italian cuisine but rarely comes out at the start of a meal. More commonly it accompanies the main course as a side dish (*contorno*) or comes after it and before dessert. The idea is that it cleanses the palate and aids digestion. Some people think the vinegar in a salad will clash with wine drunk during the meal, so they wait until after the main course. This is entirely up to you. In any case, the ingredients should be fairly simple: lettuce and radicchio and bitter greens like dandelion, arugula, watercress, and chicory. Celery is fine, as are small tomatoes, but too many ingredients misses the point of a bitter green balanced with salt, acid, and the roundness of oil. Make sure the greens are impeccably dry as well. Put some red wine vinegar in a bowl with some salt. Slowly drizzle in three times the amount of excellent extra-virgin olive oil, whisking continuously. Then toss the salad in the bowl and serve immediately.

STIR-FRIED LETTUCE/CHINA

Interestingly, lettuce is usually cooked in China rather than being eaten raw. Choose a sturdy variety with a little bitterness, like romaine. Wash a head of lettuce and thoroughly dry. Chop into bite-sized pieces and discard the base. In a superheated wok, add a few tablespoons of peanut oil and immediately throw in the lettuce. Let it fry just a minute without disturbing, then toss with the spatula. Add to the wok a few tablespoons of oyster sauce, sprinkle on some black sesame seeds, if you like, and a dash of sesame oil and mix thoroughly.

ENSALADA DE NOPALITOS (CACTUS SALAD)/MEXICO

Nopalitos are sliced cactus paddles. You can buy marinated cactus in a jar, but they are fun to make yourself. Cut off the entire edge of the cactus paddle. Remove all the spines meticulously, scraping with a paring knife, and being careful not to get any in your skin. Slice the paddles into thin strips. Cook these gently about 15 minutes in a covered pan with a little oil. You'll notice all the mucilaginous juices seeping out. Remove these and let cool. Add a lot of lime juice, salt, cilantro, and some diced jalapeños and sliced onion. Let this marinate in the refrigerator at least an hour or until all the flavors come together. Put some lettuce leaves on a plate and top with the *nopalitos*.

Mixed Vegetable Dishes

CAPONATA/ITALY

Caponata is a Sicilian dish that always includes eggplant in a sweet and sour sauce but usually includes a wide variety of other vegetables such as peppers, onions, and celery as well as olives, capers, and sometimes pine nuts. Oddly enough, in its earlier versions it also contained beans and fish and may be a form of an even earlier Catalonian recipe, which is possible as the Aragonese ruled both Catalonia and Sicily after the sixteen century and there are definitely culinary connections between the two.

Cut up the eggplant into cubes, leaving the skin on. Salt the cubes generously and let them drain in a colander for about half an hour. Squeeze out remaining water. Fry these in olive oil in a large pan. Add chopped tomatoes, sliced red and green bell peppers, and a sliced onion and continue frying on low heat. Add the olives and capers, and, if you like, a few anchovies and a crushed clove of garlic. When everything is well melded, add ½ cup of good red wine vinegar, a tablespoon of sugar, some salt and pepper, and a dash of cinnamon. Some chili flakes are also good if you like heat. Garnish with pine nuts. This can be eaten hot but is even better at room temperature.

SAUTÉED CHAYOTE/MEXICO

Chayote is a greenish oval fruit, often referred to as a squash. Peel the *chayote* and cut into chunks; about four or five make a substantial side dish. Fry this in a pan with some oil. Meanwhile, put a few tomatoes over an open flame, hold with metal tongs, and char them all over. Do the same with a fresh, green chili. Remove the seeds from the chili and scrape off some of the char if you like. Put the tomato, skin and all, and chili in the blender and process or pound into a fine paste in a mortar. Add this to the *chayote* and taste for salt. Let cook until the *chayote* is just barely tender. You can grate some fresh, white cheese on it before serving.

CASSEROLE OF ZUCCHINI/ITALY

When you have a bounty of summer zucchini, and this recipe uses many, they cook down to a manageable proportion. A plastic grocery bagful makes a decent side dish serving or dinner for two as a main course. Remove the tops and bottoms of the zucchini. Then with a sharp knife or a *mandoline* (a sharp blade set into a track in a rectangular frame), cut them lengthways into thin strips. These can be salted at this point and treated like pasta. For this dish, pour a little olive oil in the bottom of a ceramic dish and lay down a layer of zucchini, overlapping each strip. Sprinkle with salt, some oregano, and a splash of olive oil. Continue until all the zucchini are used up. Either place this in an oven and bake slowly, uncovered, or, if you are in a rush, cover with plastic wrap or a lid and microwave on high for about 10 minutes. It is one of the few dishes that really works well cooked in a microwave. It will take about 45 minutes in the oven, but in the microwave, you can pour off some of the excess liquid and cook it a further 5 minutes uncovered.

Now, if you have decided to spend some time in the kitchen, there is a more complicated and more interesting way to do this dish. Take the slices of zucchini and salt and season them. Dip them in a beaten egg and then into very fine breadcrumbs. Fry these in a pan in a little olive oil and set on paper towels to drain. (This can be done with eggplant as well.) Then lay the slices in your casserole and bake, if you like, with a little tomato sauce between each layer and a little mozzarella on top, melted. This definitely serves as a main course, sliced into thick wedges.

BUDDHA'S DELIGHT (LUOHAN QUANZHAI)/CHINA

Traditionally this recipe should contain no fewer than eighteen ingredients, recalling the eighteen different forms Buddha took. A vegetarian dish, it is also particularly appropriate for monks and for devout Buddhists at the start of the New Year as a purifying food. Familiar vegetables such as carrot, Napa cabbage, snow peas, and water chestnut go in, as well as lotus seeds, ginkgo nuts, day lily buds soaked in water, wood ear mushrooms and soaked black mushrooms, reconstituted bean curd sticks, bamboo shoots, peanuts, fried tofu, and cellophane noodles. But also add bean sprouts, baby corn, and bok choy. Almost any vegetable works well, but some of these have specific symbolic associations. For example, carrots cut in rounds symbolize money; ginkgos, silver ingots; the noodles, long life; the peanuts, birth. Cut everything into bite-sized pieces, keeping in mind that the pleasure of the dish is in varying textures and shapes. Stir-fry the tougher vegetables first, then add the more delicate ones. Once they are cooked through, lower the heat and braise them gently in soy, rice wine, and a little sugar another 10 minutes or so, adding a few drops of sesame oil at the end. Serve.

BRAISED GREENS/ITALY

Any type of fresh, tough, leafy green works here, preferably with a little bitterness. Kale, chard, escarole, or broccoli rabe are all excellent. Remove the tougher ribs, and then chop up the leaves finely and sauté in a pan in some olive oil with a pinch or two of salt. You can stop here if you prefer simplicity. Or add a little crushed garlic and a few chili flakes, or if you want something remarkable, add some chopped almonds and golden raisins. Perhaps an anchovy and a pinch of sugar. Add a little water and let this cook slowly until the greens are tender. It is especially good in the springtime as a blood purifier. This is served as a side dish (*contorno*).

Stuffed Vegetables

CHILES EN NOGADA/MEXICO

This recipe was created in Puebla to commemorate Mexico's independence from Spain in 1821. The green chilies, white sauce, and red pomegranate seeds recall the Mexican flag. With the exception of the chili itself, it is positively medieval in its combination of ingredients and technique.

Soak a bowl of shelled walnuts the night before in water. The next day, roast some fresh green *chiles poblanos* over an open flame until blackened and blistered all over. Put them into a large paper bag and let them steam about 15 minutes. Then scrape off the skin, but keep the stem intact. Make a slit down one side of each chili and remove the seeds. Then take some chunks of pork shoulder and simmer in water with onions and aromatics for about an hour until tender. Shred the meat finely. Next, brown chopped onions and garlic in a pan and add the shredded meat. To this add pepper, salt, cinnamon, and ground cloves. Then add peeled and chopped fruit, a firm peach, apples, and pears. Add raisins, pine nuts, almonds, some candied fruit peel, and if you can find it, *biznaga* (candied cactus). This mixture is called *picadillo*. Place it inside the chili peppers.

Make a batter by beating egg whites and yolks separately and then folding them together gently. Dredge the chilies in flour, dip in the egg mixture, and then fry them in a pan in some oil. It will be very messy until the batter solidifies. Set these aside.

Make the sauce by peeling the brown, papery skin off the walnuts, and pound them in a mortar with a crustless piece of bread, some goat cheese, a splash of sherry, and a little sugar. It should be a smooth, white sauce. Cover the chilies with the sauce and to serve, sprinkle on some pomegranate seeds and maybe a flourish of parsley.

MELANZANE RIPIENO (STUFFED EGGPLANT)/ITALY

You can use the mixture for the stuffing for this eggplant dish in virtually any vegetable—a cored tomato, bell peppers, zucchini cut in half vertically and hollowed out, or little rounds of zucchini work very nicely as well. Use the insides in the stuffing, too. Brown some ground beef in a little oil. Add to it chopped onions and garlic, some oregano, parsley, and salt and pepper to taste. Cut the eggplant in half vertically and scoop out the insides, leaving just a little flesh on the skin. Many recipes will have you salt and drain the eggplant or even parboil it; this is unnecessary as rarely are eggplants very bitter nowadays. Add in the scooped-out eggplant, finely chopped, to the meat mixture. Cook a few minutes until everything is amalgamated. Remove from the heat and add in some breadcrumbs and an egg to bind. Stuff the eggplant and then cover with tomato sauce and some grated Romano cheese. You can top with some mozzarella, too, if you like. Bake at 350 degrees for about 45 minutes to an hour.

FRIED ASPARAGUS OR BABY ARTICHOKES/ITALY

Nearly every culture has a method of batter-coating and deep-frying vegetables. Some batters use starch and just a little water and are light and airy; some use eggs and are heavier but sturdy. Using alcohol in the batter, such as beer, causes a rapid evaporation of moisture, giving you a very crunchy coating. These are merely random examples, but any technique can be easily combined with another vegetable.

Break off the lower end of the asparagus and peel the lower half. The fatter the asparagus the better; thin ones tend to be stringy and have a higher proportion of skin to flesh. Put these in a plate sprinkled with lemon juice so they do not discolor. Meanwhile, heat a large pot filled halfway up with vegetable oil. You definitely can use olive oil for this, if you can afford it. Make a batter using regular flour and a dry beer; Prosecco (an effervescent wine similar to champagne), or even vermouth is interesting. Season with herbs such as marjoram. There is no need to add salt yet; it sometimes makes the batter soggy. Your oil should be about 350 degrees. Dip each asparagus spear into the batter and quickly slide into the oil, just a few at a time. Let cook until golden brown, remove, salt well, and set on paper towels to drain. You can serve this with grated Parmesan cheese, too, if you like. For artichokes, choose the smallest ones, peel off the outer few leaves, peel the stem, cut in half, and remove the bit of fuzz in the center if you like, or you can leave them whole if small enough. Dip in the same batter and fry the same way (figure 4.8). This batter also works well with squash blossoms or any vegetable that is not too wet or soft; broccoli florets are especially good.

SCALLION PANCAKES/CHINA

These are a popular street food, especially in northern China. The dough is unusual in that it is made with boiling water. Put all-purpose flour in a bowl. Drizzle in a little oil, and pour in boiling water just enough so it comes together as a dough when you stir it. Knead this dough. Roll into a snake and let rest a while. Cut this up into small nubbins. Roll each out into a circle on a floured board, then top with some green scallions, cut into little rounds. Roll this up into a cigar shape, and then roll that up into a wheel. Set these aside. When done with them all, flatten each one and roll out into a thin pancake. The scallions will all be inside. Fry these in a shallow pan of oil until golden and crisp. You can serve these with a dipping sauce of soy or something more complex, perhaps with a little vinegar and ginger.

Beans

FRIJOLES REFRITOS (REFRIED BEANS)/MEXICO

So as not to hide the flavor of the bean, it is best to season this refried bean dish as lightly as possible. Soak a cup of beans overnight in water. You can use pinto beans but any variety of *Phaseolus* will work fine, even red or black beans. Drain the water. Bring the beans and a good pinch of *epazote* (an aromatic herb used widely in Mexican cooking) to a boil in a pot of fresh water, just covering the beans, and skim off the froth. Lower the

Figure 4.8. Baby artichokes are best battered and fried whole.

heat and let simmer gently until tender, about an hour or longer. Do not worry if some fall apart. Much of the water will have been absorbed; if it gets too dry you can always add more, but do not start with too much water or you will end up losing much of the flavor if you have to discard some. Add salt to taste, pepper, and oregano. Next, fry some finely chopped onions and some *chiles de arbol* (small, thin, red chilies). Add them to the beans and mash them into a fine paste with a potato masher, or you can use a blender, though the mixture will need to be more liquid. Then, heat a pan with oil or lard and fry the bean puree until it becomes thickened. This can be served as a side dish or as a filling in tacos or other dishes.

FRIED TOFU/CHINA

In large kitchens in China, tofu is often deep fried until crispy and then seasoned, but there is a much easier way to do it using much less oil. Cut a block of extrafirm fresh tofu into cubes. Let them drain on paper towels for a few minutes. Heat a flat, nonstick pan and add a few tablespoons of oil. Add the tofu and cook on very low heat, turning the cubes occasionally until browned on all sides. This will take a while. When brown and crispy, add soy sauce, some Shaoxing wine, a pinch of sugar, and let the tofu absorb the liquids. You can serve as is with some chopped scallions, or add the seasoned tofu to other dishes. For example, the tofu will maintain its shape and become chewy if braised with vegetables. It can also be added to stir-fries and just about anything. You can also buy deep-fried tofu for these same uses, but it tends to be oily.

FRESH STEWED FAVA BEANS/ITALY

If you have really young, tender fava beans, you can cook these in the pod or use whole pods of Romano beans, which look like string beans but are wider. Regular fresh fava beans are equally good shelled, and if you like with the skin of each bean removed as well. It is laborious but well worth the effort. Start by simmering the beans in a pot of water with salt, oregano, and olive oil. You can also add a sliced onion. Then add some fresh tomato puree. You can do this by peeling the tomatoes, crushing them, and adding them, or simply cut them in half and grate them into the pot, leaving the skins behind. The idea is to gently stew these; shelled beans will take less time, whole pods longer. Hint: the more olive oil you use the better. Some people cook these in half oil and half water.

Vegetable Soups

COLD AVOCADO SOUP/MEXICO

Some recipes call for frying chilies or other ingredients, but the flavors are purest when nothing is cooked. Simply take a few avocados, add light chicken broth and some cream or milk, and process until smooth. Chill and serve with some finely chopped cilantro on top.

MINESTRONE DI VERDURE (VEGETABLE SOUP)/ITALY

There are infinite ways to make minestrone, but this is a simple classic typical of Milan and Lombardy (figure 4.9). First, soak some dried beans overnight, and the next day drain and boil them in fresh, unsalted water until just cooked through. Save the boiling liquid if you like. Then make a vegetable broth. Take potato peels, carrots, celery, onions, the green tops of leeks, mushroom stems, and any woody stems from herbs you have been using. Anything, including celery leaves or turnip greens, is fine. Add in the bean boiling liquid, too. Simmer this in a pot for 30 to 40 minutes and then strain, pressing on solids to remove all the flavor possible. Salt to taste.

Dice some pancetta (unsmoked, rolled Italian bacon) and melt slowly in a pot, leaving the bits of pancetta in the pot. Or, if you prefer, heat some olive oil. Gently cook finely diced carrots, celery, and onion in the fat. Add some chopped sage leaves. Pour in the vegetable broth. Next, cube a potato and add that, as well as two or three seeded, peeled, and chopped tomatoes. You can do this by cutting an *x* into the base of the tomato, dropping it in boiling water for about 10 seconds, and then slipping off the peel. Or just chop the tomatoes and add if you do not mind the peels. Add in the beans and continue to cook. Then add in a cup of short-grain rice. Continue to cook until the rice is tender. Serve as a first course with good, hearty bread.

Figure 4.9. A minestrone or Italian vegetable soup.

RIBOLLITA (BEAN SOUP)/ITALY

This classic Tuscan soup is better the simpler it is prepared. Tradition has it that this bean soup has an ancient pedigree, but considering that it is always made with New World beans, it cannot be quite that old. The bean species may have changed over the years, but the basic procedure has remained the same. The name means "reboiled"; the beans are cooked first and then made into soup. Start with dried cannellini beans, which should be soaked overnight. Drain off the water and add fresh water to the beans in a pot—just to cover. Bring to the boil, then lower the heat to a gentle simmer. Cook covered, about an hour. When they just begin to soften, add salt. Cook until tender. Then gently sauté finely chopped onion, carrot, and celery in olive oil. You can also add a few peeled tomatoes if you like. Add the vegetables to the beans and season with some oregano and a bay leaf. Continue to cook very gently until some of the beans begin to fall apart, perhaps as long as an hour depending on how old your beans are. The older the beans, the longer it will take to cook, but there is no way of knowing their age, so this is completely variable. If necessary, add more water. Add some finely sliced *cavolo nero* (black kale). Collard greens work fine as well, but you will have to add them earlier in the cooking process. Toward the end of cooking add in a handful of breadcrumbs made from saltless Tuscan bread. Serve with a generous drizzle of olive oil. Some people also like grated cheese.

⋅{ 5 }⋅
Fruits and Nuts

Learning Objectives

+ Appreciate the diversity and importance of fruit in the three world cuisines.

+ Understand how these plants have been historically shared among various continents and why some still remain obscure.

+ Consider the ways acidic and sharp fruits serve as a flavor accent in many dishes.

+ Understand the nutritional importance of fruit and nuts in the human diet through the evolution of our species and their role in traditional medicine.

+ Reflect on the many culinary uses of nuts in sauces and for textural contrast in historic cookery.

ALTHOUGH FRUITS and nuts might appear peripheral to the core of culinary practice, they are actually very important beyond eating out of hand. They figure prominently in many sauces, and as a source of essential nutrients, they have always been important in the human diet. Many anthropologists even speculate that our ability to see colors evolved as a way to recognize ripe fruit. The surprising ways fruits and nuts figure in Italian, Mexican, and Chinese cuisines reminds us of the many inventive ways our forebears dealt with ingredients, many of which have been lost (figure 5.1).

Figure 5.1. Produce market, North Gate, Dali, Yunnan, China. © Rod Porteous/Robert Harding World Imagery/CORBIS.

Berries

Berries are without doubt among the important foods that sustained our hunter-gatherer ancestors, and the very act of picking berries probably played some role in the evolutionary development of our nimble fingers and keen, concentrated vision, and even our ability to discern colors. Most berries require no processing and are typically easily accessible to any person, or animal, who can grab them. Their primary importance is as a source of vitamin C, an essential nutrient especially in winter when fresh, green vegetables are unavailable. Many berries can be easily dried, processed, or cooked into preserves or jam, which is most common for the more sour species, or even fermented into alcoholic beverages, as with elderberry wine. Most were also, from prehistoric times to the present, gathered wild, and berries only came to be bred and commercially cultivated in the past few centuries. This is probably because most berries are delicate and are crushed in transport, or spoil after a few days. Thus, they are best picked directly off the bush and consumed immediately. The best example of this is the luscious inky mulberry, which is so fragile that it has rarely been successful in commerce—apart from silkworm culture. It is also difficult to harvest since it grows on a tall tree. White mulberries, however, can be dried, and this is the species used in silk culture. As well, practically all parts of the tree have been used in Chinese medicine. The absence of commercial viability and trade in berries means that they are largely missing from the historic record and do not move in fresh form across continents as do other foods, until recently. Within the past few centuries, they appear in cookbooks, are baked into pies and tarts, or are cooked as preserves.

Berries fall into several distinct families. There are those in the *Rubus* genus, native to Europe, Asia, and North America, including raspberries, blackberries, loganberries, boysenberries, tayberries, and many other related cultivars, many of which were developed in modern times. In the kitchen they are more or less interchangeable, usually sweetened to balance their acidity, and often cooked and sieved or passed through a food mill to remove the tiny seeds. In Italy, these berries are most commonly found as flavors for gelato and *sorbetto*. They can also be used to flavor liqueur, and in the Renaissance they were all commonly used in sauces as a condiment with meat and fowl (*sapori*). Actually, turkey with cranberry sauce is an odd rudiment of a now lost, essentially medieval, flavor combination. Apart from common raspberries, there are many less familiar types native to China and the Himalayas, including the yellow raspberry, the central China raspberry from Hubei, the Yiching raspberry, and the Chinese blackberry of Guangxi and Yunnan. These are usually eaten fresh but may be used dried and in tea or fermented into wine. In Mexico, which now has its own flourishing industry, raspberries are most commonly found in desserts, sweet empanadas, or candy. Occasionally, one does find raspberries in a nouvelle salsa, chili sauce, or in the once popular raspberry vinegar in restaurants globally.

Cranberries, a member of the genus *Vaccinium*, are native to North America, northern Europe, and Asia. They are related to the blueberry and bilberry—the latter known in Italy as *mirtillo*—and are gathered wild in the alpine regions. These are likewise used to accompany game meats or to flavor desserts and ice cream. Huckleberries, lingonberries, and whortleberries are all relatives.

It is strange that berries do not figure conspicuously in Chinese cuisine, as most of these species are found wild there, though they do play a major role in Chinese medicine. It may be that most berries are merely snacked on by passersby and, because not cultivated or a regular part of trade, never came to be used in the kitchen. These berries are not an important part of Mexican cookery either, though again, this may because they are eaten locally and gathered wild without any preparation.

A number of lesser berries have also been gathered wild and made into jams or have been used in alcoholic drinks and medicine. The barberry is one example, and the bearberry (uva ursi or *Arctostaphylos*) is common in the Alps and in mountainous parts of China. The elderberry is used to make wine or sambuca liqueur, and its flowers are also used in fritters. In China there is the wolfberry (*goji*), in the *Solanaceae* family, which has been used in teas and medicines for thousands of years and is mentioned in Shennog's ancient medical text (figure 5.2). Shennog was the celestial emperor said to have introduced agriculture some five thousand years ago. Goji berries have recently gained much attention outside China as a powerful antioxidant. They are usually sold dried

Figure 5.2. The goji berry is used in traditional Chinese medicine as well as modern medicine.

and can be eaten like raisins, cooked in soup, or sprinkled onto rice porridge. The fruits of the black nightshade and huckleberry are in the same family, though the latter term is also used in the United States for a kind of blueberry in the *Vaccinium* genus. When nightshade berries are green they contain a deadly poison, but when black and ripe they are used for jams and in folk medicine and sometimes eaten fresh. Similarly situated somewhere between food and medicine is the saw palmetto berry, used by the Maya in a tonic drink and today a staple of the alternative medicine industry.

The *Ribes* is a genus of berry, including currants and gooseberries, relatively unknown in the United States until recently. This is because they host the devastating white pine blister rust, a fungus that was introduced accidentally early in the twentieth century, and so cultivation of *Ribes* was halted in an effort to control the fungus. They survived in northern Europe and northern Asia, though. Also, strangely, the word *currant* or *zante currant* has come to be used for a type of tiny, dried raisin, totally unrelated to these, and its name may even derive from the city of Corinth in Greece rather than from any superficial similarity to true currants. True currants can be white, red, or black and are generally quite sour, so they mostly figure in sauces and jams or as a garnish. Black currants are used often as a flavoring in candy as well as the British drink syrup Ribena, popular in

Hong Kong. The gooseberry is known in Italy as *uva spina* or *uva marina* and was in the seventeenth century a popular garnish and source of sourness in sauces. It is, incidentally, unrelated to what is called the Chinese gooseberry, better known as the kiwi fruit, which will be discussed below. Regular gooseberries, introduced from Europe, are grown in northern China, and a Chinese black currant is made into wine in Hubei.

The best strawberries (*Fragaria spp.*) in Europe are a small, wild fruit gathered in forested mountains (*fragolina di bosco*). The large types common in the United States and Europe are a cross between a North American and Chilean species. The name in English derives from the practice of placing straw between the rows to keep moisture in and prevent weeds. Since they are a low-lying perennial herb, this also makes picking them easier—usually done on the hands and knees. Strawberries have been seen on Italian tables, usually in a closing dessert course in banquets, at least since the sixteenth century, but they are also, remarkably, drizzled with balsamic vinegar, the dense sweetness of which tempers the fresh acidity of the fruit and makes for a splendid combination. Strawberries figure most often in pastries and tarts, unsurprisingly. The leaves, too, were once used in salads and medicinal cordials. There are also species of wild strawberries in China, but most remarkable is an unrelated plant, the *yangmei* or strawberry tree (*Myrica rubra*)—not to be confused with the European arbutus, also called a strawberry tree. *Yangmei* is a deep red, round, and sour fruit, eaten fresh as the quintessential thirst quencher and is often preserved in sugar, canned, used to flavor *baijiu* liquor, in soft drinks, or dried for snacking. They are now being marketed in Europe as a "yumberry." In Mexico strawberries figure most conspicuously in confections, as well as *helados* (ice cream), popsicles sold by street vendors (*paletas*), and in the now ubiquitous frozen strawberry daiquiri. There is a more traditional drink of strawberry-flavored *atole* (a corn-based drink), too.

Grapes are most commonly used to make wine or other condiments such as verjus (the juice of the unripe grape), vinegar, and sapa (concentrated grape must) and grape-seed oil, which will be dealt with in another chapter. As a fresh fruit, wild species were eaten around the world, and grapes were among the first plants to be cultivated in the Middle East from about 8000 BC. Grapes spread quickly throughout southern Europe and became a principal crop of the ancient Romans. Grapes were brought to China in the Han dynasty (206 BC–AD 220) by the legendary traveler Zhang Qian, but it was not until Tang times (AD 618–907) and extensive trade contacts with the Middle East that they were grown on a large scale in China. They were also introduced to Mexico by the Spanish, though with limited success. There are also species native to North America, such as the muscadine (*Vitis rotundifolia*), scuppernong, and *V. labrusca*, one type of which is the Concord, familiar to jam enthusiasts. There is an Asiatic species, too, from the Amur Valley, which extends into China. But in most culinary contexts, *Vitis vinifera* is used most often, and there are hundreds of varieties specifically bred for sweetness, size, and seedlessness. Grapes are, of course, usually just eaten as is, but they can be preserved in sugar syrup or even brine or made into jam. In cooking, the raisin is also widely used, not only in desserts but also as a sweet accent or garnish in rice dishes, stews, or baked into breads, especially in southern Italy where there is lingering influence of Middle Eastern cookery. Fresh grapes are also used in cooking, especially stewed with wild fowl, or baked into cakes and tarts. The *schiacciata con l'uva*, a flat grape bread of Tuscany, may very well go back to ancient times. Another use for grapes is as *agresta* (verjus). It is astringent and sour and is used in spring and summer, generally when citrus fruits are unavailable. It was used extensively in medieval Italian cuisine and sometimes still today.

Melon Family

Melons (*Cucumis melo*) are sweet relatives of the cucumber and squashes. The fruit is basically a huge berry with seeds inside, with either green or orange flesh and a smooth or netted durable outer skin. The exact origin of melons is probably Africa, and they were domesticated there from a very early date, but they also range wild through the Middle East. They were domesticated in China some two thousand years ago, and in ancient Rome not long thereafter. The watermelon is in a different genus (*Citrullus lunatus*), but it, too, has African roots. There are over eight hundred species of melons, the most familiar of which are the European cantaloupes, green-fleshed honeydews, and the muskmelon types with familiar netting, including what in the United States is called a cantaloupe as well as the horned melon. In culinary terms, all melons are best eaten ripe and fresh with no adornment. But they have, of course, been adorned, with prosciutto in Italy, or equally commonly in the past, with hard cheese or salt, which was believed to cut through and counteract the corruptible humors of the fruit. Incidentally, the cantaloupe is named for a papal garden outside Rome where they were cultivated, and the melon was unquestionably the most fashionable of fruits on Renaissance tables. Melons in both Western and Eastern medical systems were also believed to be of therapeutic value. They were classified as a cold food and were thus considered best eaten in summer or by those of hot complexion. The coldness of the melon was thought to counteract the heat in the body, restoring the body's humoral balance. Melons do figure in cooking as well, either the juice or the seeds, which are toasted and eaten as a snack, much like pepitas (pumpkin seeds); the flesh cooked into refreshing soups; or the rind, which can be pickled. In China, there are duck and chicken dishes cooked in a watermelon rind. Muslim traders introduced melons to China, and they are mentioned in herbal texts from the fifth century.

Prune Family

The genus *Prunus* gives us several distinct and important fruits worldwide. These include cherries, plums, peaches, and apricots (as well as the almond). All are considered drupes; that is, a fleshy fruit with a single pit containing one seed. Most are native to southeast Europe and western Asia, except for the peach, which is from China (despite its name *P. persica*, meaning "from Persia," where it was taken probably around the time of Alexander the Great). Likewise, the apricot spread from China to Armenia (hence *P. armeniaca*). The fruits in world cuisine are primarily eaten fresh as a dessert or snack, but they are also easily dried for storage, cooked into jams and preserves and sometimes in savory dishes, and are used to flavor various drinks. A good example of the savory cooking is a *minutal* of Apicius in ancient Rome, which is a dish of pork cooked with fish sauce, wine, vinegar, honey, herbs, and fresh apricots.

In China the peach is among the most cherished of fruits, both for its fruits and flowers, and was regarded as a symbol of immortality in classical Chinese art and literature. Daoist sages used peaches in their elixirs for longevity. In the kitchen peaches have been made into preserves and a sweet soup and used to flavor alcohol or vinegar. Peaches have also been enjoyed in Italy since antiquity, but the usual custom was merely to serve them fresh or perhaps cut into cubes and served in wine. The origin of this practice stems from the early modern fear that peaches, because so corruptible outside the body, were also likely to corrupt within. The wine would act as a kind of antidote. A perhaps accidental descendant of such drinks is the bellini, made of white peach purée and bubbly Prosecco.

The use of apricots is much the same as peaches. Its name in English is a corruption of the Latin *praecox*, meaning "precocious" or "early," as it is the first tree to bloom in the early spring. In Italy they are used in desserts and pastries or are candied. The kernels are used to make amaretti biscuits and amaretto liqueur. Although apricots were a common ingredient in Italian cuisine centuries ago, especially dried and cooked with meats, today there are few remnants of this practice. However, amaretti (made with apricot kernels) are used crumbled in pumpkin ravioli, for example. The nectarine, by the way, is merely a peach without fuzzy skin, rather than a cross between two *Prunus* species, but there are lately numerous such crosses given fanciful compound names such as pluot and plumcot and aprium.

Plums are the other close relative, including many different varieties ranging from greengages to yellow mirabelles to little red myrobalans to purple damsons and everything in between. There are also both Old World and New World varieties. Some are only used for flavoring, such as the sloe or blackthorn (*P. spinosa*), used in sloe gin. Others are dried into prunes and sometimes cooked in savory dishes. Sauces made of plums or prunes (*sapori*) were once quite common in Italian cuisine. Today in China plums are still cooked with meat or used to make plum sauce, itself a cooking ingredient or a dip. The Chinese dried plum is distinct in that it is usually flavored with licorice, salt, or some other ingredient or even smoked, though what is translated as a plum in English is actually a different species, *P. mume*, from the Chinese *mei*, similar to *ume* (as in umeboshi plum) in Japan. It is smaller, rounder, and more tart than regular plums. It can also be pickled and served with meat dishes or pressed to make a sour drink.

Cherries come in an equally bewildering array of species, both sweet (*P. avium*) and sour (*P. cerasus*). In Italy, cherries are eaten fresh, candied, preserved, included in sauces for game, or made into liqueurs like maraschino. The bright red maraschino cherry is actually a modern version of a much more interesting original, which has an alluring, slightly bitter, almondlike aroma. The so-called Chinese cherry is yet another species (*P. pseudocerasus*), mostly grown for its ornamental flowers, but it also bears an edible fruit, as does the Mandarin cherry, yet another species. It should be mentioned that the prune family was introduced to Mexico by the Spanish, and these fruits are widely eaten there fresh, though most are imported from more temperate northern climes or from Chile out of season.

Apples and Pears

Wild **apples** are native to what is now Kazakhstan and western Asia, but they spread both west and east, hybridized with other species such as the crabapple in ancient times, and are now the most important cultivated fruit on earth, with China supplying a good proportion of the global supply. Traditionally in China, apples have been a symbol of peace because one of the names for apple sounds like the word for peace (*ping*). Their lore in the Western tradition is quite different: there were golden apples of the Hesperides, which conferred immortality, and also apples of discord. In the *Iliad*, an apple of discord was tossed into a wedding party for "the fairest" among the goddesses, and Paris, a shepherd, was asked to decide. He chose Aphrodite, who rewarded him with Helen of Troy, kicking off the entire set of wars recounted in the epic. Apples are also, of course, the fruit of temptation in the Garden of Eden, though in fact there were no apples yet in the Holy Land in biblical times.

Botanically what makes apples so interesting is that they need two different trees to reproduce, which leads to wild genetic variation. The solution for cultivated apples, if one wants the fruit with the same characteristics, is to graft cuttings from one tree onto another rootstock. Thus, all Granny Smith apples are descended from a single progenitor, cut and grafted ad infinitum. All Red Delicious apples likewise are exactly the same genetically as the original cultivar, and they do not change over time unless consciously crossed with another apple, which of course breeders do all the time, hence the proliferation of close relatives: Fuji, Pink Ladies, Winesap, Jonathan, Empires, Macouns, and so on. Both the Chinese and Romans mastered grafting techniques in ancient times, and thus the apple is truly a classical fruit, and the basic techniques are the same today as centuries ago: a slip is cut from the parent tree and inserted into another tree as a host and bound until the join heals. The limb will still bear fruit genetically identical to the parent.

In world cuisine apples are usually just eaten out of hand, but they can be dried, made into jams or alcoholic beverages like cider and calvados, or cooked as fritters, which are slices of apple fried in a golden batter. As with other fruits, there are some vestigial dishes from a bygone era still cooked in Italy, such as goose stuffed with apples or roasted pork loin with apples (*filetto di maiale alle mele*). Baking apples was meant to counter the fruit's harmful qualities, and recipes involved coring and filling with honey, raisins, spices, and wine. Apple cakes and apple tarts are also common. Most apples in

Italy come from Trentino and Alto Adige, where practically all production is controlled and marketed by the Melinda consortium. Unfortunately, most of the varieties grown today are much the same as elsewhere: Red and Golden Delicious, Gala, Granny Smith, Winesap; they are nice looking and sweet but often lacking in flavor and texture.

In China, there is also a kind of apple fritter typical of banquets in which wedges of the fruit are fried in a batter and then placed in a very hot sugar solution and sprinkled with sesame seeds. The fritters are then quickly dipped in ice water, creating a hard-spun sugar shell. When done properly, each slice also trails silken threads of sugar, which is what the name of the dish means in Chinese (*ba si ping guo*). A distant cousin in Mexico would be the apple empanada, a pastry crescent of fried dough with cooked apple filling.

Pears can be cooked in exactly the same way as apples but are often eaten uncooked, because cooking can accentuate the gritty texture of the individual cells. Some varieties take very nicely to being poached in red wine. The pear-shaped European pear (*Pyrus communis*) is different from the Asian pear (*Pyrus pyrifolia*), which is round and usually has a much crunchier texture and a distinct perfume. These are sometimes used ground as a sweetener in soy-based sauces or in marinades for beef. Interestingly, Asian pears, along with melons and other fruits, are also used to flavor dried beef, a kind of sweet jerky. A different species of white pear from China (*ya li*) is pear shaped and is said to have a pineapple- or roselike fragrance.

The quince (*Cydonia oblonga*) is another related fruit, which is usually inedible raw because of its intense astringency and thus is always eaten cooked, which brings out its aroma and sweetness. Although golden yellow, and as good a candidate as the real golden apple of the Hesperides, if cooked long enough the flesh turns from white to red. Quinces were baked with spices in Europe centuries ago, cooked into a jam (the original marmalade, which was reputed to have medicinal qualities), or cooked into a solid block as quince paste (*membrillo*). This was brought to Mexico by the Spanish, but interestingly similar pastes are more often made there with guava or other local fruit. In Italy this same quince paste is called *cotognata* and is eaten after a meal with cheese. There are similar, though only distantly related, fruits in Asia in the *Chaenomeles* genus, often grown in the West as an ornamental, called "flowering quince." These can be cooked and used the same way as Western quinces.

The medlar is another applelike fruit, but which strangely must be bletted; that is, left on the tree until partially rotten, when it turns soft and sweet. They can be eaten out of hand or made into jam. In Italy, these are called *nespola* from the Latin *mespila*. Used in similar ways, the sorb apple or rowan resembles a tiny apple and is also used in Italy, where it is fermented into a refreshing alcoholic drink.

Among the ancient domesticated fruits native to China, the persimmon (*Diospyros kaki*) has long been valued for the beauty of the growing tree and its bright orange fruit, which remains hanging on the tree after the leaves have fallen. There are two main types of persimmon, known in the United States by their Japanese names: the fuyu, which is firm fleshed and can be eaten when still hard, and the hachiya, which is only palatable when on the verge of collapsing under its own weight, practically rotten, though utterly ethereal in flavor. There is also a *D. virginiana* native to the southern United States. In China the many cultivars of persimmon have been stored outside in the winter frozen and then eaten like a kind of sorbet. They can also be steamed, squashed flat, and then dried, yielding a leathery, sweet snack. There are many other ways of preserving persimmons or ridding them of bitterness, involving soaking, steaming, or smoking. Reconstituted persimmons are also included in cooked dishes, as they counterbalance the flavor of duck and pork very nicely.

Pomegranate is a fruit first cultivated in Persia and adjacent areas. Its name either derives from the city of Granada in Spain or from the many grains it bears (*granata*), although its Latin name *malum punicum* or *Punica granatum* in modern taxonomy comes from the Roman name for Carthage in northern Africa. Pomegranates are eaten by breaking open and removing the seeds from the white pulp. They can be pressed for the bloodlike juice and are used to season meats. Surprisingly they have gained wide attention in recent years as an antioxidant, and the juice is now found everywhere on grocery shelves. In China the pomegranate has been used primarily as either an ornamental tree or in medicine, centuries before such use in the West. Because of the many seeds, the pomegranate is also a symbol of fertility in art, both in the West and East.

Figs are another classical fruit originating in the Middle East and eastern Mediterranean. They come in black and green varieties and can be eaten fresh, dried, or processed into jam. They can be served with prosciutto, in the same way melon is used, or stuffed with cheese and nuts. Though there have been figs cultivated sparingly in China since the Tang dynasty, more important is the bo tree (*Ficus religiosa*), under which Buddha achieved enlightenment. Thus, a fig tree is often planted on temple grounds.

The olive, native to the eastern Mediterranean, though today not considered a fruit, was in the Italian Renaissance era served at the end of a meal with other fruits. Today it features at the start of a meal as an antipasto along with other salty foods. Olives are inedible raw and must be preserved either in brine or salted. This process leaches out the bitter oleoeuropein, and the olive also slowly ferments. Olives have also been soaked in lye and then pickled, which is similar to the modern processing for canned olives, though the texture is definitely inferior. The color of the olive depends entirely on the stage of ripeness: green olives are young, and purplish or black olives are fully ripe. Olives can also be preserved in oil with herbs and garlic, chopped into a relish called tapenade, and even cooked with game like rabbit or other meats, placed on a pizza, or stuffed with strips of marinated bell pepper, cheese, or tuna. They can even, remarkably, be breaded and deep fried. They are absolutely ubiquitous and indispensable on Italian tables, and each region boasts its own celebrated varieties, from huge green Cerignolas in the south, to the diminutive Ligurian Taggiasca, to the purplish Gaeta of Naples. Of course, most olives are pressed for olive oil, a subject that will be treated at length in chapter 8, Fats and Flavorings.

The olive has also flourished in Mexico, brought by Spanish missionaries along with grapes, partly for liturgical use as holy oil for funerary rites but also as food. The varieties such as Manzanilla, Mission, and Sevillano are still grown here and as far north as California, which was once part of Mexico. Green olives in particular are used in Mexican cooking, especially with fish in dishes that originally hail from Spain. Although olives were brought to China in Tang times, and there have been recent attempts at cultivation, what is called an olive in China (or Chinese olive) is actually an entirely different species, *Canarium album* and *pimelea* or *gan lan* and *wu lan*. These, too, must be salted to remove bitter flavors and sometimes dried in the sun, but strangely enough they are also sweetened and flavored with licorice or spices. The final flavor is nothing like the European olive but is appealing nonetheless. Characteristically, they are found among the confections in Chinese shops but are sometimes used in cooking savory dishes. The interior of the kernel is also used as a nut, somewhat reminiscent of pine nuts. It should also be kept in mind, though, that dried and flavored jujubes are also often labeled in English as a preserved olive. The way to tell the difference is that jujubes have a single, large, inedible, football-shaped pit with very pointy ends.

Along with olives, mention must be made of the caper, which is the bud of a Mediterranean bush pickled and used much like olives. The fruit of this plant is also used the same way, and they resemble olives with a fleshy, seed-studded interior and long stem. Actually, stems and leaves are also pickled and make an exquisite vegetable condiment. They can be brined or simply preserved in sea salt. The best are said to come from the island of Pantelleria, southwest of Sicily. In cooking they are used with marvelous effect in tomato sauces, on pizzas, combined with fish, or in sauces descended from Arab cookery combined with raisins and pine nuts.

Citrus

The **citrus family** hails from Southeast Asia and includes the fragrant, thick-skinned citron, lemon, lime, orange, and the many permutations of tangerines or mandarins. They range from the small piquant kumquat to the grotesque Buddha's hand to the self-proclamed ugli fruit or pomelo to the grapefruit. Citrus was used in China since ancient times, and there was even a specialized book devoted to them written in 1178 by Han Yanzhi. It is difficult to imagine that such a universal flavoring in cooking could have spread only in distinct historical waves, but it is true. The lemon only arrived in Europe in late antiquity, the orange with the Moors, the sweet orange only in the sixteenth century from China, and citrus in the Americas only with the arrival of the Spanish. Even the so-called West Indian or Key lime was an import, and grapefruit is said to have been a spontaneous cross breed of the pomelo and sweet orange that occurred in the Caribbean.

Some citrus fruits are meant to be eaten fresh as a fruit or squeezed as juice, whereas others are best as a cooking ingredient. The juice of sour or Seville oranges was introduced to Mexico via Spanish cuisine, as were limes. These provide a signature sour note to cooked dishes or sauces. Citrus juice can be used as a marinade for meat, poultry, or fish. The peel can be used freshly grated, dried, candied, to flavor liqueurs, and of course in marmalade (which was originally based on quince). In China the dried peel of orange and tangerine is used in rich, flavorful dishes such as orange beef (*chenpii niurou*) from Hunan. Even citrus leaves are used in cooking, as with kaffir lime leaves in Southeast Asia. The flower too, especially of oranges, with an entirely different fragrance from the peel, is distilled into an aromatic water that has been used in confections and medicine for centuries.

In Italy lemons and oranges are a universal flavoring as well, and are mostly grown in Sicily, though Amalfi is said to produce the best. They have been used sliced as a plate garnish, where their acidity was said to cut through the gluey humors of fish. They were used in salads with fennel. They were even considered a sun symbol in Platonic philosophy, which confers wisdom and solar energy to the consumer. In modern Italian cookery one finds citrus fruits in many surprising guises. For example, the flavoring of Earl Grey tea comes from the bergamot orange. The blood orange (*sanguinello*) has recently received attention for the color of its lurid juice, in drinks and sauces. Limoncello, a popular alcoholic drink, is based on lemon. And lemon sorbet is perhaps the finest thing that has ever happened to a citrus fruit, with a lemon granita, made with shaved ice, following closely behind.

The curiously named *wampee* from Huangpi (*Clausena lansium*) is a distant relative of the citrus fruits. It is a little, round, aromatic yellow fruit that grows in dense bunches, with a mucilaginous pulp. It originates in southern China and has now spread throughout Southeast Asia. Some types can be eaten like a fresh fruit, but the sour types are usually combined with meat.

The mamoncillo (*Melicoccus bijugatus*) is a coincidentally similar green fruit native to the Caribbean and Mexico and is sometimes called *quenepa* or *limoncillo*—or oddly Spanish lime. It is either eaten fresh or the fruits are boiled to make a refreshing drink. This fruit is actually related to a whole group of Asian fruits in the *Sapindaceae* family, including the lychee (*lichi*), longan, and rambutan. These fruits all have a delicious sweet, white flesh and a single large seed in the center. The lychee's texture is sometimes compared with that of a grape, but it is firmer, juicier, and more perfumed. Fresh, they are eaten merely by peeling the tough outer skin, popping the eyeball-like fruit in one's mouth, and spitting out the shiny pit (figure 5.3). Lychees can also be dried and come to resemble little intensely fragrant plums, which are also used to flavor tea. They have always been considered one of the most prized of Chinese foods and were celebrated in classical poetry and art. There was even a treatise on lychees written in 1059 by Cai Xiang. The rambutan, which is imported to China (*hong mao dan*) (*Nephelium lappaceum*), is a similar tasting fruit, but with a fearsome spiked appearance to the shell, which is actually easily peeled. Likewise, the longan (*Dimocarpus longan*), with its white shell, is also quite similar and eaten the same way.

Ackee (*Blighia sapida*) is a related West African fruit in the same family brought to the Caribbean with the slave trade, but it is entirely different from its relatives. The flesh is slightly sweet but resembles scrambled eggs or brains more than anything else, and in

Figure 5.3. Fresh lychees are sweet, juicy, and slightly crunchy.

fact it is cooked, usually with salt fish in the Caribbean. The seeds, as well as the unripe flesh, are extremely poisonous.

Indigenous to southeastern China and cultivated since ancient times is the loquat (*Eriobotrya japonica*), a small, round, yellow or orange fruit with a bright acidic flavor, usually eaten fresh whole, minus the shiny brown pits. It also now grows well in the Mediterranean and California. Because loquats bruise easily, they are mostly canned in syrup and used in desserts in Asia. They can also be made into candy or used to flavor liqueur. A loquat syrup is also popular for coughs.

The kiwi, which is actually a name invented as a marketing device by a grower in New Zealand, is nonetheless native to China, where it is known as *mihou tao*, meaning "monkey peach," or *yangtao*, though that name now refers more often to the starfruit. Its botanical name is *Actinida deliciosa*. Once known as a Chinese gooseberry, kiwis are still gathered wild in hilly regions of the southwest. Kiwis are most often eaten fresh. Today they are grown extensively in Italy, which is the leading producer globally. There is also a variety of tiny kiwifruit that can be eaten skin and all.

In Asia there is a small ovoid fruit called the jujube (*Ziziphus jujuba*), not to be confused with the hard gummy candies of the same name. It is sometimes called a Chinese date, though its flavor more resembles a fragrant apple. Jujubes are ripe when turning from green to brown and can also be dried, preserved with sugar, or even smoked. There are actually related edible species in the Mediterranean, one of which is said to have provided the thorny branches for Jesus's crown as he was approaching Golgotha. The common jujube also spread to ancient Rome and has been a well-known fruit in Italy.

Proper dates grow on a palm tree native to the Middle East and northern Africa. They only fruit well in very hot, dry conditions, so in Italy they have been a dried imported item. The same is true of China after contact was made with the West via the Silk Road. As in Italy they are only grown in the southernmost regions. Dates are eaten as a snack, made into confections, or sometimes used in sauces and savory dishes.

Tropical and Subtropical Asian Fruits

There are a number of fruits that although not extensively grown in China have nonetheless been important in cuisine and obtained through overseas trade for centuries with adjacent Southeast Asia. Durian is perhaps the most notorious of these, originating in Malaysia (figure 5.4). It is a large fruit weighing up to ten pounds with a hard, spiked

Figure 5.4. Difficult to open and remarkably aromatic, the durian has a creamy, white flesh.

exterior. On the inside is a creamy, white flesh whose odor some have compared with rotting flesh and excrement, but in fact it is considered one of the most exquisitely delicious fruits on earth. The carambola or starfruit originates in the same area but is now grown throughout the tropics, including the Caribbean. It is yellow and oblong with five deep ridges, so when sliced it makes a decorative star, which is ideal for garnishes. The flavor ranges from sour to sweet, and it has a pleasant, crunchy texture. The mangosteen (*Garcinia mangostana*) is another tropical exotic from Malaysia and Indonesia, with creamy, white segments encased in a thick, round, purplish-black rind (figure 5.5). It has been hailed as the most delicious of all tropical fruits, with a fragrant sweet and sour flavor. It is only just beginning to appear in North America.

There is also the mango (*Mangifera indica*), also from the Malay Peninsula and India, but mangoes are grown extensively in China in southern Guangdong Province and since the sixteenth century as a major crop in Mexico, probably introduced via the Philippines. In Mexico and China they are eaten fresh, juiced, or in desserts. The luscious, sweet flesh of the mango has an alluring piney or turpentinelike aroma, and on the streets in Mexico one finds slices of fresh mango on a stick, with a sprinkling of chili powder and salt. This sweet/sour/spicy/salty combination is also used in candies, lollipops, and the like, both in Mexico and perhaps coincidentally in Indonesia.

Figure 5.5. These canned mangosteens are much less flavorful than fresh, which are only beginning to appear in U.S. markets.

The jackfruit (*Atrocarpus heterophyllus*) is another of these tropical global travelers, introduced first to China perhaps from India in the Tang dynasty. It is a huge, green fruit weighing up to sixty pounds or more. In fact, it is the largest tree-borne fruit on earth, with a skin made up of hundreds of tiny, soft bumps. It is also difficult to cut up because it exudes a sticky, white sap. The flavor is by most accounts rarely appreciated on first encounter, but it is very sweet and pineapplelike. It is eaten fresh as a fruit, immature as a vegetable, and fried into chips, and even the seeds are cooked. In the United States it is usually found canned in syrup, either in Asian shops or in Caribbean stores, as it was also transplanted there. So, too, was a close relative, the breadfruit, infamous as the fruit being taken aboard the HMS *Bounty* from Tahiti to the Caribbean by Captain Bligh, inciting the mutiny when precious water was used on the saplings rather than for the sailors.

Among all these exotics we must also include the all-too-familiar banana (*Musa acuminata*), which followed the same routes as those fruits mentioned, but of course has achieved worldwide fame. Bananas are primarily imported in our three world cuisines and eaten as a fresh fruit, but they can be cooked in fritters and pies. The plantain is a starchy, green, related species eaten cooked, in stewed dishes or fried, and quite popular throughout the Caribbean. Plantains are most often used in Mexican cooking on the east coast, perhaps accompanied with rice (another Asian import) and black beans. *Tostones de plátano* are made by frying slices of plantain, then squashing them and refrying, resulting in a crunchy, salty snack or side dish. Plantains are even boiled and formed into a dough that is stuffed with meat. The leaves, too, are used to wrap tamales, line cooking pits, or serve as disposable plates in coastal areas.

Tropical and Subtropical American Fruits

There are an equal number of important fruits native to the Americas, which have also traveled the globe, though in the opposite direction. Again, like the aforementioned fruits, these are mostly tropical imports and have been incorporated into world cuisines for the past five centuries. Some are more recent introductions elsewhere and may still be relatively unfamiliar outside their homeland.

This is certainly not the case with the pineapple, a Caribbean fruit from South America that spread to Central America and the Caribbean in pre-Columbian times. In fact, Columbus was the first European to have tasted it. Its name in botanical Latin is *Ananas comosus*, the genus name coming directly from the Tupi Indian name used in Brazil. Its English name comes from the resemblance to a green pinecone, formerly called a pine apple. In Mexican cuisine pineapples (*piña*) are eaten raw, juiced or cooked with spices and fermented into a drink called tepache, or cooked with savory dishes. Pineapples contain bromelain, a protease enzyme that breaks down proteins in meat and thus is often used as a meat tenderizer. Pineapples were probably brought to China by the Portuguese via their colony in Macao. They were rapidly assimilated into the cuisine, particularly in sweet and sour dishes, probably because they hold their form nicely when cooked. Pineapple is also commonly grilled on skewers with chicken or shrimp in Chinese American restaurants. Pineapple appears in all our cuisines in sweet desserts. Pineapple peels can also be soaked, fermented, and left to sour into a form of vinegar. Pineapple is also used in Dai (an ethnic minority in Yunnan) cookery a lot.

Other American fruits that have found a welcome home around the world are the guava, papaya, and cherimoya or custard apple. Guavas (*Psidium guajava*) are little yellow, pear-shaped fruits with red mucilaginous flesh and seeds. Extremely high in vitamin C, they are commonly juiced or made into jelly or guava paste or eaten for dessert. There is also a relative called the strawberry guava (*P. littorale*) and several other South American relatives. These are not related to what is sometimes called the pineapple guava (*feijoa*, *Acca sellowiana*), which is green and oblong with an odd, astringent taste. The papaya is much better known globally than these others. Native to Central America, it is also used in marinades and acts as a tenderizer. It contains papain, an enzyme that breaks down the proteins in meat, and actually if left too long in a papaya-based marinade, the meat will disintegrate and become unpleasantly soft. It can also be dried, candied, and squeezed into a thick juice or nectar. In Southeast Asia, the green, unripe fruits are shredded and made into spicy salads. They feature in Chinese cuisine like other tropical fruits, are made into sweets or desserts and even savory soups, or are merely eaten fresh. They are sometimes used as a garnish or ingredient in savory dishes as well, a descendant of which in the United States are sweet and sour stir-fries with pineapple chunks.

There are also a number of other tropical American fruits that are practically unknown in the United States. Such are the sapote (*Pouteria sapota*), which has an orange, pumpkinlike flesh and a large, almond-shaped pit. The flesh is scooped out of the rind and eaten, and the pit is mixed with corn, sugar, and cinnamon to make a drink in Oaxaca. This was the fruit that was said to have kept Cortés's troops alive as they marched southward from Mexico in the sixteenth century. The canistel (*P. campechiana*) is a close relative, with green skin and a mealy flesh that is usually eaten with salt, pepper, and lime juice. The sapodilla (*Manilkara zapota*) is unrelated but similar in flavor and use as these others. It looks like a little potato, with dark, beanlike seeds inside. Aside from the fruit, the latex exuded from a closely related species, known as chicle, has been the prime ingredient in chewing gum (as in Chiclets), though today most gum is synthetic rubber. The mammee (*Mammea americana*) is another Central American fruit like these others, and often confused with it, though it is actually more closely related to the mangosteen. They are rarely shipped far, and there seems to be little interest in it outside those regions where it grows.

Nuts and Seeds

Tree **nuts** are probably along with berries among the earliest foods that sustained our prehistoric ancestors, as well as hominids and other animals. This is because most nuts provide abundant calories from protein and fat for relatively little energy, beyond knocking them down or collecting before animals find them and breaking the shell. Few nuts require cooking. But many can be roasted or briefly heated for a deeper flavor, a technique possibly discovered by accident while trying to open more intractable species like pine nuts. These and other types can also be ground into a sauce, or ground and soaked to form a kind of milk. When sweetened they are the base for numerous sweets, and historically they have been used to garnish dozens of different dishes. And of course all nuts make fantastic snacks.

Pine nuts are certainly among the earliest nuts enjoyed around the world, as species of pine are indigenous to the Old and New World and still grow wild. In the Americas, the piñon pine bears a large cone with a sweet kernel; in Italy it is primarily the stone pine, but there are other species that bear nuts. The pine apple (not to be confused with pineapple, which is named for its resemblance) is a large cone, picked ripe but green, and heated until it breaks open and the seeds can be removed. The **seeds** in turn have their own shells, which are gently crushed to remove the whole nut kernel. Pine nuts are used in such classic sauces as the Genoese pesto with basil, oil, and cheese, but this is merely one of a number of pounded, nut-based sauces typical of the Middle Ages. Pine nuts are also used in cookie doughs and to garnish individual pastries. Pine nuts have been used in China since prehistoric times, too, and pine needles have been used to smoke meats and to give *lapsang souchong* tea a resinous, smoky flavor. They are also common in Lu cookery of Shandong in the north. There is also a very small variety of pine nut from China, sold widely in the United States, which has been implicated in a condition sometimes called pine nut mouth, which leaves an unpleasant bitter taste in the mouth for several weeks. In general, pine nuts are very expensive, and cheap ones should be held suspect.

Almonds, unlike pine nuts, are primarily a cultivated tree native to the Middle East. They spread in classical times to Italy, during the Tang dynasty to China, and after the Spanish conquest to America. There are two kinds of Chinese almond (*bei xin* and *nan xing*), which are really the kernels of an apricot variety used in much the same way as almonds, though they are somewhat toxic if eaten in great quantities. Almonds are actually closely related to the peach, and peach root stocks are often used to bear almond trees today. They look very much like peaches when immature and green and can even be pickled in this state. A bitter almond is also used mostly in flavorings and medicines. It is different from the sweet almond because it is poisonous if eaten whole and unprocessed.

Sweet almonds are perhaps the most versatile of all nuts. They can be ground finely and used to thicken sauces or soaked in hot water to create an almond "milk." Sweet almonds were commonly used in medieval cooking as a substitute for animal milk, which was forbidden during Lent. Ground almonds are also pressed into oil, made into a butter, and when mixed with sugar to form marzipan were used not only to make little fruit-shaped confections but also as a cooking ingredient in their own right. They are used in cakes and pies, and in the past even in savory dishes. Almonds can also be candied in pastel colors as Jordan almonds, which are traditional for weddings in Italy. Almonds are, of course, a staple ingredient in the confectionery industry today worldwide and are frequently combined with chocolate. All these uses were adopted in Mexico. Almonds are also finely ground and mixed with egg whites to make macaroons, though in the case of amaretti, apricot kernels are used. There is also a sweet almond soup used for sore throats in China and a cool, refreshing drink made from almond syrup in Italy called *orgeat*.

Though etymologically related to *orgeat* (derived from the Latin *hordeum*, meaning "barley," from which the original drink was made), *horchata* in Mexico is made from tigernuts or *chufas*. These were introduced by the Spanish, who got them from the Moors, who originally planted them in Valencia. Strangely, the plant (*Cyperus esculentus*) looks like a weedy grass, and its nut, sometimes called an earth almond, is the tuber. This is soaked, ground, and made into a sweet drink, though the more popular version sold on the streets in Mexico today is made with rice.

Native to southeast Europe and western Asia, walnuts (*Juglans regia*) have been cultivated since antiquity and were a common snack food in the past, just as they are today. The nut when growing is encased in a green fruit, which can actually be pickled unripe, and in Italy these are used to flavor alcohol, called *nocino*, which is drunk as a digestive. Normally they are left on the tree until black, then the walnut casing is removed, leaving the hard-shelled nut with the nutmeat inside. There are species of black walnut from America, the related butternut, and another species from Manchuria. So this is one of the few trees with representatives on most continents, with dozens of species, though the one most commonly eaten is the Persian walnut or English walnut. They can be eaten whole raw or toasted and are often incorporated into baked goods, used as a garnish, or ground as a kind of thickening agent in pounded sauces related to pesto. In China, walnuts are used in mixed dishes to provide an interesting contrast in texture, with perhaps chicken, or finely ground to make sweet walnut soup. The nuts also yield an aromatic oil, which will be discussed below. The pecan (*Carya illinoinensis*) and hickory are related nuts and are used in similar ways. They are native to the southern United States but also grow in Mexico.

The pistachio (*Pistacia vera*) is native to Iran, Turkmenistan, and Afghanistan. It spread from there both east and west in late antiquity, but significantly only in the Middle Ages. Today pistachios are grown in Italy and Mexico, though they are of lesser importance in China, and most of those eaten in Asia and Europe are grown in California. Pistachios, like other nuts, are eaten simply roasted and salted, or flavored with citrus and other flavors. But they can be used as a cooking ingredient, with best success as a flavor for ice cream, both in Italian gelato, and a lurid green version once popular in Chinese restaurants in the United States. A relative of the pistachio is the mastic tree, grown originally only on the island of Chios. It was the source of an edible gum, which was widely chewed in ancient times and from which we get the word *masticate*.

Hazelnuts (*Corylus avellana*) or filberts are native to southern Europe and were cultivated since ancient times. Typically they are just eaten as a snack, but they also make a seductive flavoring and are used in liqueurs such as Frangelico or mixed into chocolate to make what is called *gianduia*, which has a low melting point and pleasant mouthfeel. Hazelnuts are the base for the popular Italian spread Nutella. They can be used toasted and ground as a garnish on savory dishes, especially fish or chicken. There is also a rich, unctuous hazelnut oil, used almost only on salads. The Chinese also have a native hazelnut species common in the north.

Not a nut at all in any sense but culinary, the peanut has followed a long and twisted path around the world. The bean, which like some other relatives that reinsert themselves in the ground (the Bambara ground nut, for example), comes originally from South America. It was brought in one path to Africa, where it replaced similar species ground in soups and stews, and from there to the American South. It also traveled eastward, or perhaps more likely westward from Mexico to China, where it is a common ingredient in stir-fries. Perhaps its most common use in Asia is in the form of peanut oil, which has a fragrant flavor and high burning point. Finely ground peanuts can also be used as a sauce base or as the modern health/convenience food peanut butter. Although peanuts were domesticated in Mexico several thousand years ago, they seem to have been somewhat displaced in cooking. However, some peanut-chili sauces used, for example, with chicken (*pollo encacahautado*) that are surprisingly similar to a Thai peanut sauce, are popular in Veracruz, which historically was influenced by African cooking. In Italy peanuts are a fairly recent introduction, used mostly in sweets, though ironically Planters Peanuts was founded by two Italian immigrants to the United States.

Ginkgo, despite its recent popularity in the West as a dietary supplement, is actually an ancient tree, unchanged after millions of years—a living fossil, as it were. It has unmistakable yellow, fan-shaped leaves and was often planted in Chinese temples. The fruits, or technically a kind of cone, are found only on female trees, whose odor is reminiscent of smelly cheese and sewage. It is inedible, but it bears a large, smooth, almond-shaped seed that is delectable (figure 5.6). Ginkgo nuts are shelled, which requires wearing gloves because the flesh is a caustic alkaline, and then boiled or roasted, grilled on a skewer, or added to more complex dishes, especially those for vegetarians. They are also eaten as other nuts are, salted as a snack.

In China the seed or nut of the lotus is also eaten as a dessert or snack and is sold on the street as such. These can also be candied in syrup, pickled, ground into a paste used in pastries, or used as an ingredient in mixed dishes. They are eaten to promote fertility as well because their name in Chinese sounds like the word for "many sons" but are also considered a medicinal food for strength and virility. In Asian markets, they are sold dried and should be soaked before use, such as in lotus seed soup. In the West the seed heads of the lotus are best known in dried flower arrangements.

Green pumpkin seeds (pepitas) are a New World snack eaten much like nuts are, toasted and salted. They are also ground after toasting and are used as a thickener, an ingredient in sauces, and other dishes. Pumpkin seeds can also be pressed as a rich, dark oil. They have been eaten in Mexico since pre-Columbian times. Sunflower seeds are

Figure 5.6. The nuts of ginkgo trees are used widely in Asian cooking.

native to North America and can be used in similar ways, usually salted and eaten as a snack, though their greatest use is as a source of oil.

Several other tropical nuts bear mentioning, though they are used minimally as exotic imports. The candlenut (*Aleurites moluccana*) comes from Indonesia and is used cooked as a snack and also as a source of oil. Macadamias, although usually associated with Hawaii, actually come from Australia and have an incredibly hard, round, woody shell, and so are usually sold already shelled. Their use in Europe is primarily in chocolate or as a snack. Their introduction globally has been only in the last century. Cashews, a nut native to tropical South America, will not grow in temperate climes. Thus, they have always been an import to both the West and China, where interestingly, the perishable red fruit from which the nut is appended is completely unknown. The nuts are used mostly as a snack or in confections. Likewise, the Brazil nut (*Bertholletia excelsa*) only grows in the wild in the Amazon jungle. None of these played a major role in the traditional cooking of our three world cuisines, but they are increasingly doing so nowadays. But there is one tropical nut that has enjoyed great prominence: the coconut.

The coconut is borne on a palm tree and is widely used in Asian cooking, both its milk (that is, shredded, soaked, coconut flesh, not to be confused with the water inside the coconut), as well as its flesh in countless other savory and sweet dishes. Coconuts can be grown at the extreme southern edge of China and Hainan Island but have mostly been imported. Coconut is used to flavor coffee and candy. The use of coconut milk is mostly in Southeast Asian curries. The flesh can be shredded and dried and incorporated into other dishes (for example, coconut shrimp) or candied in syrup or jellies for dessert. There is also a gelatinous coconut pudding that is cooked into squares and commonly served as dim sum. The sap can be fermented into palm wine. Coconuts, of course, have also spread widely across the Pacific to Central America and in the other direction to Africa. In Latin America, coconuts are now used in a wide array of cocktails such as the piña colada, made with sweetened coconut syrup, pineapple juice, and rum. Coconuts also figure in Mexican candies and cakes, and there is also a flan flavored with coconut, the *cocada imperial*. In Italy one is most likely to find coconut as a flavoring for gelato.

Other edible seeds such as sesame and poppy will be covered in the chapter on fats and flavorings because they are not eaten alone as a snack. In that chapter, there will also be coverage of oilseeds such as sunflower and canola.

Study Questions

1. In what inventive ways did cooks in the past use nuts and berries, and how might these be revived today?
2. How and why do berries and other fruits straddle the line between food and medicine in the past and today? In what ways did people in the past correct what were perceived to be harmful qualities of certain fruits?
3. How has modern agriculture and commerce actually reduced the number and types of fruits and nuts regularly consumed, despite the global trade in such ingredients?
4. Describe the ways fruits are used to balance flavors with their acidity and sweetness.
5. How and when do many fruit and vegetable species spread from their origin in the Middle East both to Europe and Asia?
6. How are some fruits used as symbols in art, folklore, and religion? Speculate why this happened.
7. Speculate why certain fruits and nuts are quickly adopted around the world but not others. Is this a matter of flavor, perishability, or ingrained cultural preferences?

Recipes

Fresh Fruits

CROSTATA DI FRUTTA (FRUIT TART)/ITALY

Most fruit is consumed fresh after a meal or as a snack in Italy. However, in the past there were a number of recipes and sauces that contained fruit as a kind of sweet-and-sour counterpoint to meat. This is a recipe that merely accentuates the Italian love for the best fruit in season.

Start with the freshest and most unblemished fruit you can find: any combination of peaches, strawberries, apricots, kiwi—any soft stone fruit that requires little cooking.

Next make your pastry dough, a *pasta frolla*. This is quite different from a standard French pastry dough and is both sweetened and uses egg yolks. Its roots go very far back in Italian gastronomic history, and for a touch of that past flavor, you can also incorporate rosewater and a hint of cinnamon. Either way, find what in Italy is called farina 00, or regular fine pastry flour, which has a low gluten content and will remain tender after baking. Use 4 cups of this flour. Then cream together 2 cups each of butter and sugar with a wooden spoon until light colored. Whatever proportions you use, the volume of sugar and yolk should be equal to the volume of flour. Incorporate 4 eggs into the creamed butter and sugar, much the same as when making cookies. Then gently incorporate the flour as quickly as you can and knead into a dough. Wrap the dough and put it in the refrigerator.

Next, make the pastry cream. Heat a pint of milk on a low flame. Add in a few slices of lemon peel, without the bitter white pith, and a split vanilla pod, with the seeds scraped into the milk. Let this steep on very low heat for about 15 minutes, and then strain into a bowl. In another bowl, beat together 4 egg yolks and ¾ cup sugar. When the mixture is light, add ¼ cup flour. Temper this mixture by adding a little of the hot milk at a time and beating vigorously with a whisk. Then pour the egg mixture into the hot milk and continue beating. Place the bowl over a pot of gently simmering water and continue beating until thickened. Let this cool.

Preheat the oven to 350 degrees. Roll out the dough into a circle larger than the pie pan you will be using. It should be about an inch thick. Roll the dough around the rolling pin halfway, and transfer to the pie pan and unroll. Without stretching the dough, ease it into the pan. There should be a little overlap at the rim, which you can cut off evenly by rolling your pin across the top of the pan. Decorate the edge if you like with the tines of a fork or by pinching a fluted pattern with your fingers. Poke holes in the bottom of the dough with a fork. Line the crust with a circle of parchment paper and put a few cups of beans in and prebake the crust. The beans keep the dough from rising up. (You will want to keep the beans for this purpose; they are not good for eating anymore.) Bake the crust until golden, about 20 to 30 minutes.

While this is baking, remove the pits and peels if necessary, and cut up your fruit into even slices. Gently toss these in a pan with butter and a little sugar, just to glaze them. You do not want to cook them through or too long so they fall apart. Remove the crust from the oven and let cool, remove the baking beans and parchment paper, then fill with the pastry cream and arrange the fruit on top in a decorative pattern. Sprinkle with powdered sugar. Slice into wedges and serve.

PEAR MINESTRA (ANONIMO TOSCANO, FIFTEENTH CENTURY)/ITALY

Peel fresh pears and place in water to soak. Then throw away this water and let them boil in new water with salt and oil, then fry some onions with spices and saffron with a little water and add it to cook. And when it is done serve with a little of the spices in a soup plate. And similarly you can make this in a little almond milk without oil and without onions, adding a little sugar and a little salt.

While the flavor combinations are certainly unusual to the modern palate, this kind of savory fruit minestra was typical in the Late Middle Ages. The absence of meat also means that this would have been suitable for fast days and Lent. The spices are left to the discretion of the cook, but normally these would have included some combination of pepper, cinnamon, and ginger, with perhaps nutmeg. They are fried with the onions, then moistened with water and poured into the poaching pears. Almond milk is made by pounding blanched almonds with hot water, soaking them, and then allowing them to drain in a sieve. It often served as a milk substitute. The soaking of the pears may have been to keep them from oxidizing while they are being peeled, but throwing away the water seems to be a precaution against ingesting the crude, noxious juices of pears, widely believed by physicians then to be dangerous to health.

PICKLED PEACHES/CHINA

Take firm, small, slightly underripe peaches and put them in a large, sealable glass jar. Mix together equal parts rice vinegar and water and add sugar to taste, about ½ cup,

Figure 5.7. These crunchy preserved plums are sweet and salty.

and a few spices such as cinnamon and cloves. Add a pinch of salt as well. Pour the liquid over the peaches and leave at room temperature for about a week, then refrigerate. These can be eaten sliced as an appetizer or added to stir-fries. This recipe also works well with pears and any other firm fruit.

PICKLED PLUMS/CHINA

What is called a plum in English is actually more closely related to the apricot (*Prunus mume*), the fruit of which is called *meizi* in Chinese. The trees are especially revered for their blossoms, which bloom first after the winter. They are also made into plum wine and plum sauce, which goes with duck and a variety of other condiments. *Suan meizi* is a kind of pickled plum, made much like the peaches in the previous recipe (figure 5.7). The underripe fruits are soaked in vinegar, a lot of salt, sugar, and flavorings such as licorice root. There are also dried, preserved versions, which are salted and left in the sun to dry. Most people buy these in stores, and they come in an infinite variety of flavors.

CHERRIES IN CONSERVE/ITALY

Although most Italians purchase maraschino or amarena cherries in syrup, they are easy to make at home and much less expensive. These are made with particular cherry types, but it is fine to use any sweet dark cherry such as Bings or Morellos. If you use sour cherries, simply add more sugar. Take half your cherries with the pits and boil them in water and sugar with some cinnamon, cloves, and similar spices for about 30 minutes. Drain the solids, and cook down the liquid into a syrup. Add a little almond extract. Put your fresh cherries with stems attached into clean jars and pour over the hot liquid. Let sit several months until thoroughly infused. These are fabulous with ice cream or in drinks. Other fruits can also be preserved in a syrup or with alcohol. Anything firm enough to keep its shape, such as whole apricots and plums, is submerged in hot syrup and then strong, clear alcohol is added and the fruit is left to macerate for several months, sometimes with whole spices. These are eaten for dessert, with some cream drizzled over or with *panna cotta* (cream cooked with gelatin and solidified).

MOSTARDA DI FRUTTA (FRUIT MUSTARD)/ITALY

This is a traditional product of Cremona, served as a condiment with other dishes, though there are variations in nearby cities as well. It is not a yellow mustard like in northern Europe or in the United States but rather a preserved fruit conserve. It does incorporate mustard, the seed, or sometimes mustard oil. They key is to not let this become a gloppy jam but to keep all the fruits separate and whole. You can use any combination of the following: cherries, crabapples, small pears, apricots, figs, as well as orange and lemon peel, which have been blanched. Remove the pits from the stone fruit, but keep whole if possible. Heat a thick sugar syrup made with equal parts sugar and water or white wine. You can also use honey if you prefer. Add freshly ground yellow and black mustard seed—you can do this in a spice grinder or mortar—plus cinnamon sticks. Then plunge in the fruit, while the syrup is still hot. This will just barely cook it through without making the fruit mushy. Next, add vinegar and a touch of salt. Leave undisturbed in a glass jar as long as you can—at least a month for the flavors to come together, but it should last indefinitely if stored in a canning jar and processed in a water bath. Once opened, refrigerate. This goes wonderfully with *bollito misto* (boiled meats), salami, or cheese.

FRIED BANANAS/CHINA

Though one sees these often in Chinese American restaurants, served with ice cream, there is an original prototype made with a fragrant banana. If you can find a small red banana or fingerling type, they are usually more interesting than the large yellow variety. Heat a wok of peanut oil to 365 degrees. Make a thin batter of all-purpose flour, a little baking powder, and a tablespoon or so each of sugar and water. Cut the banana into bite-sized pieces, dip into the batter and fry, a few at a time, then put on paper towels to drain. In a small pot combine sugar and water and boil, without stirring, until you have a thick syrup. Serve the bananas with a drizzle of syrup. Or, you can bring the sugar syrup up to the hard crack stage—test a droplet in a bowl of ice water and it should be crunchy and hard. Take each fried banana, dip into the syrup, and then plunge immediately into the ice water. If you are feeling very daring, you can also try to spin the sugar into a kind of floss. This is done with a chopstick, flicking long threads of hot caramelized sugar over the bananas. Just be sure not to burn the sugar or yourself, as this is very hot.

TOSTONES DE PLÁTANO (FRIED PLANTAINS)/MEXICO

Not a sweet but a kind of snack made from green, unripe plantains, these are popular throughout the Caribbean and on the west Gulf Coast of Mexico. Cut the plantain into chunks and remove the peel. It is easier than trying to peel it whole. Fry these chunks in a few inches of olive oil and set aside. Then with a board or other hard, flat object, squash each chunk into a flat disk and fry them again. They will be crispy from all the starch inside. Salt them, put on a little chili powder if you like, and serve as a side dish.

GUAVA PASTE (GUAYABATE)/MEXICO

This is a direct descendant of the Spanish *membrillo* made with quinces, and they look similar. Remove the seeds from guavas and soak in water. Boil the guavas in water with sugar and a cinnamon stick until they break down, stirring constantly. Drain the liquid from the soaking seeds and add it to the fruit. Then pass the cooked guavas through a food mill or strainer. Return to the pot and continue cooking on very low heat until it becomes a solid mass. Set out on a piece of parchment paper and place in a low-temperature oven to thoroughly dry. The paste is sliced and eaten with cheese or for dessert. It is also used as a filling in pastries.

TANG HU LU (CANDIED BOTTLE GOURD)/CHINA

These are not really gourds at all but candied haws from the hawthorn tree, on a skewer, which resemble the bulbous gourd. Haws are a kind of sour medicinal fruit, but other fruits are also used, such as crabapples, grapes, and strawberries. First the fruit is cooked in a sugar syrup and left to dry. Then it is dipped several times into a clear rock sugar syrup, until it has a hard, clear glaze. They are sold by street hawkers.

Frozen Fruits

PALETAS (POPSICLES)/MEXICO

Paletas are a wide variety of Mexican popsicles, some of which are based on fruit, others on milk with flavors such as chocolate, coconut, and rice pudding (figure 5.8). Even cucumber with chili and other exotic flavors can be found. The name comes from *palo* (stick). They are sold from street carts as well as *paletarías*, shops specializing in them. All you really need is a popsicle mold, or any shape will work. Simply puree the fruit, straining out seeds if you are using raspberries or similar fruit. Add a little sugar, if you like, and lime juice for added acidity if necessary. Pour into the molds and freeze. Mangoes

Figure 5.8. A *paleta* (popsicle) made from watermelon and sugar.

work very well, as does pineapple, cantaloupe, and strawberry, but pretty much any fruit can enter into a *paleta*. Coconut milk with shredded coconut works well also. To these you can also add a little regular milk for a creamier version.

GELATO AL LIMONE/ITALY

Mix ½ cup lemon juice with ½ cup sugar and a little grated lemon peel and heat until the sugar dissolves. Let cool thoroughly in the refrigerator. Add to this 2 cups milk or half and half (or a cup of milk and a cup of heavy cream) and put in an ice cream maker. When nearly solid, put in a sealed container and leave in the freezer to firm up. You can also serve this in a hollowed-out lemon peel, frozen with the gelato inside. Some recipes also call for a stiffly beaten egg white added in, which makes the texture considerably lighter.

⁂{ 6 }⁂
Meat, Poultry, and Dairy Products

Learning Objectives

+ Speculate why some societies but not others become heavily dependent on meat for protein, and why certain species predominate due to environmental factors.

+ Learn how various cooking methods have been used to render meat more palatable, nutritious, and tasty.

+ Consider why many societies use every part of the animal, whereas others waste the so-called lesser cuts.

+ Think about the true costs of animal rearing and its impact on the animals themselves, humans, and the environment.

+ Consider why these three cuisines are relatively free of food taboos concerning meat.

THIS CHAPTER discusses the dominant sources of animal protein, including meat and dairy, as well as poultry and other edible creatures, in the three world cuisines. The focus will be on why certain species became prevalent due to agricultural or environmental factors, as well as how these ingredients are prepared and incorporated into the cuisines. In some cases, the dominance of a particular form of protein was a simple matter of the accidental confluence of the right terrain and economic forces, such as a demand for meat in burgeoning cities. In others, it was specific aesthetic choices dictated by the culture and its values. Certain meats were favored for their flavor and texture or their ability to be preserved. Conversely, perfectly nutritious animals are sometimes avoided for purely cultural reasons, such as the horse in the West historically.

Wild Game

As hunters and gatherers, and for most of human existence on this planet, the principal source of animal protein came from hunted wild game. These could range from the smallest rodent to large elk and the now extinct woolly mammoth. Even after the domestication of cattle, hunting provided an important food resource and, in many cases, became the exclusive privilege of aristocrats in Europe, who may have been the only people legally allowed to hunt and often were the only ones with horses and weapons to take down larger animals. *Venison* once referred to any hunted animal, though the word is now normally used for deer species. Wild deer are prevalent in the Old and New World, but understandably their populations have severely declined in densely populated areas. This means that today in China they are very rare as food, although they are sometimes used as medicine (especially the antlers), and the same applies to Mexico, where deer were overhunted in previous centuries, though there are still populations of white-tailed deer in northern Mexico. In Italy, deer populations are also very small, but there are still roe deer (*capriolo*); fallow deer (*daino*), with their big, shovel-shaped antlers; and huge red deer (*cervo*), which survive only in the Alps; as well as the chamois (*camoscio*), which is a kind of wild mountain goat. These are all best roasted or stewed with strong spices and perhaps flavored with vinegar, as was common in the Middle Ages. One can still find salami or dried *bresaola* made from venison from deer that are today semidomesticated, that is, captured and kept in pens. Otherwise, these meats are cooked much like beef, with the caveat that they are much leaner and prone to dry out much quicker.

Wild boar still roam the mountains of Italy, and they are cooked much the same way as other pork products and are made into salami and prosciutto, which one sometimes sees in central Italy hanging from butcher shop ceilings with the fur still intact. Its taste is much denser and gamier than that of domestic pork. Equally important, in the past and today, especially on refined tables, is wild fowl, which will be discussed with poultry. In this class of food also belongs the wild hare as well as smaller rodents. Despite our antipathy toward the idea of eating rats and similar creatures, they must not be discounted as an important food source throughout history. The ancient Romans kept dormice in little perforated clay pots called *gliraria*, and there they fattened them up for the table. This practice persisted into early modern times. The Chinese have consumed rats, which were common around rice paddies. In Mexico, there are guinea pigs, the *paca*—a fat, spotted creature—and its smaller relative, the *agouti*, as well as anteaters and tapirs, all of which suggest that there was hardly a lack of animal protein in the past. Today, consuming such animals is largely restricted to rural areas, but such animals are not greeted with the kind of horror that might meet them in the United States or Europe. In fact, Western culture is the anomaly, and the consumption of domesticated animals meant that many perfectly edible species remain neglected.

Insects

Although the modern diner in most Western cultures rarely has recourse to eating **insects**, largely because of culturally ingrained prejudices and an aversion to insects in general as unclean and unfit for human consumption, it would be remiss to neglect them as a historically important source of protein, particularly in China and Mexico. Insects were an integral part of pre-Columbian cuisine and remain to a great extent delicacies in many locales (figure 6.1). For example, the eggs of small water insects called water boatmen and backswimmers or *ahuatle* are considered Mexican caviar and eaten on crackers. The larvae of *mecapale*, a kind of diving beetle, as well as those of dobsonflies,

Figure 6.1. The diet of the indigenous peoples of Mexico, before the Spanish Conquest, consisted of many edible insects and other exotic foods. Today, food scientists recognize the high nutritional value of these original organic products and are attempting to promote their consumption as an alternative source of proteins, carbohydrates, vitamins, and minerals. A selection of edible insects from Mexico: foreground, *cocopachis*; middle, a large beetle; in the back, *jumiles* (stink bugs). © Keith Dannemiller/CORBIS.

damselflies, and dragonflies (*padrecitos*), are also a traditional ingredient. These as well as grasshoppers (*acachapoli*) or *chapulines*, particularly from Oaxaca, are cooked much like crustaceans or salted and spiced and sold as a snack in the spring. Ants (*chicatana*) are also considered delicacies, and in Mexican markets they actually cost more than many mammalian meats. Nostalgia presumably plays a role here, as well as perhaps the kind of machismo we see in daring diners who find strange foods an interesting challenge. However, local people hardly consider these strange. In the old silver mining town of Taxco, there is an entire festival dedicated to the consumption of stinkbugs. The maguey worm in a bottle of mezcal may be the only place one finds them outside of Mexico, but these *gusanos del maguey* can also be deep fried or seasoned and served in a tortilla. We should also note that cochineal, a red, edible dye, comes from an insect, as does honey of course, both of which will be covered in subsequent chapters.

For purely nutritional reasons and as a natural way to keep plant predators down, entomophagy deserves serious consideration as a food resource, as well as a gastronomic experience. The Chinese have always and continue to consider many insects rare delicacies. Locusts are both collected in rural areas and sold in urban shops, and even in the West they were once a valued food source. By Levitical law, they are deemed kosher. In Guangzhou, silkworm pupae, beetles, fly larvae, waterbugs, and stinkbugs are also consumed. These may be boiled, pickled, baked, deep fried, or cooked any number of ways for particular dishes or for use in medicine. At banquets one particular rare delicacy is jellied bees. The real question, by way of comparison, is why insects were reviled in Italy and most of the rest of Western culture. The few isolated examples, such as *casu marzu* cheese from Sardinia, crawling with maggots, are the exception that proves the rule.

Domesticated Mammals

Dogs were probably the first domesticated animal on earth, as they happily followed bands of nomads in exchange for food and served as pack animals willingly and acted as guards in return. Dogs, in fact, came with humans in their migration from Asia to North America in prehistoric times. Dog flesh has been commonly consumed historically in all three of our world cuisines. The practice declined gradually in Europe as dogs were used extensively in hunting and were brought into the house as pets. Even as late as the sixteenth century there are reports of certain people in Italy who continued to eat dog as a regular food; that is, not in times of famine. Today, of course, eating dogs is considered abhorrent in Western culture. There was a hairless dog native to Mexico, used extensively for food by the Aztecs. The breed has since gone extinct in the wake of the Spanish Conquest, partly because the Spanish salted their flesh and used them to feed sailors. It is in China, however, that the practice lasted longest, down to the present, and dog meat there has been used since ancient times. Guangdong Province, according to reports from outsiders, was the most common region to find dog meat for sale, though it is more common in the north. There were many people who avoided dog meat for various reasons, particularly Buddhists. Nonetheless, dog has been prepared much like pig, made into hams, stir-fried, or "red-cooked." In many places today, such as Hong Kong, it is officially banned, but dog meat is still surreptitiously sold or served in private households. Some people even tell stories of pet dogs stolen by neighbors and eaten. The story is much the same for cats, though in China they were used more for medicinal purposes than cuisine.

Horses are also among the earliest domesticated animals. In prehistoric times archaeological evidence suggests that they were systematically butchered and eaten, but this practice gradually died out and was even officially banned by the Catholic Church. Horses were naturally much more valuable as traction animals than for food, pulling carts and plows as well as people. Strangely, consumption of horsemeat returned in the nineteenth century, particularly in France and Italy. Not coincidentally this happened at the same time that mechanical means of transport and traction were invented, and it initially took a conscious campaign to convince Europeans to eat horsemeat. There are still today specialized horse butchers in Italy, and horse flesh is considered a healthy and easily digested food. It is dark, sweet, and extremely lean, so the meat is normally quickly sautéed as a steak or finely ground and cooked. There is also a spiced, dried, and finely shredded horse product called *sfilata di cavallo* found in the Veneto, which is used in various cooked dishes and as a pizza topping. Horsemeat is also made into salami and *bresaola* (a dried, cured cold cut usually made from beef). In China there is mention of horsemeat as a delicacy in Han times, and there are reports in more recent centuries. But the scarcity of horses, especially in the south, meant that it could never become a common food. It is, however, eaten in Inner Mongolia, and mares' milk is fermented into a sour drink called *kumis*. In Mexico, horses were brought by the Spanish, and they were so highly valued for riding that the idea of eating them, as elsewhere in the Americas, was highly repellent.

Sheep and Goats

Sheep and **goats** were also early domesticates, roughly ten thousand years ago in the Fertile Crescent. They are eminently useful animals, providing wool and skins for containers, milk for cheese, meat, and natural fertilizer for growing crops. Most important, sheep and goats can be fed grass that grows naturally on mountainsides, thus converting what is otherwise unusable energy (humans cannot digest grass) into food. In traditional Italian culture, the meat most commonly consumed would have been lamb and kid (baby goat) for the simple reason that it is difficult to rear animals through the winter when mountain ranges become inaccessible. So most younger animals, usually males, are slaughtered in the autumn, and the reserved fodder is given to the adults, who can still be milked and will rear the next generation. The practice of transhumance, or moving sheep en masse to high pastures, is historically important not only for husbandry but also for gastronomy in general. Shepherds would have to survive in small huts, devising their own makeshift cuisine, and of course they made cheese in the mountains as well.

Lamb is one of the favorite foods among Italians, and in Christianity it is also a symbol of Jesus, the "lamb of God," whose sacrifice took the place of animal sacrifice in the remission of sins. Butchered cuts of lamb look quite different from those found in the United States, but there are comparable cutlets that are grilled or sautéed: joints such as the leg for roasting, coarser chunks for stewing, as well as odd misshapen cuts used in dishes like *scottadito* (meaning "burn finger," presumably because of how it is eaten in anxious haste), made from young, spring-milk-fed lamb (*abbacchio*). *Agnello* is usually the name reserved for an older lamb, and *castrato* for a young, neutered, male sheep. Fully grown sheep or mutton, more rare, is *pecora*, or *montone* for mountain ram. Kid is also a specialty around Easter time, usually simply roasted whole, and one will see them in butcher shops being skinned whole. Lamb organ meats are also highly appreciated. *Coratella d'agnello con carciofi*, for example, is a dish of a lamb's "pluck," which is lungs, windpipe, and heart stewed with artichokes. This dish can still be found in the Jewish quarter in Rome.

Sheep are most common in the northwest of China where there are open plains for grazing, and particularly among China's Muslim population. Here lamb is commonly barbecued, braised, or thinly sliced and quickly cooked in hot pots. The Manchus of the north are also very fond of lamb, and specific lamb dishes have spread elsewhere in association with these groups. For example, marinated lamb stir-fried with leeks is now enjoyed throughout much of China.

Sheep and goats were also imported to Mexico, but their use here is relatively limited. This is surprising given that mutton was among the favorite meats of Spanish conquerors, and most likely it was simple environmental factors that limited extensive pastoral operations, or perhaps they destroyed crops when allowed to roam freely. There are exceptions, though, such as roast kid (*cabrito*), which is common in the north in and around Monterrey. The kid is split and flattened, then roasted slowly for hours over a slow, charcoal flame. But it can also be cooked much like beef, grilled or stewed. Actually it is probably the importance of beef in northern and western Mexico that accounts for the relative scarcity of sheep. But this is not really the case in the mountainous center of Mexico, where sheep are reared and where one is more likely to find mutton on a barbecue than beef, and this would include the head and the stomach stuffed with other organs.

Water Buffalo

Water buffalo must also be mentioned as an important food source, especially in China where they have been a principal draft animal, as well as in Italy, where their milk is used to make mozzarella. But the meat has also been consumed, especially in marshy places, such as around Rome. These should not be confused with what is called buffalo in the United States, which is properly a bison.

Beef

The cow also belongs to the complex of animals domesticated in the Neolithic Revolution, descended from the once-wild roaming aurochs. They gained great importance as a source of traction, pulling plows and carts, especially fully grown and castrated oxen, which are both powerful and docile. It is for this reason that **beef** was not a regular food, especially in places with high population density, little room for pasturage, and intensive agriculture. It had been reserved for special occasions, though.

This is especially the case in China, where Buddhist vegetarianism and a need to preserve cattle for plowing and fertilizer made beef a very expensive food, until recently. Either oxen in the drier regions or water buffalo in the wetter rice-growing areas were normally eaten only after they ceased to be able to pull a plow. Because dairy and cheese have been practically nonexistent historically, there was not the constant need to keep cows calving, thus there was no regular supply of veal, either. There were actually laws prohibiting slaughtering cattle in various periods as well in an effort to keep an adequate supply of plow animals. Cattle were not offered as sacrifice.

The ubiquity of beef on Chinese American restaurant menus more closely reflects American tastes than Chinese. In Chinese cuisine, a small amount of beef might be used to flavor a large dish of other ingredients. Beef is nonetheless today one of the most highly appreciated meats in China and readily lends itself to being sliced thinly and quickly cooked, far more so in fact than pork or chicken, because beef is best rare or slightly undercooked. It is also used to make rich broths, long-simmered stews, cold salads, and dried as a kind of Chinese beef jerky, surprisingly with fruit flavors. The panoply of beef dishes today in all likelihood use traditional techniques and flavorings applied to this now more-abundant ingredient. Thus, we find beef stir-fried with black bean sauce or oyster sauce, red-cooked in a stew with soy where one might have found pork, in a Mongolian hot pot where lamb would formerly have been used, or even steamed. The major difference in Chinese beef recipes, as compared with those in the West, is that one never finds a large slab of beef cut with a knife or sliced at the table. Beef is typically cut into small, thin pieces raw, perhaps marinated, and then cooked. The strong flavor of beef is also usually accompanied by other strong ingredients, such as five spice powder, garlic, orange peel, or chili. Sometimes a larger cut is simmered and then cooled and sliced and seasoned. Or ground beef may be used in poetic dishes, like "ants climbing a tree" from Sichuan, in which cellophane noodles swathed in a sauce with meat seem to resemble trees, up which ants are climbing when one lifts a strand to the mouth. Beef, of course, exacts a large toll on the environment, as cattle are very inefficient sources of protein, requiring a great deal of grain in modern industrial husbandry and producing a great deal of pollution in the form of manure and methane gas. For this reason, environmentalists have been especially concerned with the rapid increase in consumption of beef in China recently.

In Italy, veal is more common than beef, for the same reasons that lamb and kid have predominated. A cow provides renewable food resources only when alive. Cow's milk products naturally are of central importance in the Mediterranean and will be discussed below. Only recently has beef been eaten on wider scale, the most famous being the huge T-bone cut of Chianina beef found in Tuscany, called appropriately *bistecca* after the English term "beef steak." Strangely, beef is not typically roasted but boiled, or rather slowly simmered in a *bollito misto*. This contains a large cut of beef and any number of other ingredients, such as vegetables and sausages, perhaps *cotechino* (a sausage made of pig skin) or *zampone* (a stuffed pig's foot) to give it body. The meats are removed and sliced, and the broth drunk separately. The dish probably arises from the fact that normally older animals only would be used, and they need slow cooking to tenderize. The younger cuts of soft, lean meat raised today would yield only a pallid, bland dish. Interestingly, medical literature of the past always recommended boiling beef as well, to temper the dry flesh and make it more digestible. Ground beef is used too, sometimes in combination with veal and pork, to make meatballs and pasta fillings and as the base for *ragù Bolognese*, which is a meat sauce with only a little tomato and enriched with milk. Italians also make use of lesser cuts: intestines are used for sausage casings, and sometimes beef is used for *salumi* (salted, cured meats), often in combination with pork. The tail is stewed (*coda alla vaccinaria*), and practically all the organ meats are used in a wide variety of contexts.

Again, veal is still more common and is best appreciated in thinly sliced cutlets that are served with a marsala (wine) sauce, or rolled in prosciutto and sage in a *saltimbocca*, which means "jump in your mouth," or in the deservedly popular veal *alla Parmigiana*, or the intriguing *vitello tonnato*, which is a slice of veal covered in a tuna fish sauce. Larger cuts of veal are also common, such as the stuffed breast of veal of Genoa, which is slowly simmered and served cold and sliced. There is also the classic *ossobuco* of Milan, one of the great triumphs of Italian cuisine. It consists of sawed veal shanks braised in a *battuto* of carrots, celery, and onion with broth. The meat becomes unctuous and tender, and the marrow can be scooped from the "eye" of the bone. A sprinkling of *gremolata* (finely chopped garlic, grated lemon peel, and parsley) hits the dish before service.

Mexico is a strange exception to these rules about the preference for veal, mostly because cattle were reared on huge *haciendas* in the flatter open ranges of the north and west and could be kept outside grazing year-round. This was partly to supply leather. But it meant that the price of beef was relatively lower than back in Europe through the colonial period, and consequently there are more beef recipes in Mexico, perhaps even than in Spain. Beef (*carne de res*) is roasted in the ubiquitous *carne asada* (normally sirloin or rib), or it is stewed and perhaps shredded, as in a taco or in *gorditas* (little, fat-stuffed circles of corn masa). Dried beef is also common, called *tasajos* or *cecina*. Both these are not native kinds of jerky applied to beef but rather taken directly from Spanish traditions. *Cecina* is similar to what the Italians call *bresaola*, an air-dried beef, eaten much like prosciutto, thinly sliced and raw. Ground meat is also used in the various dishes that in the United States would be called chili, as well as meatballs (*albóndigas*), which also come directly from Spanish cookery. The flavorings and sauces (like those based on chipotles) in which they are cooked, however, are native.

Milk, Butter, and Cheese

The historical importance of dairy products in the human diet has largely been limited to those cultures in which herds of cattle and the rich grasslands to feed them have been available. This has meant that until fairly recently southern China made practically no use of **milk**, apart from human breast milk for infants. In the far north and West, among Turkic populations, and in Tibet where yak **butter** is mixed with tea, this has not been the case, and for brief periods the Chinese experimented with products like *kumis* or yogurt, especially in the Tang dynasty. But in general, milk as a food for adults is virtually unknown, and this has been the probable cause not only of a cultural aversion to milk but also even widespread lactose intolerance. The Chinese, as have many other peoples on earth, never developed the ability to digest milk in adulthood. It is only very recently that Asians have begun to consume milk, and it is given to children in particular to foster growth.

The exact opposite is the case in Italy, which adopted the domesticated cattle of the Middle East and has always made use of dairy products, primarily in the form of **cheese**. Aged cheese, apart from the practicality of keeping a food source through the winter when cows would often stop producing milk, also made sense because dairy products, along with meat, were forbidden during Lent, the forty days preceding Easter. Technically, cheese is merely milk that has been curdled by means of rennet or, in some cases, vegetable extracts like cardoons or fig sap (which are both vegetarian and kosher), and they are usually heated. The curds are then separated from the whey and pressed, traditionally in a little wicker mold, from which comes the name *formaggio*. These young cheeses can be aged, and many receive piquancy from naturally forming lactobacillic cultures as well. The leftover whey can then be fed to pigs or further processed into ricotta, which means "recooked." Ricotta is a smooth, fresh cheese used in lasagne, manicotti, and other fillings.

There are a few world-renowned cheeses native to Italy based on either cow, sheep, or goat milk, or some combination of these (figure 6.2). An exhaustive list would be impossible, but here are some of the most important. Mozzarella is probably the best known of cheeses, thanks to the popularity of pizza and other dishes topped with or incorporating melted cheese, such as lasagne or baked ziti. Most of this today is made industrially with cow's milk, but the finest mozzarella comes from water buffalo milk. It is normally sold very fresh, in small balls (*bocconcini*) or larger globular pieces, which are formed by stretching and forming the cheese in very hot water. Mozzarella is pale yellow and actually does not melt the way the rubbery product familiar in the United States does. It tastes primarily of fresh milk and is thus best simply adorned in a salad.

The king of all Italian cheeses, however, having been produced for centuries in the Po Valley, is Parmigiano-Reggiano, which is nothing like the cheaper, industrially produced hard cheeses or what is labeled Parmesan in the United States. And, of course, the dried product shaken out of cardboard can is a travesty. The true cheese is a strictly controlled and legally protected artisanal product, with a hard, grainy interior and a complex, buttery aroma that develops during the long aging process (figure 6.3). In the past, cheeses from Piacenza, Lodi, and other cities were equally famous, but today these are all lumped together as *Grana Padano*, most of which are produced industrially. They can be an affordable substitute for Parmigiano. The best way to appreciate the real thing

Figure 6.2. Cheese shop in Milan, Italy. © Michael S. Yamashita/CORBIS.

is an entire half wheel, which is carefully separated with a series of wedges, and then served in hunks pried from the exposed surface. Parmigiano can also be grated and incorporated into ravioli fillings, sprinkled on pasta, or as a flavoring in meatballs. Its use is practically ubiquitous. A similar but quite distinct cheese is called Pecorino, which as the name suggests is made from sheep's milk. It is lighter in color, saltier, and without all the nutty subtleties of Parmigiano. Nonetheless, excellent versions come from Sardinia and exquisite aged versions from Tuscan cities like Pienza. What is found in the United States is normally Pecorino Romano.

Gorgonzola is a white, soft cheese streaked with blue veins with a pungent aroma and flavor. The blue mold is actually penicillin, today injected, which is why they appear in regular columns. Although it can be melted into cream sauces, it is usually simply eaten after dinner in a cheese course. There are both sweeter and drier versions; the former is usually eaten after a meal.

Other popular cheeses include provola, made similarly to mozzarella with the *pasta filata* method of pulling strings and rolling them into a ball, but it is aged and sometimes smoked and hung in pear shapes surrounded by netting. Provolone is a bigger, sausage-shaped version that is sliced and has a pronounced sharpness. There are also regional cheese specialties such as *taleggio*, a flat, square, soft cheese covered in a white mold. There is *asiago* from the area of Trento and Vicenza, which may be eaten young

Figure 6.3. A cheesemaker smells a sample of cheese bored from a round of Parmesan to determine its quality in the storeroom of a cheese factory in Parma, Italy. © Owen Franken/CORBIS.

or in a powerful aged version. From the northeastern Alps and the Val d'Aosta comes *fontina*, which when ages and is melted makes the ideal *fonduta* (or as the French call it, a fondue).

Cheese was introduced to Mexico with the Spanish, though Mexican forms tend to be unique. Most are fairly fresh white cheeses such as *queso blanco*, a kind of firm cottage cheese, which is used for fillings, although it does not melt. *Queso fresco* is normally crumbled on top of other dishes in texture something like feta, but it can be aged and hard, in which case it is called *queso añejo*. *Requesón* is another fresh cheese similar to ricotta and also used as a filling, and *queso Oaxaca*, which is more like mozzarella, works nicely in quesadillas. There are also aged cheeses like that from Chihuahua, which is more like cheddar and was introduced by German Mennonites, and *manchego*, which is a hard, pungent cheese introduced directly from La Mancha in Spain. Last is *cotijo*, a hard, crumbly goat cheese, used much like Parmesan in that it is crumbled on other dishes. The familiar melted nacho cheese one encounters in the United States is for all practical purposes unrelated to these.

Today in all three cuisines one can find other dairy products, including cream and butter used in cooking and yogurt as well as cheeses imported from all over the world. These have not been used extensively in traditional cookery, though, for the very simple reason that they spoil easily without refrigeration. Thus, in dairying regions they would have been used on the ranch, in northern Mexico, in the Alps or Apennines, but it is only in modern times with refrigerated transport or with ultrapasteurizing technology that they have become common. All the same, butter is not essential in a well-stocked kitchen of any of these three cuisines, though northern Italians, especially those from Bologna, might well disagree.

Pork

Pork is one of the most ubiquitous and versatile meats on earth, and it features centrally in all three of our world cuisines. Pigs are native to Eurasia and still run wild in many forested areas, feeding on nuts and acorns. They have also been domesticated as long, if not longer, than sheep, in many separate locations independently. The wild boar is genetically similar to the domesticated pig, and in fact if left in the wild pigs will revert to feral status within a few generations, growing longer tusks and thicker bristles. They were introduced to the New World after 1492, though there are native species of piglike animals such as the javelina or peccary, but these are too feisty to be domesticated.

The usual quip that every part of the pig from tail to snout (except the oink) is used applies equally to all three cuisines. This is simply a matter of economy, and as every part of the pig does have many uses and can be easily preserved, there is no reason to waste anything. The so-called lesser cuts, including ears, snout, and organ meats, can be used in stews or sausages; the intestines provide casings. Pork fat is a typical cooking medium, either in liquid form or semisolid lard, or rendered from a preserved hunk as with pancetta, *lardo*, bacon, and the like. The caul fat, or fatty lining of the viscera, is also used wrapped around chopped meat or other cuts to baste while they cook. The pig's legs are cured as hams. Even its head is boiled down and molded into a gelatinous headcheese in Italy (*coppa di testo*). Its blood is used in soups and sausages, and blood scrambles when cooked, like an egg, as in the Italian *migliaccio*, a sweetened pancake of blood, traditionally made with millet, or the *sanguinaccio*, made mostly of blood (though oddly today made of chocolate). It is also used in black sausages such as *buristo*, or the southern Italian sweet *biroldo* with pine nuts. The pig also provides chops, ribs, feet, and roasts. Its skin is cooked down into cracklings (*ciccioli* in Italian, *chicharrones* in Mexico) as well as *cueretos* (skin marinated in vinegar); and various kinds of cracklings in China. Skin can also be cooked stuffed into a sausage casing, in Italy called *cotechino*. Pigs, unlike cattle, provide few services on a farm. They cannot be used for traction or milking, and they compete with humans for food resources because they are not ruminants and cannot digest grass. However, no animal provides more varied and delicious ingredients for the kitchen.

Fresh pork can be simply roasted or fried in a pan, and the loin, the most delicate part, is almost always cooked over quick, high heat and then finished in an oven. This is also true of chops. The majority of cuts available in U.S. butcher shops must be cooked this way, primarily because they are nowadays bred to be leaner and butchered with less visible fat, so the flesh dries out quickly. A breadcrumb coating on chops and cutlets helps retain moisture. Lean pigs are not as highly valued elsewhere in the world, though modern breeds and industrial-scale rearing practices are catching on outside the United States and Europe. This does mean that many traditional recipes are hard to make well with available cuts of pork. There are, however, numerous recent efforts in Europe to revive traditional breeds that mature slowly and have good flavor and texture. The Cinta Senese, a black pig with a white sash around its shoulders, bred in Tuscany, is a good example of these.

Larger cuts of pork such as the shoulder or butt are best either roasted or cooked slowly in a braise or stew. Each of our three cuisines has a unique way of treating these cuts, with very different flavorings, but there are essentially comparable procedures. Simply slowly stewing pork in Mexico, with onions and garlic, and then letting it cool and shredding it, provides one of the most common fillings for tacos or stuffing chilies. With chili peppers and hominy (corn that has been nixtamalized and freed from its outer seed coat, but still whole) added to the cooking liquid directly, the pork is cooked until it falls apart, a classic rendition called *posole*.

In Italy, pork is sometimes boiled or braised in an enclosed pot set in the oven, strangely enough in milk, which is then cooked down and provides a sauce. Pork is usually roasted, though, and simply sprinkled with salt and fennel or rosemary. Norcia in Umbria is said to have the best pork butchers in Italy, and a butcher everywhere is even called a *Norcino*. The typical pork butcher's shops also cure their own preserved meats or salumi. But fresh sausages are equally important, and have been since classical times, when several of the basic forms were already being made. *Lucanica*, named for Lucania in the modern province of Basilicata, is made with spices and sometimes smoked. Today they are made all over Italy, though, as is basic *salsiccia* with infinite regional variations. These can be grilled or fried whole or crumbled into sauces or over pasta or on top of pizza.

Roast pork (*Shao zhu*) is a staple item in the Chinese kitchen but differs from the Italian in that it is never served cooked as a huge slab of meat. Rather, it is cut up and combined with other ingredients as a flavoring. Its importance in China can partly be explained by intensive agriculture, which left little land for pasturage and made pigs an ideal animal to raise in a small place, fed with table scraps. Consequently, it is the most important source of animal protein and figures in hundreds of dishes. Not only roast pork with its signature red hue but also ground pork, pork balls, shredded pork, barbecued ribs, and practically every anatomically distinct part of the pig can be found in a Chinese butcher's shop, each with its own culinary uses.

In cooking fresh pork, there is also a fairly universal method of flavoring large cuts, or whole pigs: marination. This involves setting the meat in a wet medium including salt, possibly providing the derivation of the word *marination*, from "marine." Pork lends itself extremely well to this process. As the meat marinates, the salt concentration difference between the cells in the meat and the surrounding medium cause the marinade to be absorbed through the cell membranes, along with the other flavors. Water is also drawn in, making the meat juicier. Brining is not chemically different from marination. This process is used extensively in our three world cuisines in surprisingly similar ways. For example, the Yucatan dish of *cochinita pibil* replicates an ancient process, using more recently introduced ingredients. In it a suckling pig, or today often a large piece of pork shoulder or butt, is marinated in a combination of *achiote* paste, made from crushed red annatto seeds, which are native, along with allspice (also native), cumin, pepper, and sometimes other spices. These are mixed with chilies and the juice of bitter Seville oranges, garlic, and salt. The pig is then marinated overnight. The next day it is wrapped in banana leaves and placed in a pit containing hot embers and then buried (*pibil* means "buried" in Mayan; *pib* refers to the underground oven) and left to cook slowly for five or six hours. The marination causes the flavors to penetrate the flesh in a way that cooking with them alone would not achieve.

A similar effect is achieved, coincidentally, with Chinese dishes, in which the meat is marinated in soy, ginger, garlic, and vinegar or other acid such as oranges (which are native to Asia). As in the aforementioned, the acid ingredients also begin to break down the tough fibers in the meat, which means less tender cuts can be used. There is a classic wedding dish in China as well, called red pig, in which a whole pig is rubbed with a mixture of salt, soy paste, ginger, and Gaoliang (or Kaoliang) wine made from sorghum and red peppers, and then hung, in a kind of semimoist marinade. It is then left to dry and drip slowly. More seasonings are added, and then it is roasted on a spit or placed in a pit lined with hot stones and covered in leaves. The red color, as in all Chinese food, denotes good luck.

In Italy, a marinade usually consists of vinegar and oil, herbs, and garlic. *Porchetta* is the Italian dish comparable to those mentioned: a boned and stuffed whole pig, which is slowly roasted on a spit or baked in an oven. It comes from central Italy, Ariccia near Rome being one place associated with it, as well as Monte San Savino in Tuscany. On a good hunk of bread, it makes an incomparable dish, and people travel miles to eat *porchetta* sold from a truck.

Letting the meat sit longer in brine completely changes its flavor and texture, and rather than marination, this is technically a kind of pickling because bacterial cultures begin to proliferate, lowering the pH and giving the meat a sour flavor but also killing harmful pathogens so the meat can be stored. Pork belly is commonly pickled in many cultures and used either as an addition to slow-cooked dishes or rendered for fat, a subject that will be treated later.

Beyond fresh and pickled pork, there is also an enormous range of cured, salted pork products, known in Italy as *salumi*, and where they are purchased, a *salumeria* (figure 6.4). These include hams such as prosciutto, salami (or properly *salame*) made from

Figure 6.4. Traditional food shop window in Bologna, Italy. © Atlantide Phototravel/CORBIS.

coarsely chopped pork and fat stuffed in casings from intestines or other organs, *culatello* made from the rump encased in a bladder and tightly bound, *guanciale* made from the jowls, *pancetta* from the stomach, *zampone* from the feet, and so on. These are both cooking ingredients and are eaten raw throughout Italy as an antipasto or snack. The process of curing pork products is essentially the same in all these examples. The meat is salted, which draws out moisture from the cells and prevents bacterial growth. At the same time, a kind of lactic fermentation takes place with a good kind of bacteria, keeping out the bad bacteria and giving the cured meat a sour, tangy flavor. The red color of most salami was traditionally achieved with saltpeter or potassium nitrate. Today sodium nitrite is more common. Nitrates, although there is some concern over their use today, also effectively kill the spores that can cause botulism and other diseases and do lend very distinctive gastronomic qualities to cured meat. Sometimes white mold is encouraged on the exterior casing of salami as well, also acting as a preservative. Some types are also smoked; others like mortadella are cooked. This is a huge sausage of very finely ground pork studded with lumps of fat and peppercorns or pistachios. Its American descendant is called bologna, after the city most associated with the original. In Italy it can be thinly sliced but is also often cubed and served with toothpicks as an appetizer.

There are also cured sausages in China, like *lap cheung* (*la chang* in Mandarin). Lighter versions are made only of pork, and darker types may include duck liver. These are normally sliced and cooked with other ingredients rather than consumed raw like salami. In Mexico, chorizo, flavored with chili peppers, is normally an uncured fresh sausage, unlike its namesake in Spain, and it must also be cooked. It is one of the most typical ingredients used as a filler, normally squeezed out of its casing and sautéed first. Sausages are also made with pig liver (*tomacella* in Italian), and some types of *lap cheung* use pig's liver as well.

Pork is also smoked, which also dries the flesh and prevents bacterial growth. *Speck*, a kind of smoked ham, from the very north of Italy in the Sudtirol, is a good example of this. It is first brined with pepper, juniper, and garlic, then lightly smoked and hung to dry. The result is an intense pork flavor. There are comparable smoking procedures for preserving pork in China and Mexico as well. Unlike these are cured and air-dried hams, the most famous of which is prosciutto. The best come from Parma, or the Veneto in the case of San Daniele, though there are many other regional versions. Prosciutto is merely salt cured and air dried in a complex and traditional procedure, which again, concentrates the flavor. The famous Jinhua ham of Zhejiang Province in eastern China can be surprisingly similar. Ham is often processed in association with dairying, as the whey from the manufacture of cheese is fed to the pigs and is said to lend it an unmistakable flavor. But in many other locales pigs are still allowed to roam and feed on acorns and beech mast, which lends it an entirely different astringent and nutty flavor. *Capocollo* is another kind of ham made from the neck, and today one can also find cooked or boiled hams, which are uninteresting by comparison. Most of these products are eaten simply sliced thinly as an appetizer, but they can also be used in cooking—the prosciutto bone, for example, used to flavor a soup, or cured meats finely diced in a stew or pasta sauce.

Rabbit

Rabbit meat, either from rabbits kept in captivity or, rarely, caught wild, is delectable but is gradually disappearing from American tables for reasons of squeamishness. In Italy there are no such qualms about eating small mammals, and in restaurants one might even be served rabbit freshly plucked from a cage and dispatched for dinner. Its taste is often compared with that of chicken, but it is actually quite different when prepared well. Although it is light colored and sometimes cooked quickly, in Italian cuisine rabbit truly excels when slowly braised or cooked in a rich pasta sauce. Unlike bland chicken breast, rabbit's flavor concentrates and intensifies and can stand up to tomatoes, olives and capers, and wine. *Coniglio alla cacciatora*, in which the rabbit is browned in oil or lard in a casserole and slowly cooked with white wine, whole garlic cloves, rosemary, and sage, perfectly evokes the rustic hunter's meal. Rabbits for food and fur are a fairly recent addition to the Chinese repertoire, but apparently they are reared in enormous numbers and are cooked in much the same way as chicken. Rabbits are also native to Mexico and are usually stewed with green peppers and corn, a dish that has scarcely changed in centuries, though similar dishes, *fricase de conejo* and *guisados*, were also common in Spain.

Poultry

Wild fowl have been hunted and cooked in each of these cuisines from prehistoric times through to the present. They have often been considered the most elegant and refined of foods as well, especially in times when only a few had the leisure time and extensive land to devote to hunting. Fowl include well-known birds still considered delicacies, such as pheasant, grouse, and partridge, as well as tiny birds such as thrushes, fig peckers, snipe, and quail. In Italy these tiny birds (*uccelletti*) are typically simply roasted on a turnspit, sometimes covered with a piece of bacon to prevent the breast from drying out. Incidentally, little rolls of veal or beef are also called *uccelletti* or *uccelletti scappati*—something one makes when the birds get away. With equal irony, *fagioli all' uccelletti* is just a bean dish. Actual wild fowl can also be boned and made into a sort of rustic paté (*sformato*). Small wild fowl have also always been eaten in Mexico and China, though increasingly with urbanization such birds are disappearing from the table in favor of bought domesticated fowl, though this might include domesticated quail in China, which is highly prized both for grilling and for its tiny eggs.

Although uncommon today, in the past wild waterfowl were considered an elegant dish throughout Europe. Swans, despite their dark flesh, as well as cranes were a delicacy. Their long necks could also be stuffed. Peacocks, too, were considered aristocratic fare, and the bird might be cooked and then resewn into its feathers, a common table trick used with pheasants as well. Many of these birds were captured and fattened for the table, which might be considered a form of semidomestication, and in the case of quail, they were domesticated and bred in captivity. This is equally true of pigeons, perhaps the most important of all small, wild fowl that could be kept in towers and dovecotes. This was, strangely enough, an exclusive aristocratic privilege in much of Europe, not particularly appreciated by peasants, whose freshly sown fields could be devastated by a flock of pigeons. In modern times, because of the pigeon's association with urban parks and garbage, it is looked down on as food, but it is basically the same creature as the dove—a symbol of purity—and the squab, which is what it is called when it appears on a plate. Pigeon meat can be cooked much like any other small bird: roasted whole or cut up for use in a pie. In Chinese cuisine, pigeons might be served in a tonic broth with wolfberries, which is considered a powerful aphrodisiac, or even boned and stir-fried like other poultry dishes.

CHICKEN

A domesticate of the Southeast Asian wild jungle fowl, there is perhaps no other creature on earth used for food that has become so ubiquitous or enjoyed in more places than the **chicken**. This is, of course, because the bird is tough and survives the industrial rearing practices, some undeniably inhumane, imposed upon it. This was not the case in the past, however, when birds were left in farmyards to peck and scratch at worms and bugs, making them darker-fleshed and perhaps tougher but much deeper in flavor. For this reason, many traditional cooking practices will not work with the young and relatively flavorless birds available today. Long, slow stewing, for example, is practically impossible with a modern, hyperextended white chicken breast. In the past, typically female chickens were kept for egg laying and the eggs sold, while male chickens were castrated, after which they are called capons. Castration causes the bird to grow rapidly and much bigger than normal. These large, flavorful birds hold up nicely to roasting, especially when larded with strips of bacon fat, or to simmering in a rich broth, which is considered an ideal food for invalids, or to cutting up and browning in a pan and then braising with vegetables and wine—a classic chicken cacciatore. In the past, Italian cooks were even more inventive with chicken. One classic dish, the blancmange, involved poaching the capon breast, picking it apart into minutely fine threads, and then cooking with almond milk, rosewater, rice starch, and sugar. Again, the delicate white dish was thought to be easily digested and extremely nutritious. Today, blancmange or *biancomangiare* is a sweet dessert without the chicken.

In the modern Italian kitchen one is more likely to find chicken taking the place formerly held by veal, not only in consideration of cost but also perceived health benefits as well as animal welfare concerns. Chicken breast is often pounded flat and breaded as a cutlet or grilled with herbs and olive oil. It is also cooked down into a rich broth, in which might be served little tortellini *in brodo* (in broth).

Chicken broth is also a universal base in Chinese cuisine, though the process of making it is somewhat different. Just as was common in Europe in the Middle Ages, chickens in China are usually blanched in boiling water first. Many contend that this is to remove any unpleasant odor, but in fact it firms up the flesh and helps it retain moisture. This is done both before making broth and when cooking the chicken in some other preparation, perhaps stewed in a clay pot or slowly roasted. Interestingly, chicken is also served quite differently in China. Rather than carving the meat off the bone, or serving smaller, disjointed pieces such as legs, wings, thighs, and so forth, the chicken is often hacked straight through the bones, spine and all, which is said to make it more savory, and indeed much of the flavor is close to the bone (*gu xiang* means "bone fragrance"), and this allows seasonings and sauce to penetrate all parts of the chicken. Chicken meat is also finely sliced and cooked in stir-fries with vegetables, or steamed, or marinated in wine to make dishes like drunken chicken. Another technique is called velveting, in which the chicken pieces are coated in egg white and cornstarch, then hot oil is poured over the meat or it is plunged briefly in hot oil, giving it a rich texture and flavor and sealing in the juices. It can then be incorporated into other dishes. This should not be confused with what is called velvet chicken, in which the chicken breast is finely hacked with two cleavers, and a little cornstarch and seasonings are added until it is a fine, white purée and then cooked. The result is soft and almost custardlike, strangely similar to the aforementioned blancmange. Another interesting technique used to make shredded chicken dishes involves the breast being cut finely along the grain rather than across; this prevents it from totally falling apart when cooked.

In Chinese markets one will also encounter a black chicken, called a silky or silkie, from its fine, downlike, furry white feathers. The skin and even bones are dark purple or black. These are reputed to be medicinal, good for building up strength and blood, especially when stewed with medicinal herbs like ginseng and wolfberries. It is often given to pregnant women and also eaten for other ailments. Coincidentally, in Italy, it has always been the custom to serve women chicken broth after childbirth.

Chickens have also been an enormously popular ingredient in Mexico for the past five centuries, and here they can be slowly stewed and shredded as a filling for tacos and tamales much like any other ingredient. Pieces of chicken are also cooked in a molé sauce, much like turkey. Chickens are also barbecued, seasoned with achiote paste, onions, and chilies, or sautéed with chilies and cream—a dish that appears to be a combination of Spanish, indigenous, and French ingredients. The popularity of chicken is, of course, due to its incredible versatility; practically any dish made with meat can also be made with chicken.

Equally important historically and in the present are eggs from chickens, and eggs without doubt are the most versatile ingredient on earth, lending themselves to dozens of different techniques. They can be eaten raw, soft-boiled, hard-boiled, poached, steamed as a custard or flan, coddled in an enclosed container, scrambled, baked into a quichelike pie, fried, whipped into frothy egg whites, the yolks whipped and gently cooked into a *zabaglione* flavored with marsala—the list is endless, and this does not even include eggs in baked goods. In the past eggs were even roasted, either by throwing directly into hot coals or carefully skewered and turned before the fire. They were also the preferred thickener in sauces before the invention of butter-based sauces; the whole egg or yolk was added at the last minute and very gently heated over a double boiler until thick.

In Italy the egg is also a symbol of rebirth and resurrection, perfectly suited to Easter, and colored Easter eggs even have their roots in pagan practice, long before the advent of Christianity. Hard-boiled, red-colored eggs are baked into braided dough. A classic use of eggs in Italy is the frittata, which is quite different from an omelet because it is cooked into a solid mass and then sliced at the table. It can contain shredded zucchini or other vegetables, cheese, and practically any leftovers for a quick, easy supper. Eggs are also incorporated into soups, either placed in whole as in *zuppa alla pavese*, or cooked into fine strands by vigorously beating the egg into the soup, called *stracciatella*. The exact same technique is used in Chinese egg drop soup, though here the longer strands are achieved by pouring the beaten eggs slowly into the soup. The Chinese are also extremely fond of preserved eggs, though ducks are the preferred species. Eggs are also used in a savory custard (*zheng shui dan*) that might include meat or fish. There are also tea eggs (*cha ye dan*), which are hard boiled then gently cracked and soaked in tea, which gives the egg a marbleized surface when the shell is removed.

In Mexico, eggs are also a universal food, the most familiar dish being *huevos rancheros*, in which the fried eggs are served on tortillas covered with red or green chili sauce and cheese. When the egg is scrambled with fried onions, chilies, and cheese, it is called *huevos revueltos*. Like the frittata, it can incorporate leftovers and makes for a hearty and satisfying midmorning meal called *almuerzo*—something like brunch—but not lunch as it is often translated, because it happens before noon. The logic of this meal is that breakfast (*desayuno*) normally consists only of bread and coffee or churros and chocolate. Eggs are ideal for the morning meal, scrambled with chorizo or chopped tomatoes and chilies, or nopalitos, beef, or virtually anything. Eggs are also featured in sweets and desserts, in flan, for example, or *yemas*, which are sweetened and flavored egg yolks, both of which come from Spain. Bread also is often enriched with egg yolks.

Turkey

Turkey is one of the greatest contributions of the Americas to world cuisine. Turkeys were domesticated in ancient times and sold in the Aztec markets by the thousands, according to conquistadors' accounts. They were also enthusiastically and almost immediately adopted in Europe, and there are recipes recorded in cookbooks like that of Bartolomeo Scappi in the sixteenth century. Confusingly, the turkey arrived in Italy at about the same time as the guinea fowl, and both are sometimes called *gallo d'India*, but the latter comes from Africa, and neither comes from India. Today they are called *tacchino*, supposedly from the sound they make, which in English is construed as "gobble gobble" and in Italy as "tac tac tac." Typically the turkey is roasted, but parts can also be cooked as cutlets, sautéed with other ingredients, or as in the bizarre turkey tetrazzini, laced with cream and mushrooms and baked with noodles, which is an Italian American invention. The classic cookbook of Pellegrino Artusi in the nineteenth century suggests making a soup out of turkey, with rice, cabbage, turnips, or spelt; roasting with garlic and rosemary; or the pounding thin of the breast and grilling. Today it figures in precisely the same way any other meat would. Not surprisingly, the most inventive turkey dishes are Mexican—the most famous being turkey (*guajolote*) cooked in *molé poblano* sauce, which is a thick combination of chilies, spices, nuts, raisins, tortillas, tomatoes, and a hint of chocolate.

Ducks and Geese

There are species of **ducks** native to both the Old and New World, many of which remain wild and are to this day primarily hunted, but domesticated species also come from both sides of the Atlantic and Pacific, and several modern crosses have been made among these. For example, the mulard ("mule") is the offspring of a male American Muscovy and a female Pekin and is the preferred variety in the production of foie gras in France. Most domesticated ducks are descended from the mallard (*Anas platyrhynchos*), and the process of domestication probably occurred independently at different times and places. The most important difference between these and their wild relatives is that they lost the ability to fly.

Nowhere is the duck more appreciated than in China, where they were kept in great numbers since antiquity, perhaps as long as three thousand years ago. Their importance there may be partially explained by their preference for water; that is, in rice paddies, where they eat insects and other creatures that would feed on the rice seedlings. But another consideration is that duck fat is highly prized in places where there is little other available animal fat, and in China they have been specifically bred for fatness. Duck cookery is also a prime example of the ingenuity and multiple procedures used in Chinese cuisine. Ducks are dried whole or seasoned and left to hang slowly so that the fat drips and the flavor concentrates—a familiar sight in any Chinese poultry shop—or they are roasted and then chopped through bones and all, or the flesh is added to stir-fries. They are also smoked, braised in a clay pot with vegetables, and boiled to make soup. Moreover, every part of the duck is used: the liver, giblets, as well as the webbed feet, which are pleasantly gelatinous like a chicken's.

Peking duck is the best known of duck dishes, wherein the skin has been inflated with a straw to separate it from the flesh, lacquered with barley syrup, and hung to dry. Then the cavity is sewn up, filled with boiling water to let it steam from the inside out, and roasted in a wood-burning oven. It is then ceremoniously carved at the table and rolled in little pancakes with plum or hoisin sauce and scallions. Another example of a succession of techniques used on a single duck is the camphor-smoked duck of Sichuan, which entails six separate procedures. First the duck is brined overnight with spices such as Sichuan peppercorns, ginger, orange peel, cinnamon, and garlic. Then it is blanched with hot water, by ladling the water over the bird. Then a sweetened red vinegar syrup is poured over to color and flavor the skin. Next it is hung and dried for eight hours and then smoked in a covered wok over a parcel of camphor wood dust, sugar, rice, chilies, and tea leaves. Last, this is deep fried to crisp the skin. Every step adds another layer of flavor and texture.

Duck eggs are also more highly appreciated in China than in the West. Salted duck eggs are a delicacy. They are either simply pickled raw or packed with salted charcoal, making them black. Unlike pickled hard-boiled eggs, these are first cured, which turns the yolk red and firm, but the white remains runny. They are boiled before use and then used in rice porridge, or the yolk, as a symbol of the moon, is used in mooncakes. Today they are usually sold dyed red, so they are not confused with fresh eggs. These are completely different, however, from so-called hundred-year-old eggs, which have been cured in a casing with clay, ashes, and salt and then covered in rice straw. The alkali or lye in the coating causes the white to turn gelatinous and brown and the yolk deep green. The purpose of the procedure is to preserve the egg, but it creates dramatic changes in the flavor and texture as well and has a pungent aroma of ammonia. These are not cooked further but served simply sliced as an hors d'oeuvre or on top of a simple tofu dish. They can be cooked with rice porridge, though. Also, they are not a hundred years old, but usually about one hundred days, though modern techniques have significantly shortened the curing time. Fresh duck eggs are also eaten, and their taste is similar to that of a chicken egg, only a little richer and deeper.

Ducks were also kept by the ancient Romans, though they may not have been fully domesticated until the medieval period. That is, they may have been merely captured and enclosed for fattening. In Italy too, the full range of techniques could be applied to duck—either simply roasted on a spit, or baked in a casserole with sour fruits, or braised with lentils or olives. Sautéed duck meat can also grace pasta, and the breast meat can even be cured and sliced thinly like prosciutto or cured into a duck salame.

We may not think of duck as a typical Mexican ingredient, but it is one of those foods normally overlooked by those who claim that pre-Columbian Mexico was bereft of significant sources of animal protein. A dish in which duck is slowly braised in a green sauce of pumpkin seeds, for example, is completely indigenous. The Muscovy duck is also native to Mexico.

Geese are cooked much like ducks, and though of less importance in China, geese have always been an important food in Italy. The ancient Romans apparently used them as guard animals, as they will honk vociferously at strangers. Goose liver is also important in patés, and the fattened goose liver common in France (*foie gras*) was adopted from the Gauls in ancient times. In Italy this was also practiced by local Jewish communities, who because of their prohibition on pork, would raise geese and use their meat in sausages, prosciutto, and other dishes.

Guinea fowl are another domesticated bird, from Africa, and are uncommon in the United States but popular in Italy since the early modern era. They are cooked much like turkeys, though they are usually smaller.

Study Questions

1. Describe food taboos or the relative lack of them in certain cultures. What accounts for these?
2. How are animals changed through domestication and selective breeding? What advantages do these processes confer on humans, and at what cost?
3. How do environmental factors influence the preference for particular types of meat over others?
4. Explain the ways cattle are so useful, but also the negative side of modern cattle rearing.
5. Why are certain cuts of meat considered lowly in much of the modern world but not in the regions described here?
6. Why were certain animals reserved for nobility, and what has become of these today?
7. How are modern health concerns rapidly changing the range of foods commonly consumed in these three cuisines?

8. Describe the importance of dairy products in Italy and Mexico. Why were they not traditionally used in China?

9. How do food prohibitions such as Lent or kosher laws lead to innovation in the kitchen?

Recipes for Meat

Raw Meat

CARPACCIO (RAW BEEF)/ITALY

This is a modern recipe, allegedly invented in 1950 at Harry's Bar in Venice when a customer was told by her doctor to eat raw meat. The name was apparently suggested to the proprietor Giuseppe Cipriani because the deep red color is evocative of the palette used by fifteenth-century painter Vittore Carpaccio. In fact, his reds are no more or less vivid than those of his contemporaries. And why exactly a doctor would recommend raw meat is unclear. Regardless, it has become deservedly popular, even though adaptations abound using meats other than beef or even thinly sliced vegetables.

Use the best and freshest beef you can buy, preferably filet, whole and not divided into steaks. Slice it thinly across the grain, and pound each piece gently and patiently with a *batticarne* (meat pounder) until large and paper thin. It is easier if you do it between two sheets of plastic wrap. Arrange a few slices on each plate. Toss a few leaves of arugula with a drizzle of olive oil and place on top. Shave some Parmigiano-Reggiano on top and add a squeeze of lemon juice. Perhaps add a few grains of sea salt and a turn of black pepper. But resist the urge to doctor this up any further.

Boiled Meats

BOLLITO MISTO/ITALY

This is a classic dish of the Piedmont region in the north, meaning a "mixed boil." It will probably strike American cooks as odd to boil beef and other ingredients together, but as a grand and festive dish it offers diners a wide variety of meats to choose from, and it is not actually boiled but rather simmered, technically a slow braise (*lesso*). Traditionally it is made with seven different kinds of meat, seven vegetables, and seven sauces, but there is no reason not to use whatever you have on hand or prefer. The ingredients should at least include a large cut of beef shoulder or brisket, a whole chicken, and then perhaps a tongue, a breast of veal, tail, a sectioned head if you can find it, and most importantly *cotechino*, which is a sausage made of pig skin, which gives the broth a glossy sheen from the collagen in the skin. This should be pricked with a knife before cooking. As you want the meat to cook without losing all its flavor to the broth, it is best to bring the salted water to the boil first, then add the meat. Then quickly lower the heat. A very large stockpot is required, or several smaller ones. The order that these meats enter the pot depends on what you are using. The head will take at least three hours, the beef and tongue two, and the chicken only about one. Ideally, you want each simmered to tenderness but not falling apart. Vegetables such as onion, carrot, celery, turnips, and potatoes should be added only in the last hour or so.

The most intriguing part of the dish is that the meats are removed and placed on a platter, covered to keep warm while the broth is reduced. They are ultimately served separately, or perhaps with a little broth to moisten the meat. The meats themselves are served with a variety of sauces. The carrots and other vegetables can be served separately as well.

A *bagnet vert* (or salsa verde) is the most important of the sauces. It is made by chopping parsley coarsely, then adding a crushed clove of garlic, a few anchovies, some capers, a little stale bread, a cooked egg yolk, and oil and vinegar. It is somewhat like a green salad dressing. Several different kinds of mustard are also necessary, both mustard as known in the United States as well as *mostarda di frutta*—a fruit mustard. Red sauce made from tomatoes, peppers, and onions is typical, as is a simple white sauce of onions and breadcrumbs. There might also be a quince sauce (*cotognata*). The more variety the better. To serve, the meats are sliced, moistened with broth, and the sauces offered on the side. The rest of the broth can be served separately.

VITELLO TONNATO (VEAL WITH TUNA SAUCE)/ITALY

This dish means "tunified veal," and how it could have originated is a mystery, though it appears to have come from Piedmont and Liguria. It is certainly no older than the nineteenth century as it is always made with canned tuna. One also suspects that it was invented from leftovers, but the combination of flavors is both surprising and delicious. Begin with a veal roast, from the round, which has very little connective tissue and practically no fat. This should be tied and simmered in water flavored with carrots, onion, celery, salt, pepper, a bay leaf, and parsley. About an hour at a gentle simmer is ideal. Remove, let cool, wrap tightly, and chill in the refrigerator, preferably overnight. The next day, remove the strings and slice thinly.

The sauce is essentially a mayonnaise of tuna. Most people today add commercial mayonnaise, but there is no reason to do so. Start with a can of the best imported tuna in oil, drained. Either in a food processor or a sturdy mortar, puree this with a raw egg yolk, a few anchovies, and some capers. Drizzle in about a cup of olive oil until thick. Add the juice of a lemon and mix thoroughly. If you like, thin with a little of the chilled poaching liquid. To serve, arrange the slices of veal with the tuna sauce, covering the meat completely, and garnish with some capers and slices of lemon in a pattern on top of the sauce.

Fried Meats

COTOLETTA ALLA MILANESE (VEAL CUTLETS)/ITALY

This is essentially a simple breaded veal cutlet. It can be used as the base for a number of dishes, though. Veal parmigiana is simply this cutlet topped with tomato sauce and mozzarella and baked until the cheese melts. But served plain, perhaps with a wedge of lemon and an anchovy, or a poached egg on top, it is far more subtle and delicate. In Milan these normally come with the bone attached, which adds to the flavor. When there is no bone and they are pounded flat, they are called *orecchie di elefante* (elephant's ears). In either case it should be a cutlet of milk-fed veal, which is pale and white in color.

If you are using boneless veal, use a *batticarne* (meat pounder) to flatten the meat. A meat mallet will often tear the meat, so use very gently with the flat side if this is all you have. Season lightly with salt and pepper. Then dip into beaten egg and immediately into a bowl of breadcrumbs, pressing on the cutlets to make sure the crumbs adhere. The breadcrumbs are best from good, thoroughly stale bread pounded in a mortar or ground in a blender, but not too finely. You can also simply grate the stale bread. Shake off the excess and put on a plate while you do the remaining cutlets. Do not, despite all the directions you may have read, dredge these in flour before dipping in egg. It not only makes a thick, overbearing crust but also often causes the breading to separate from the meat and fall off after cooking.

These should be fried gently on both sides in butter until golden brown. You can substitute pork loin cutlets here with excellent results or even chicken breast similarly pounded thin. These are best without a sauce, which only makes the breading soggy.

PEKING BEEF/CHINA

There are dozens of different dishes that are known by this name in the United States. Some use thinly sliced beef, others use shredded beef—supposedly reminiscent of the Mongolian influence on northern China. This is a simple, straightforward rendition that can certainly be altered to suit one's own preference. A cheaper cut of beef is preferred, either skirt steak cut thinly across the grain or flank. Marinate the beef in soy sauce, rice wine, minced ginger, garlic, a chopped red chili, and sesame oil. Add some cornstarch to bind this mixture, and set aside for at least an hour or longer. When ready to cook, heat a wok to the highest possible temperature. This will take some time. Add in a few tablespoons of peanut oil or other neutral oil and immediately add the beef, letting it cook undisturbed for a few minutes. This will allow the meat to brown and will seal in the juices. With a thick dishtowel, grab one of the short wok handles in your left hand and use the long-handled metal spatula in your right to scoop up the meat and toss it over several times to stir-fry. This whole process should take only a few minutes. Scoop the meat out and set aside in a bowl. Meanwhile, with the wok still hot, add a little more oil and stir-fry some strips of green and red bell peppers, thickly cut onion, and any other vegetable you like, such as broccoli or asparagus if in season. When nicely colored, but not overcooked, add the beef back in with the vegetables, toss well, and sprinkle with sesame seeds. Serve.

As quickly as you can, take the wok to the sink and run under very hot water. Throw in a sponge, and with your spatula holding down the sponge, scrape out any remaining food residue and drain. The wok will still be hot. Place it back on the burner, and it will soon be ready to use again. In professional restaurants there is always a hot-water spigot positioned right over the cooking area as well as a drain for this very reason. It not only keeps the wok in action but also prevents leftover food particles from sticking to its surface. In other words, at home, do not just leave the wok to cool while you eat, or it will be very difficult to clean later and you may end up removing the slick, seasoned surface.

Roasted Meats

COCHINITA PIBIL (ROASTED PIG)/MEXICO

For this recipe you must start with a whole suckling pig or alternately a large pork shoulder roast. The *pibil* in Yucatec refers to a pit oven dug in the ground in which the meat is very slowly baked. The pork is first marinated in the juice of bitter Seville oranges, though a combination of lime and orange juice is a rough approximation. To make the marinade, finely grind a few tablespoons of *achiote* seeds (annatto), or use a ready-made paste. This will stain everything yellow, including your skin, so be careful. Add to this finely ground cumin, oregano, cinnamon, and a few whole cloves. Add a few tablespoons of salt and pepper. Add to this an entire head of garlic, crushed, and a handful of finely chopped chilies. Use habañeros if you like it very hot, or a milder chili. In a large, sealable plastic bag or bowl, add enough orange juice to this mixture to make the marinade, add the pork, and let marinate overnight in the refrigerator.

The next day, dig a large hole twice the size of the pig. Make a hardwood fire in the pit, and when it reduces to hot coals, add some rocks (lava rocks intended for the barbecue are best) and let them heat. Remove the pork from the marinade and wrap completely in banana leaves, which are available from Asian and Latin grocery stores, usually frozen. Bind this securely with wire and set on top of the hot rocks. (It helps if you also make a wire handle so it can be easily removed later.) Lay down some more banana leaves, and cover the hole with dirt. This should cook for at least four or five hours, preferably even longer. To serve, dig out the parcel, brush off all the dirt, discarding outer leaves if necessary, and place in a large baking pan and carefully unwrap. The meat should be tender and falling off the bone. Serve with a salsa made of chopped onions, habañeros, and the same bitter orange juice. You can also cook the banana leaf parcel in the oven, but the flavor is not really the same.

PORCHETTA (STUFFED ROASTED PIG)/ITALY

For this recipe you will need a young pig about a year old, weighing about 75 to 100 pounds. A suckling pig or a full-grown hog will not work. The meat will be light and tender, in color and texture somewhat like turkey. Most recipes substitute regular pork shoulder or some other cut, but you really do need the entire animal with the skin and head attached. Have your butcher singe the hair and remove the viscera cavity, but do save the liver and other organs such as heart and spleen as well if you like. First remove the legs, cut the meat from the bone with the fat, and grind finely. This will be your stuffing. Add to it grated cheese, garlic, and enough eggs to bind. Add herbs such as fennel seed and rosemary. Then chop the organ meats coarsely and add to the stuffing.

Next, spread apart the cavity of the pig and carefully remove the ribs with a sharp boning knife. Remove the shoulder blades, spine, and all other bones so you have a single large flap of skin and meat, still attached to the head but boneless. Do not worry if you end up removing some of the meat in the process, just place it back on the skin. On this flap, season very well with salt, pepper, herbs, and a lot of chopped garlic. Spread the stuffing evenly over it. Then fold in the sides and bind tightly with string from top to bottom. Place the trussed pig in a roasting pan. It will fit into a conventional oven, but ideally it should be a wood-fired oven. Roast for about 5 hours until the skin is browned and crisp. Remove from the oven and let cool thoroughly, at least an hour.

When ready to serve, remove the strings as necessary and slice straight through the entire pig into thin slices. You will see separate rings of skin, meat, and stuffing. Eat as is, or serve on bread with condiments.

If you care to experiment with a scaled-down version, you can use a large section of pork belly for the outer layer, a loin of pork and some sausage meat, rolled and tied in the same way with the same flavorings, but again, mature pork has a very different flavor and texture.

Grilled Meats

CANTONESE ROAST PORK (CHAR SIU)/CHINA

This is typically not eaten on its own but sliced and used as an ingredient in other dishes. The characteristic red outer coating is achieved using red fermented bean paste or sometimes red food coloring or a special red powder made just for this dish, but it is not absolutely necessary. Simply take a piece of pork shoulder and marinate it in a combination of dark soy sauce, honey, a little hoisin sauce, and rice wine—preferably overnight. Then roast this in a 400-degree oven on a roasting rack set in a deep pan so all sides become crispy and glazed. Turn over halfway through and baste with the accumulated juices. Or, place it right on the barbecue and grill it. For even better results, skewer this and properly roast over an open flame, turning slowly. When done, let it cool and slice and either serve or use in a stir-fry or other dish. The same marinade also works very well for barbecued baby-back ribs.

CARNE ASADA (ROASTED MEAT)/MEXICO

Not really roasted as the name (*asada*) suggests but grilled, this is best made with flank steak or skirt steak, marinated lightly and grilled, then sliced thinly and placed inside tacos or burritos (figure 6.5). To marinate, combine lime juice, chopped green chilies, chopped onions, and garlic, cilantro, salt and pepper, and a dash of olive oil in a plastic zip-top bag. Place the meat inside and let marinate at least 4 to 6 hours or longer. You can also add a little chopped papaya, the enzymes of which will break down the meat

Figure 6.5. Carne asada in a soft taco.

and tenderize it, but do not add more than a few slices or leave it too long or the meat will disintegrate. Bring the meat to room temperature before grilling, and scrape off the solids from the marinade. When ready, heat a grill and place the meat on, not moving it around too much. After about 5 minutes, turn over. If you have flames flare up, you can move the steak to another part of the grill, or squirt with a little water to douse the flame. You want the meat a little charred but not burnt. Touch the meat with your finger; it should be about as firm as the fleshy part of your hand just beneath the thumb. There are no hard and fast rules about this; all will depend on the heat of your coals or gas jets and the thickness of the steaks. Most important, do not overcook. Let rest at least 10 minutes, loosely covered with foil. Then slice on a wooden cutting board and serve with tortillas and condiments, such as shredded lettuce, guacamole, or tomato salsa.

BISTECCA ALLA FIORENTINA (STEAK FLORENTINE)/ITALY

Using an extra-thick-cut T-bone of Chianina beef (about 3 inches thick, weighing about 2 or 3 pounds) or the best prime beef you can find, simply sprinkle with good, coarse sea salt, leave at room temperature for about 20 minutes, and grill over a charcoal grill using fragrant wood. The coals should be superhot. This must be cooked rare or medium rare, with a good char on the outside but still very pink inside. No further adornment is required, except perhaps a drizzle of olive oil. Some people like lemon. Some also contend that the meat should not be salted until after grilling, but it does taste better if done before. As a compromise, you can add a little salt to flavor the meat and then finish at the end with coarse sea salt to give it a pleasant crunch. Incidentally, this is rested and served as slices, as no one could eat a whole steak of this size.

Braised Meats

OSSOBUCO (BRAISED VEAL SHANKS)/ITALY

Veal shanks sawed across so the "hole" of the bone is showing and served upright so the marrow can be scooped out is one of the triumphs of Milanese cookery. It is a simple dish, basically just braised shanks with vegetables. It can be made a number of ways as well, in a light wine base, or deeper with tomato and browned veal stock. This version is the latter.

Tie each shank around the middle with twine to make sure the meat does not fall off in cooking. Season with salt and pepper and dredge lightly in flour. Then brown each 2 to 3 inch segment in melted lard rendered from diced pancetta (unsmoked bacon). Remove and add to the same pan diced onion, carrot, and celery and cook a few minutes. Add about a cup of fresh peeled and seeded tomato puree, and then add the veal back into the pan. Add about a cup of white wine, and scrape the bottom of the pan to dislodge any browned bits, and bring to a boil. Add to this a cup of veal stock. (Chicken stock will work fine as well.) Also, add a tied bundle of herbs: thyme, parsley, and bay leaf. Turn down the heat, cover and simmer gently for about an hour and a half. Add extra stock while braising if necessary. The liquid should come about halfway up the pan, and do turn the pieces of meat over from time to time.

To serve, remove strings and place meat on top of vegetables, perhaps with polenta on the side, or serve with risotto. Just before service, garnish with a light sprinkle of *gremolata*: a mix of finely shopped garlic, lemon peel, and parsley. Be sure to provide people with small spoons for scooping out the marrow.

This is how the great cookbook author Pellegrino Artusi describes the dish in the nineteenth century:

This is a dish that must be left to make by the Milanese, being a specialty of Lombard cuisine. I intend therefore to describe it without any pretension, or fear of being mocked.

Osso bucco is a piece of muscular bone with a hole in the end, from the shank or shoulder of a suckling calf, which is cooked in a braise in a way that it becomes delicate and tasty. Place on the fire enough pieces, according to the number of people who will eat it, on top of raw aromatics such as chopped onion, celery, carrots and a lump of butter, seasoned with salt and pepper. When it has become flavorful add another lump of butter rolled in flour, to give it color and to bind the sauce, and let it cook with water and tomato sauce or conserve. Strain the sauce and degrease and return to the fire, giving it aroma with lemon peel sliced in fine shreds, mixed with a pinch of chopped parsley before removing it from the fire.

MAO'S FAVORITE RED-COOKED PORK FROM HUNAN (HONGSHAO ROU)/CHINA

Apparently Chairman Mao's favorite recipe from his home in Shaoshan, this is made with fatty pork belly. Cut the belly into 1 inch batons, leaving on the skin. Put it into a pot and add rice wine, soy sauce, some whole star anise, a cinnamon stick, a few dried red chilies, a few slices of ginger, and some pieces of scallion, perhaps a pod of black cardamom as well. Add water almost to cover. Some caramelized sugar gives the dish a deep-brown color. To do this, heat the sugar in a dry pot until it melts and turns brown, then add some of the cooking liquid to dissolve the sugar and add it back to the pot. Braise for about an hour, covered, then remove the lid and reduce the cooking liquid until thick and glossy. Serve with plain white rice. There is no need to remove the spices because they are all still whole.

Steamed Meats

BIRRIA (STEW)/MEXICO

This dish from Jalisco is frequently just stewed goat meat or lamb, but properly it should be steamed in a pot in the oven. Take a handful each of dried red guajillo and brown ancho chilies. Remove the stem and seeds and toast them on a hot griddle (*comal*). Then soak them in water about 20 minutes and puree in a blender with garlic cloves, salt, a dash of vinegar, ground cumin, and pepper. Spread all but 1 cup of this marinade on large pieces of goat, shoulder or leg, and refrigerate overnight. The next day place a steaming rack in a large pot, add some water, and put the meat onto the rack. Cover the pot. Place in an oven at 300 degrees and slowly steam about 3 hours. Check every now and then to make sure the water has not boiled off. When done, remove the meat, brush on the reserved pureed chilies, and place briefly under the broiler to brown. Meanwhile, to make the broth, skim off any fat from the pot and add water with tomatoes pureed in it. Simmer for about half an hour. Serve the meat and broth separately, mixing the two in a bowl at the very last minute, and garnish with chopped onions, cilantro, and a lime wedge. Serve with warm corn tortillas. A quicker and simpler version can be made just by stewing everything together, but the texture and flavor are somewhat different.

STEAMED BEEF/CHINA

Steaming is among the most favored cooking methods in China, said to preserve the pristine flavors of foods. This dish can also be done with chicken, pork, or other meat.

Take a pound of beefsteak and slice it into bite-sized pieces. Marinate this in soy, ginger, rice wine, and a little chili paste for about 30 minutes. Then take glutinous rice flour and toast it in a dry skillet. Add a touch of five spice powder. Then drain the marinade from the beef and toss the pieces in the rice flour. Place these in a bamboo steamer, either in a wide, flat bowl, on a cabbage leaf, or atop a piece of parchment paper with holes poked in it. Steam this for about 15 to 20 minutes. Many Chinese dishes are steamed directly in a bowl set in the steamer, so feel free to add vegetables and other flavorings with this technique instead of stir-frying.

Meatballs

POLPETTE/ITALY

In Italy a meatball is not served on a bed of pasta with sauce, as it is in the United States. That method is the legacy of southern Italian immigrants. In Italy, it is a dish in its own right as a second course, varying in size from tiny meatballs added to soups to large, fist-sized balls, or even the *polpettone*, a kind of meatloaf. Cooks will disagree vehemently about exactly what should go in them, so here various options will be offered (figure 6.6). Begin with ground meat. This can be beef alone, or as many contend, a combination of beef, pork, and veal. The former adds flavor; the latter, tenderness and delicacy. Into the ground meat add chopped parsley, salt and pepper, Parmigiano, and if you like some cooked onion. Breadcrumbs are also requisite, and these can be added directly to the mix, or you can use stale bread soaked in milk and then squeezed out. The former tends

Figure 6.6. Meatballs can be made with a variety of different meats.

to keep the meatballs together. You can also add an egg, but if long cooked, these tend to make the meatballs too firm. If you are searching for historic flavors, then add some raisins and pine nuts.

Mix the ingredients well, and roll into balls about the size of a walnut. You can then either gently sauté these in a pan with olive oil, or you can drop them directly into gently simmering tomato sauce, without stirring. Cover and let them simmer until cooked through, about half an hour. For the tomato sauce recipe, see the Fats and Flavorings chapter recipe section.

ALBÓNDIGAS/MEXICO

A meatball in Mexico is similar to the Italian version. Tiny ones are called *albóndiguillas*. Both words derive ultimately from Arabic *al-bunduq*, meaning "hazelnut" or "small, round ball." Here, they are most commonly added to a soup—*caldo* or *sopa de albóndigas*. Mix about a pound of ground beef with a cup of uncooked rice that has been soaked for about 20 minutes in water. Add an egg, salt and pepper, and some chopped herbs such as cilantro and a pinch of cumin. Roll this mixture into good-sized balls. In a skillet, sauté some diced onions, garlic, some jalapeños or other chili peppers, and some crushed tomatoes. Blend this with water to form the soup base. Add some carrots cut into chunks, and simmer for a few minutes. Then add the meatballs and simmer gently. Taste to make sure the rice is cooked through. Toward the end of cooking, add in chunks of diced zucchini.

LION'S HEAD MEATBALLS/CHINA

This dish originates from Yangzhou, though is most commonly associated with Shanghai today and traditionally cooked in a clay pot. The meatballs are large, like a lion's head, and the greens evoke its mane (figure 6.7). Start with a pound of ground pork. If you like, use pork shoulder with a good proportion of fat, and chop very finely with a cleaver. Add an egg, a tablespoon of soy sauce, some sesame oil, minced ginger and scallions, white pepper, a splash of rice wine, and enough cornstarch so the mixture will hold together well. You are supposed to stir these only in one direction to mix. With wet hands, roll these into large meatballs and flatten a little on each side. Fry these in several inches of hot oil in a wok, just three or four at a time, until lightly browned and not yet cooked all the way through. Place these in a clay pot with chicken broth and shredded Napa cabbage and let them cook through. This can also be done in a casserole in the oven, if necessary. Serve one meatball per person in a soup dish, surrounding the meatball with the greens. There is also another version that is red-cooked, or braised in soy sauce.

Fermented Sausages

SALAME FELINO/ITALY

Of the three fermented sausage recipes, only the Italian is commonly eaten raw. The name Felino does not refer to cats, but to a city of that name. Traditionally it was made

Figure 6.7. The lion's head meatball is named for the "mane" of green vegetables that surrounds it in the soup.

in the part of the large intestine (*budello gentile*), but you can use hog or beef middles or another wide sausage casing. The proportion of meat to fat should be about 70/30, and the fat should be cut into large knobs, which are apparent when the *salame* is sliced. There are an infinite variety of types, large and small, with a fine texture or coarse and seasoned many different ways (figure 6.8). A five-pound batch is easy for one person to handle. If you are using a meat grinder, it is important to keep everything cold, and perhaps even semifreeze the meat. With just a cleaver, this is not necessary, in which case just coarsely chop pork shoulder and fat. Add 4 tablespoons sea salt, 3 tablespoons superfine sugar (this is to aid fermentation; it will not be sweet in the end), and season with garlic, oregano, whole peppercorns, and, if you like, other flavorings like fennel. A ¼ cup red wine also adds to the flavor.

You must also add 1 teaspoon Instacure #2. This is a combination of sodium nitrite and sodium nitrate, the latter of which is slowly converted into the former in the course of curing and is slightly different from standard pink curing salt. It is indispensable, not only to prevent botulism but also to give the finished product a pink color and the right texture. In the past saltpeter (potassium nitrate), would have been used, and in the distant past mined salt probably contained naturally occurring nitrates. The only other

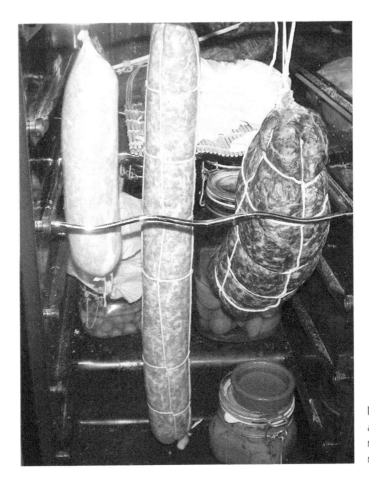

Figure 6.8. A typical *salame* from Italy, *felino*, is named after a city of this name.

option is a cure derived from celery juice, which is also high in nitrates. It works well with quickly cured and cooked sausages, but less so with these. If you see products like bacon, salami, or hot dogs labeled "uncured," they are actually cured with a celery juice–derived form of nitrate.

Tie one end of your casing tightly with string. Using a wide-mouth funnel, or if you are nimble fingered, just use your hands, stuff the casings with the chopped meat. Each *salame* should be two or three feet long, but you can always make them smaller by tying off sections. Make sure you press all the air out, but not too firmly or the casings will split. Tie the other end, and prick with a clean pin all over. Hang these in a cool, moist (around 55 degrees) place for about three months. A cellar works, as does a wine fridge. Within a few days you will see the casing tighten and the color turn red. Periodically check them; they should smell sweet and fragrant. To achieve a powdery white mold on the casings, you can buy an inoculant for this purpose, or simply hang a store-bought salami nearby and it will share with your homemade ones. When they are firm, slice on a bias thinly, remove the casing, and enjoy with some olives, cheese, a crusty bread, and a glass of wine.

LAP CHEONG/CHINA

This is the most commonly used Cantonese name for what are actually a variety of sausages. Sometimes they contain pork or duck liver; sometimes they are smoked. They can be dried until completely hard. In any case, they are always cooked and usually added to other dishes. Like Italian *salame*, the principal reason to process them in this way is preservation, and these will last without refrigeration through the whole year. Few people make these at home anymore, but there is no reason not to do so. The basic technique is similar to that for Italian *salame*, though you definitely want a more narrow hog casing for these. The proportion of fat is also higher, roughly 60/40, and the fat should be cut into rather large knobs. Also essential is that this must be cut by hand. Fine, evenly ground meat will not give you the characteristic knobby appearance. The only principal difference is the seasoning. For a five-pound batch, use both soy sauce (4 tbs) and salt (1 tb), rice wine, and five spice powder, and add a little cayenne pepper if you like. Five tablespoons of sugar are used here, and the final product should be somewhat sweet. Commercial versions are often dyed red as well, but this is not necessary. Traditionally they do not use nitrates, but it is advisable, in the same proportion as in the previous *salame* recipe. Hang and dry as in the previous recipe, and use them at any time you like because they will be cooked. They can be sweated in a pan or steamed before adding to other dishes. Like *salame*, they should have a characteristic lactobacterial sourness, paired with the salt and sweetness. The flavor is nicely balanced.

CHORIZO/MEXICO

This is a Mexican descendant of a hard-cured *salame* of the same name in Spain. In Mexico it is soft and must be cooked. In fact, it is so finely ground and soft that it is squeezed out of the casing and fried before using in other dishes. The basic stuffing technique is the same, but the flavoring includes a lot of ancho chili powder, red wine vinegar, cumin, oregano, and several cloves of garlic. A pinch of ground cloves is also nice. Also, as this will not be long cured and dried, Instacure #1 works fine. Although it can be used immediately, in which case there is no need to use nitrates, it is much better if left to hang at least a week or two to allow the meat to ferment. There should be a noticeable sour tang playing off the heat. If you prefer, you can also use fresh chilies, very finely chopped. These are also excellent lightly smoked, in which case you will want to hang them in a smoker for a few hours over mesquite or other hard wood.

Organ Meat

It is really only in Western culture that organ meats are derided, discarded, or fed to pets. They are often the tastiest part of the animal, and they have recently come into vogue among serious chefs and gastronomes. In Mexico, China, and Italy they never went out of fashion.

FEGATO ALLA VENEZIANA (LIVER VENETIAN STYLE)/ITALY

In the Venetian dialect livers are called *figa*. It and the word *fegato* descend from the Latin word for fig, which was fed to animals to fatten their livers (*iecur ficatum*). The key to this dish is cooking thinly sliced onions in butter and olive oil until they are unctuous and soft and very sweet but uncolored. It will take an hour or longer. Add salt to taste. Then slice calves' liver into little pieces, and in another pan, quickly cook these on very high heat, no more than a minute a side, in butter. You want them still a little pink inside. As they are done, put them into the pan with the onions. When done, splash in a little dry white wine and scrape up any bits stuck to the pan, and pour out into a serving dish. Salt the liver after cooking, and toss on a handful of chopped parsley and a drizzle of vinegar if you like, preferably one very mild, or lemon juice.

LENGUA (TONGUE)/MEXICO

The tongue is perhaps the most delicate and flavorful of all parts of the cow, mostly because it is nearly always in action, curling around grass before incisors cut it off. As pure muscle, it also needs to be slowly simmered a long time. Place a 3- or 4-pound tongue in a big pot of water with aromatics such as carrots, celery, bay leaves, onions, and salt. Simmer for about 3 hours, then remove from the pot and let cool. First run a knife just down the center of the bristly part to cut barely through the outer skin. Peel this off all around the tongue and discard. Now, slice it into even pieces and cut these into long strips. You can reheat them if you like in a pan to brown a little. Put a generous spoonful of the meat into a corn tortilla and top with a little salsa verde, made with tomatillos and green chilies, and perhaps a little avocado.

BLOOD SAUSAGES/MEXICO

Asian grocery stores are the easiest places to find raw blood for sale in the United States. It congeals on its own, is cut into squares, and is used in restorative soups or just cooked in blocks called red tofu. It can also be used in various sausages, similar to those found everywhere in the world there are pigs. The *biroldo* is simply a casing with pig's blood, onions, garlic, pine nuts, and raisins, poached and then lightly sautéed. Or it can be eaten sliced and cold. The *moronga* or *morcilla* is a pig's blood sausage descended directly from Spanish versions, though usually flavored with hot chilies and large chunks of fat. It is usually squeezed from the casing, pan fried, and served with onions in a *gordita* (a little fried, corn-based stuffed cake) or crumbled and put in a taco.

Skin: Cicciole, Chicharrones, Crackling

In cultures where no part of the pig is wasted, it is no surprise that pig skin is appreciated as a delicacy. It is prepared virtually the same in these three countries, though the seasoning might be a little different. Salt is requisite. Start with skin, with a little fat attached. It can be cut into thin strips or a large piece, which is common in Mexico. The key is first to fry slowly to render the fat. You can do this in oil, or if you are doing a large batch, in rendered lard alone. This first stage should be about 225 degrees and will take about 20 or 30 minutes, depending on how many you are frying. Incidentally, if you can do this outside, all the better, as the fat tends to splatter and the aroma of cracklings will linger in the air a long time, which is not necessarily a bad thing. Next raise the heat up to about 425 degrees and plunge the skin in briefly, just a few minutes, until it puffs up and is very crunchy. Salt immediately, and perhaps season with chili powder or pepper.

COTECHINO (SKIN SAUSAGE)/ITALY

In keeping with the dictum that nothing ever gets wasted, skin is also used in other imaginative ways. Although most Italians buy *cotechino* today, and it is common all across northern Italy, it is very easy to make from scratch. It is classically used in a *bollito misto* to lend glutinous body to the broth, as the high collagen content of the skin has an unctuous gummy texture, evident in the sausage itself. It is also served on New Year's with lentils. As with the sausage recipes given here, it is naturally fermented.

The casing for *cotechino* is the beef bung or caecum, one of several large stomachs that is closed at one end. The entire casing holds well over 10 pounds of meat and is also used traditionally for mortadella and other large sausages. One casing will make two or three individual sausages weighing 3 to 5 pounds each. Begin with 5 pounds of pork skin. You can take this from pork belly or buy it separately. Chop very fine, putting it in the freezer for an hour to help with chopping, or finely grind with a meat grinder. Take 2½ pounds pork shoulder and 2½ pounds fat. If you have used pork belly, it should have sufficient fat. Grind this as well and mix with the skin. Add 8 tablespoons sea salt, 6 tablespoons superfine sugar (necessary for fermentation), and 2 teaspoons Instacure #2. Season as you like with oregano, pepper, perhaps sage, juniper, and so forth. Mix well with your hands. Rinse the casing well, and cut it into two or three parts making sure one end of each is tied with string. The opening is very wide, so it will be very easy to fill the casings by hand. Make sure to squeeze out any air pockets, pushing the meat down firmly. Tie the other end and then truss with string, making sure that the strings will bear the weight once hung. Prick the sausage all over with a clean pin and hang it in a cool, dark, and humid place for about a month. A 55-degree cellar is perfect, but a wine fridge will also work fine. In the ensuing month, the sausage will ferment and also shrink. It may need retying as well.

Once cured, remove the string, place in a large stockpot, and gently boil for about 5 hours with aromatic vegetables. You can also add other cuts of meat, such as cured tongue, brisket, a pig's foot, a whole capon, and, in its grandest iteration, a whole calf's head. Before serving, remove the casing and slice the *cotechino*. Whether simple or otherwise, it should be served with mustard of the yellow type as well as Italian *mostarda di frutta*, horseradish sauce, and green parsley sauce.

Figure 6.9. *Menudo* is a common breakfast dish made with tripe and hominy.

MENUDO (TRIPE SOUP)/MEXICO

Menudo, a tripe soup, is among the quintessential Mexican dishes, often eaten at celebrations and even for breakfast as the ideal restorative and hangover cure (figure 6.9). It takes many hours to prepare from scratch. The base of the soup is tripe, though *tripas* in Mexico refers only to the small intestine. This includes the lining of the stomach, honeycomb tripe, and other parts such as calves' feet and tendons. Some versions are clear, others use a red chili base, and some include hominy. This red version with hominy comes from northern Mexico. It is served with fresh corn tortillas, chopped raw onion, and chilies, and perhaps some chopped cilantro and lime wedges for each person to add as they like.

Start by boiling the calves' feet (split) and tendons in a pot, with the tripe cut into small squares, covered with very lightly salted water and simmered for about 2 to 3 hours. Skim the pot of impurities regularly. Add a chopped onion and a crushed clove of garlic, some oregano, and a pinch of cumin. You can remove the feet at this point and strip off the meat or serve as is. Next, take some dried red chilies and toast on a *comal* (Mexican griddle) or skillet. Remove the stem and seeds, soak in water until soft, and then pound finely. Add this to the soup. Then add the hominy. Hominy can be bought canned, frozen, or prepared from dried corn, soaked in hot water and calcium hydroxide ("cal"), which is well rinsed, and then the outer coating is rubbed off by hand. This is also known as *posole*. Salt to taste. Let the soup simmer until all the flavors come together.

Recipes for Poultry

Chicken

POLLO ENCACAHUATADO (CHICKEN IN PEANUT SAUCE)/MEXICO

The area around Veracruz, facing the Caribbean Sea, shares with the Caribbean a deep connection to Africa. Peanuts, although originally from South America, were brought to West Africa and then to eastern Mexico with the slave trade. People and recipes of African descent like this one survive in the area. Start with a whole chicken cut into serving pieces. Season each piece with salt and pepper and brown in lard or olive oil. Remove from the pan and set aside. In a mortar or blender, puree a few tomatoes, add whole peanuts, and puree the whole mixture. Add some onion and a seeded ancho chili softened over an open flame and then soaked in water. Fry this puree in the pan in which you fried the chicken, and then put the chicken back and simmer in the sauce for about 45 minutes. Add water if necessary to thin.

SALT-BAKED CHICKEN/CHINA

This recipe comes from the Hakka people of southeast China, and despite the intriguing technique, it is actually very simple. Start with a whole chicken with giblets removed, rub the inside and out with rice wine, and fill the cavity with scallions, ginger, cilantro, star anise, and tangerine peel. You can also season with ground Sichuan peppercorns and a little salt. Place on a rack and let dry thoroughly for about an hour. Then, truss the chicken tightly with twine to form a compact ball, wrap the chicken carefully in lotus leaves or parchment paper and tie the bundle up. Then pour coarse sea salt into the bottom of a steel wok and heat it until very hot. Pour out about three quarters of the salt into a bowl, place the bundle on top of the hot salt still in the wok, and then cover the package completely with the remaining hot salt. (Add more if necessary.) The hot salt will gently cook the chicken, for about an hour and a half with the flame on very low. Before serving, remove the bundle, brush off the salt, carefully unwrap and chop the chicken into bite-sized pieces, across the bones, and place the chicken pieces with all the juices onto a serving platter. This goes nicely with a vinegar-spiked chili sauce.

CHICKEN CACCIATORE/ITALY

Although most recipes call for a whole chicken cut up, inevitably the breast meat dries out when stewed in this way. It is much better to use only thighs. As it is "hunter" style, wild foraged mushrooms such as hunters would spot in the woods would be ideal, but any kind will serve, or they can be omitted altogether. Lightly salt and pepper the thighs and then dust with flour, shaking off the excess. Fry these in olive oil, until brown on both sides, and remove from the pan. You may need to do this in two batches rather than overcrowd the pan. In the same pan, fry onions and garlic and cut-up green bell peppers until cooked through. Then add the mushrooms, and arrange the chicken in between the vegetables. Add some dry white wine, about a cup or more, and let it reduce. Then add some peeled and chopped tomatoes and simmer until the vegetables come together as a sauce and the chicken is cooked all the way through. You can add chicken broth as well for a deeper flavor, but it is not necessary. Red wine also works well if you want a heartier dish for the winter.

CHICKEN STIR-FRIED WITH VEGETABLES/CHINA

There are dozens of ways to do this recipe, and even the vegetables you add are entirely up to you (figure 6.10). Begin by slicing boneless, skinless chicken breasts across the grain (not too thinly or they will fall apart). Marinate this in soy sauce, Shaoxing wine, ginger juice, and cornstarch. The ginger juice is made simply by grating ginger and then squeezing out the juice with your hands. This will give you a concentrated flavor. You can also use other forms of starch, such as arrowroot or even tapioca starch, which can be bought in any Asian grocery store. Then cut up any of the following vegetables:

Figure 6.10. A common stir-fry with chicken and vegetables.

water chestnuts, bamboo shoots, baby corn (any of these canned are fine, but fresh is better if you can find them), plus green bell peppers, onions, bean sprouts, and bok choy. Use maybe three or four vegetables, not all of them, with a good combination of colors, textures, and flavors. Heat the wok on the highest flame possible. Pour in a few tablespoons peanut oil. Throw in the chicken, and leave for a few moments to sear. Then, toss by holding one handle of the wok and scooping up the contents with a long-handled spatula. After a couple of minutes remove the chicken and set aside. Add some more oil and cook the vegetables, tossing continuously. Add the chicken back in. There should be roughly equal amounts of chicken and vegetables, but the proportion is up to you. Then season this once more with soy, rice wine, a little sesame oil, and something crunchy like peanuts or cashews. It need not be thick, but if you prefer it that way, add in a slurry of cornstarch and water and continue to cook. The key to this dish is keeping the vegetables crunchy, the chicken tender and not overcooked, and the sauce covering everything but not overpowering, so that you taste nothing but soy sauce. Put this into a serving dish, and immediately take your wok to the sink and pour in hot water, swirling an abrasive sponge inside, held down by the spatula. It will be clean, and there is no need to ever use soap on your wok. It takes about a minute, but it would be a real chore if you leave it until after the meal to clean out. Serve the chicken with bowls of steamed rice, each person putting a little directly into an individual bowl.

CHICKEN BURRITO/MEXICO

Burritos made with flour tortillas come from northern Mexico and mean "little donkey," perhaps because of the shape or the resemblance to a rolled blanket carried by a donkey. They have, of course, taken on a life entirely their own in the United States, often made with huge tortillas and packed with a wide variety of ingredients. Smaller ones are not only easier to eat but also better tasting. It is common to see fillings like shredded flank steak or *carne asada*, or just beans. But this version is simple, and the flavors meld together well. Take chicken parts on the bone and marinate them for at least 30 minutes in a mixture of salt, pepper, oregano, ground chili powder, garlic, a touch of vinegar, and olive oil. Then grill these until cooked through. Remove the meat from the bone and cut into strips. Briefly place the tortillas on the grill until charred lightly in a few places. You can also cut up the chicken first and cook it as fajitas on a hot iron *comal* (griddle). Place a few strips of chicken on each flour tortilla, and add a little fresh tomato salsa and perhaps avocado and some shredded lettuce, but not much else. Roll by first folding in the bottom half, then one side, then rolling the whole into a fat cylinder. Then roll each in paper or napkin, and eat outside, preferably standing up. You can add more salsa, if you like, or hot sauce.

UCCELLI ARROSTO (ROAST BIRDS)/ITALY

In the countryside, hunters will shoot or snare a variety of tiny little birds, including wild pigeons, larks, snipe with their long beaks, or larger birds like pheasant and partridge, which are usually hung for a few days before roasting. Domestic quail can be used as well, but cornish game hen and the like are not good substitutes. The basic procedure is the same for wild ducks, which tend to be small and relatively skinny. Traditionally the tiniest of birds are not gutted, and in the case of snipe, the insides can be squeezed out onto toast. For the squeamish, of course, the birds can be cleaned. What you will need for this is a turnspit, which can be either a manual version or a small clockwork mechanism, which is not very expensive. The key to the dish is that you are not cooking these over hot coals, but beside burning logs. The smoke rises while the heat roasts the birds, and the flavor is completely different from if you were to barbecue them. First, generously salt and pepper the birds and season inside with sprigs of thyme, sage, or rosemary. Cover the breast with a thin slice of *lardo*, or bacon works well too, and tie it securely in place, perhaps with a sage leaf tucked in. Thread them on the spit by passing the pointed end through the tailbone and the other end through the wishbone, so the birds will not slide around. Some spits come with little pins as well to secure food in place. A larger bird should be skewered on an angle so the breast meat is slightly farther from the flame than the legs and wings. If using small birds, separate each one with a small piece of stale bread. Place a rectangular pan beneath the birds to catch the juices, and set the spit beside the fire, turning slowly. As they cook, mop up the juices, perhaps adding a little lemon juice, and baste the birds. These should be cooked just until browned nicely and barely firm to the touch; overcooking not only makes them tough but also tastes awful. They must be just a little pink inside. Smaller birds might take as little as 20 to 30 minutes. They are eaten with no further adornment.

Duck

RAGÙ DI ANATRA (DUCK SAUCE)/ITALY

You will need a whole duck to make this sauce, which comes from Arezzo. It is a rich, dense sauce that goes on fresh tagliatelle or a similar-shaped pasta. The nice thing about this version is that it uses every last part of the duck. Begin by taking the duck apart completely: cutting off the legs and wings, and removing the breast meat. Remove any visible fat and set aside, remove the bones from the legs, and separate all the skin. Set all the skinless meat and giblets aside.

In a pot, first render the fat from the skin and let it brown. Pour off the fat into a bowl. Add in all the bones, neck, and wings with onions, celery, and carrot, and brown everything well. Then cover with water to make a broth. You can also add nutmeg if you like or a few leaves of sage or sprigs of rosemary. Simmer this gently for about an hour and a half, skimming the top occasionally. Salt to taste.

Next, chop all the meat from the breast and legs plus the giblets and brown this well in the reserved fat in a saucepan. You want a good fond (the browned bits) on the bottom of the pot. Add in more carrots, onion, and very finely diced celery, and some herbs tied in a bundle. When everything is browned thoroughly, strain the reduced broth into this pot. Add some tomato paste and two glasses of red wine, and simmer until everything is tender, at least another hour or more. Boil the fresh pasta and add to the sauce, toss well, and serve, sprinkled with grated Pecorino and chopped parsley.

DUCK IN PUMPKIN SEED SAUCE/MEXICO

Ducks are indigenous to Mexico, as are pumpkin seeds, and this dish certainly originated before the time of the conquest (figure 6.11). A Muscovy duck (despite the name) would be the most authentic species. First, prick the skin of the duck and gently brown in a pan, cover the pan, and continue to cook on low heat on the stovetop or in the oven for about an hour. In another pan toast a cup of pumpkin seeds with some cumin and peppercorns. Transfer to a mortar and pound until very smooth. Add some tomatillos, a clove of garlic, a few serrano chilies, and continue pounding, adding a little water from time to time. This can also be done in a blender with sufficient water. Pour off most of the melted fat from the duck, and reserve for another use. Then add the green pumpkin seed sauce, and stir on very low heat until the two come together, again adding water if the sauce is too thick. Taste for salt and serve with tortillas.

Figure 6.11. This green sauce based on pumpkin seed is called *pipián* in Mexico.

SHANGHAI BRAISED DUCK/CHINA

For this a small Peking duck (or Long Island) is ideal, as it has little fat. Plunge the whole duck into boiling water for just a minute or two and rinse. Then in a clay pot or casserole, add light and dark soy sauce; Shaoxing wine; star anise; slices of ginger, smashed; some scallions; and a little sugar. Add the duck breast side up and about a cup of water just so the liquid comes about halfway up the duck. Gently braise for about 45 minutes, spooning over the braising liquid from time to time. Let the duck cool in the liquid for several hours. When ready to serve, remove the duck, and if you like place it under the broiler for a few minutes to make the skin crispy. Then remove the legs and wings. Chop the carcass, straight through the bones, with a cleaver into small pieces. Serve at room temperature with the braising liquid as a dipping sauce.

Soups

CHICKEN SOUP WITH MUSHROOMS/CHINA

In Chinese cuisine, a chicken used in a soup is usually blanched first for about five minutes, rinsed, and then put into the pot with fresh water for soup. This is said to remove any unpleasant odors. It also means you will not need to skim off much froth, as the soup base will be fairly clear. After blanching the chicken, cut it up, using everything, including the giblets and feet. For flavoring you can add mushroom stems, peeled and sliced ginger, and if you like some pork neck bones for extra flavor (which should also be blanched as in the menudo recipe). Bring this to a boil, and then let simmer for at least 2 hours or longer. Strain stock through a sieve before using.

Next, take some dried black mushrooms, soaked until soft. Use only the tops. Take some fresh chicken, including the bones, and cut into small pieces through the bones. Marinate for about 20 minutes with soy sauce, sugar, and a little cornstarch. Into the hot broth add some sliced carrots, turnip greens (or broccoli rabe), and bamboo shoots. Then add the chicken and simmer until cooked through.

HOT AND SOUR SOUP/CHINA

This is a Sichuan soup but is now popular throughout China and among Chinese American communities (figure 6.12). Start with chicken stock. Take a skinless chicken breast, divide it in half across the grain, and then cut it up into very fine strips along the grain rather than across it. This will prevent the chicken from completely disintegrating. You want fine threads about 2½ inches long. Marinate this with soy sauce and a pinch of cornstarch and sugar. Soak a handful of dried black wood ear mushrooms. Remove any woody bits, and cut into small slivers. Fresh also works well if you can find it. Add them to the simmering stock. Then add some julienned firm tofu, some Shaoxing wine or other cooking wine, some red vinegar, a spoonful of chili paste or more depending on how hot you like it, and vegetables such as carrots, bamboo shoots, and minced ginger.

Figure 6.12. Combining heat and sourness, salt, and an array of textures, the classic hot and sour soup is a favorite on both sides of the Pacific Ocean.

Let simmer until vegetables are tender. Taste for a balance of flavors. Then add the chicken and perhaps a few shrimp if you like. Let this cook through for just a minute or two. Then add a few beaten eggs, stirring constantly. To thicken before service, mix some cornstarch and water and stir into the soup. Season with white pepper.

Chili paste or sauce (*la jiao jiang*) can be easily bought in any Asian grocery store, but it is also fun to make your own. Find some fresh, hot red chilies. Remove the stem, and finely mince with the seeds included. Fry this in oil and add some minced garlic, making sure not to burn them. Then add vinegar to taste and simmer for about 10 minutes. Add salt to taste. It is now ready to use.

CHICKEN BROTH (BRODO)/ITALY

It is fairly common in Italy to simmer a broth for a full 24 hours on the lowest possible heat. This draws the most flavor possible from the bones, and the resulting broth can be used as the base for the simplest of soups. For this you will need some chicken carcasses, coarsely chopped—wing tips, necks, plus the head and feet if you can get them. I normally save bones from two or three chickens in the freezer until ready to make broth.

In a pot add the bones and some carrots, onion, celery, plus a few bay leaves and a sprig of thyme. Cover with water. Bring to a boil and skim the flotsam that rises to the top for a few minutes, then turn down to the lowest possible setting. Leave on the stovetop or even in a low oven for 24 hours. You will need to check periodically to make sure the liquid has not reduced too much; just add water if it has. The next day, strain the broth. It will be somewhat gelatinous, which is good. There is no need to remove any fat, either. To this broth you can simply add a few vegetables and small pasta, or use in a recipe like *zuppa pavese* (below), which lets the flavor of the broth dominate.

ZUPPA PAVESE/ITALY

According to legend, this soup is said to have been invented in the wake of the Battle of Pavia (1525), when the defeated King François I was rummaging for a meal in a local farmhouse. He was served a simple soup with bread and cheese in it, and in went an egg in an effort to impress the king. In reality, he was captured in the course of the battle and made to sign a very humiliating treaty, which he later ignored. In any case, this soup is a magnificent use of the simplest of ingredients. Start with the best Italian bread, sliced and browned gently in butter. Place a slice in a hot soup dish and ladle some hot broth on top. Gently crack an egg onto the toast, and then hit it with a good blizzard of grated cheese, Grana Padano seems right, plus freshly ground pepper and maybe a sprinkling of parsley. The hot broth will cook the egg, and along with the soaked bread the texture is soft and soothing.

TORTILLA SOUP/MEXICO

A good tortilla soup should be very simple and based on the best broth. You need not spend hours making the broth, though; simmering for a few hours will still give you a fine base. The flavorings for this include chilies, tomatoes, and garlic, which should be cooked over very high heat almost until seared, and then these are pounded or blended into the broth. Next, take some corn tortillas and fry them quickly in a few inches of oil until crisp. When serving the soup, add some lime juice for sourness, a little salt if necessary, some cumin for aroma, and garnish with the fried tortilla strips, chopped cilantro, and a dab of sour cream. More elaborate versions include shredded chicken as well, and vegetables, perhaps some diced raw onion or avocado. The simpler version is excellent, too, though.

Eggs

1,000-YEAR-OLD EGGS/CHINA

Although the ammonia and sulfurous odor may be off-putting to Westerners, this is a delicacy in Chinese cuisine, originally intended as a preservative for duck or chicken eggs. The method was originally intended to preserve the eggs but is now enjoyed for its unique flavor and texture. The brownish, pellucid, white and gelatinous bluish-green yolks are themselves something amazing. Traditionally these were made by taking a combination of clay, wood ash, lime (calcium oxide, not the fruit, which is made from heated limestone), black tea, and salt, and molding it around each egg (with gloves), and then covering this with rice hulls. These are then placed in jars to age at least several months. The extremely alkaline pH of the mixture in a sense cooks the egg, or at least changes it chemically and serves as a preservative. There are simpler ways to preserve eggs in alkaline solutions today, and they can be purchased at an Asian grocery, but it is still worthwhile making them at home.

The eggs can be served as is, quartered as an appetizer, but they are also used as an ingredient in many dishes, added to rice *congee* or served on a cold platter on festive occasions. One particular flaky pastry in Cantonese cooking includes a whole 1,000-year-old egg and a slice of pickled pink ginger. In Hunan cuisine, the eggs are cut into little chunks along with salted duck eggs and slices of ham, which is served on braised pea shoots.

TEA EGGS/CHINA

These eggs are much less challenging but exquisite to look at and eat (figure 6.13). Use regular chicken eggs, or quail eggs, which are tiny and taste pretty much the same. First, fill a pot with water, 3 tablespoons black tea, some star anise, a stick of cinnamon, some

Figure 6.13. The tea egg is gently cracked and soaked in tea to create this pattern.

smashed slices of ginger, and some soy sauce. Let this simmer for about 20 minutes. Then hard-boil the eggs for about 10 minutes in another pot. You do not want to use very fresh eggs for hard-boiling as they are much harder to peel. While still hot, tap the outside of the shell all round, without breaking any of the shell off. Place them in the tea mixture and cook for another half an hour or so. Let them cool in the tea. Peel and you will find a gorgeous pattern where the tea soaked through the cracks.

FRITTATA/ITALY

This is a glorified form of omelet, into which you can really put anything you like. It is the technique and presentation that make it unique. First, start with at least six eggs or as many as will fit in your pan according to the number of people you are feeding. Beat the eggs well. Into this put any of the following: sautéed onions and peppers, chopped cooked greens such as spinach or broccoli rabe, grated cheese, or small cubes of provolone. Ricotta is also very nice. Even julienned *salame* works well. Mix everything together, and pour into a hot pan with a little olive oil. Let cook very gently, covered, without disturbing. Shake the pan, and when the bottom will slide in the pan and the whole is fairly solidified, take a large plate and put it on top of the pan. Flip the entire thing upside down, return the pan to the heat, and slide the frittata back into the pan to continue cooking the other side. Then slide it back onto the plate. This is usually served at room temperature, sliced into wedges, with a little tomato sauce if you like.

HIGADITOS (LIVER OMELET)/MEXICO

The name of this recipe means "little livers," and the more you add the better. It is also called *higaditos de boda* because it is served on the morning of weddings. It should be made with gizzards as well and chicken. Place a whole, cut-up chicken with its giblets, but not the livers, in a pot barely covered with water, some chopped onions, and a little salt. Simmer for about an hour. Remove the chicken and shred the meat, discarding the skin and bones, and finely chop the gizzards. Gently sauté the livers until barely cooked through. Chop coarsely and add to the rest of the meat. Strain the broth, and let it continue to simmer uncovered, reducing. Then, gently fry some onions with some ground cumin, cloves, and a pinch of saffron threads. Add some of the broth and five or six chopped tomatillos. Cook this until the tomatillos have broken down. Add the meat and more broth. It should be quite soupy. Beat a dozen eggs with some salt and pour them over without stirring, but make sure they seep into every corner of the pan. Continue to cook, covered, until the eggs are set. Serve with a chili sauce, if you like.

EGGS AND TOMATOES/CHINA

The combination of eggs and tomatoes certainly does not seem particularly Chinese, but it is extremely popular. Heat a wok or nonstick frying pan. Beat some eggs with a pinch of salt and a few drops of toasted sesame oil. Cut ripe tomatoes into bite-sized pieces, and, if you like, a few scallions. Add some oil to the wok and immediately throw in the tomatoes, not stirring them for a full minute or so, to make sure they sear. If you stir them immediately the liquid will exude out and they will only boil. Then toss the tomatoes for a minute or so, push them to one side of the wok and add the eggs, stirring and tossing and mixing them with the tomatoes. Serve.

HUEVOS MOTULEÑOS/MEXICO

Similar to huevos rancheros and its various cousins in the United States, this version comes from Yucatan. Begin by toasting corn tortillas over an open flame. You can also fry them in a little oil if you like. Next, make a red chili sauce with fresh habañero chilies (or just one with seeds removed if you prefer it less hot), onion, and tomatoes, all finely chopped together with salt and some lime juice. Put some black beans in a pan and mash them coarsely, then set aside. In the same pan brown cubes of ham, and set aside. When ready to serve, make a layer of black beans on each tortilla, fry eggs sunny-side up, and place one on top of each tortilla with beans. You want the yolk to remain uncooked. Then pour over some of the sauce and top with the ham and some green peas. Sprinkle on some *queso fresco* and serve.

SAVORY EGG CUSTARD/CHINA

Any number of ingredients can go into this dish, which is popular in Cantonese cooking, where it is called *soi tan* (watery egg). The only requisite parts are a large bamboo steamer and small soup bowls. These proportions are for four servings. Take four eggs and mix with a cup of good chicken stock. Slowly mix together, as you want a smooth, silky texture, not a fluffy, aerated one. Divide the mixture into the four bowls and add other ingredients such as sliced mushrooms, shelled raw shrimp chopped up and marinated in some Shaoxing wine, and some chopped scallions. The dish is also delicious with crab. Cover the bowls with their lids or use tin foil and place in the steamer, covered, for 10 to 15 minutes. They should be very gently set, not overcooked, or they become rubbery.

FLAN/MEXICO

Flan is directly inherited from Spanish and French cookery but is a common dessert in Mexico, too. It can be made in small individual ramekins or in one large, shallow pie plate. An earthenware dish is very nice, but a springform pan will make turning the dish out at the end easier. Start with ½ cup sugar. Melt it in a pan until dark and caramelized. Put this into your pan or ramekins and swirl around. Then mix 6 eggs with 3 cups full-fat milk, a few capfuls vanilla extract, and ½ cup sugar. Some recipes call for a small can of condensed milk and evaporated milk, which works nicely as well, but if you use a sweetened version, do not add sugar. Pour this into your pan, place it into a larger roasting pan, and set in the oven, preheated to 325 degrees. Pour boiling water into the larger pan so it comes up just an inch or two. Bake for about 45 minutes, or until just set. It will take much less time if you are using small ramekins. Remove from the oven and chill for several hours in the refrigerator. When ready to serve, run a knife around the edge of the pan, place a plate over the pan, and flip it over. It should turn out as one solid, jiggly mass with its own caramel sauce. You can decorate with fruit if you like or just slice and serve with some of the caramel sauce.

STRACCIATELLA ALLA ROMANA/ITALY

In this recipe, we are referring to the soup rather than the gelato flavor of the same name, which is analogous to chocolate chip or ripple in the United States. The name means "shredded in the Roman style," which is what the eggs look like in the soup, rather like egg drop soup in Chinese cooking. You should start with an absolutely clear chicken broth, which is a base for countless other recipes. This recipe has a very long history and is related to the *panada*, which was a simple broth, bread, and egg mixture, usually given to invalids. Consequently, this is excellent comfort food, considered easy to digest and very nutritious.

Start by chopping carrots, celery, and onions. Heat a deep pot and brown chicken parts. Use wings, backs, necks, giblets, and other parts you have on hand. It makes sense to buy whole chickens, cut them up yourself, and save these parts in the freezer until you want to make the broth. When lightly browned, remove and add the vegetables with a bay leaf and other herbs such as thyme and parsley, and lightly brown these. Add the chicken back to the pot and cover with cool water. Bring to a boil, and then turn down the heat to low. Gently simmer, uncovered, for about 1½ hours. Skim the top occasionally. Then strain the contents through a fine sieve into another pot; press on the solids to extract all the liquid. If you like, clarify the soup by gently stirring in a beaten egg, with the heat on. This will collect any stray particles. Pass through the sieve again to remove this egg. This step is not strictly necessary as this will not be a clear soup in the end. Salt to taste, only at the very end.

When ready to serve, take a cup of cool broth and add 3 eggs and a good handful each grated Parmigiano and finely ground semolina flour. You can also use fine breadcrumbs. Grate in a touch of nutmeg. Drizzle this mixture into the hot broth, and stir constantly until you see shreds appear in the broth, about 5 to 10 minutes. Serve at once, sprinkled with a little chopped parsley.

CAPIROTADA (BREAD PUDDING)/MEXICO

This bread pudding is a popular dish for Lent in Mexico, and its ingredients are said to symbolize various parts of the Passion: the bread the body of Jesus, the cloves the nails, cinnamon the wood of the cross. Interestingly, it did not start out with these associations as the recipe from the colonial period below attests, and it contained meat and would not have been suitable for Lent. Neither would the modern version below, as it contains eggs and cheese. In any case, comparing the two versions shows how much a dish can change over time, though its name remains the same. Incidentally, most of Diego Granado's book is actually a translation of Scappi's Italian cookbook of a few decades before, but this recipe is nonetheless typical of those brought by Spanish conquistadors. The lights (*livianos*) here are the lungs and trachea.

A COMMON CAPIROTADA (DIEGO GRANADO, 1599, 77–78)

Take two pounds of Pinto cheese, and another pound of rich cheese that is not too salty, and mash in a mortar ten cloves of garlic that have been first boiled with the meat from a capon breast that has been roasted. When it is all pounded, you add ten raw egg yolks and a pound of sugar, and infuse everything with cold chicken broth, because if it is hot it will not infuse, nor can you pass the composition as is required through a grinder or sieve, because of the cheese. Being sieved, put it in a tin-lined casserole, and put it on a bed of coals from the fire, and add an ounce of cinnamon, half an ounce of pepper, half of cloves and nutmeg, and a good bit of saffron. While it is cooking, shake it enough to help it take shape. When it is cooked, if you want the best flavor, next take lights of veal or kid fried, mix it all together, and serve with sugar and cinnamon on top. In place of the lights you can also put breast of veal cooked and then fried, in the same way you can make a capirotada with any kind of roast.

MODERN CAPIROTADA/MEXICO

Start by tearing up some *bolillo* rolls (see recipe on p. 138) or other light, eggy bread into small chunks. Put them into a casserole, and pour over some melted butter. Put this into an oven to toast lightly. Next mix eggs and milk with a generous sprinkling of sugar (use *piloncillo* if you can), cinnamon, and cloves. Grate a couple of handfuls of cheese, such as cheddar or Colby, or a lighter Mexican cheese like cotija if you prefer, and mix everything together with the bread. Return to the oven and bake until solidified and browned on top.

TIRAMISÙ AND ZUPPA INGLESE/ITALY

There is absolutely nothing traditional about tiramisù; it probably dates no further back than the 1980s when it first became popular, but it is so well loved that omission would be a pity. There are earlier forms of the dish, in particular the zuppa inglese ("English soup" or "sops"), which is a relative of the English trifle, and it allows for more freedom and creativity. It can be made with layers of sponge cake, alternating with fruit such as strawberries or raspberries, and then layered with custard. The custard is made simply by heating milk with a strip of lemon zest. Meanwhile, beat egg yolks with sugar until thick and pale yellow. Add in a little of the milk to temper, and then pour the egg-milk mixture back into the pot of milk and whisk vigorously until thick on very low heat. Let cool and add to the zuppa in layers with the fruit and spongecake.

If you are adventurous, the spongecake, or as it was called *pan di Spagna*, is very simple to make at home. Here the proportions do matter. Separate 3 eggs. Beat the yolks, gradually adding in ½ cup sugar, and beat until thick. Flavor with 1 teaspoon vanilla extract (rosewater is also very nice). Then beat the egg whites with a clean whisk in another bowl until a meringue of stiff peaks is formed, beating in 1 tablespoon sugar toward the end. Add a dollop of the meringue to the yolks to loosen, and then add the yolks into the meringue a little at a time, gently folding. Take a generous ½ cup cake flour and sprinkle in also, gently folding with a rubber spatula. Continue folding all three together, and at the end quickly mix in 3 tablespoons melted and cooled butter. Place the batter in a parchment-lined baking pan and bake at 350 degrees for about 25 minutes. Let cool, then run a knife around the outer edge, and turn out of the pan. Divide into two even layers using a serrated bread knife or a taut string. Proceed with the layering instructions above. You can also make two such cakes for a large *zuppa*.

Tiramisù is similar, except Savoiardi biscuits or ladyfingers are used and mascarpone cheese is used instead of the custard. Dip the biscuits into the strongest espresso coffee. Line the bottom of a glass bowl with the biscuits. Drizzle in a dash of rum. Add a layer of mascarpone cheese, sweetened with sugar, thinned with a little whipping cream so it is spreadable and flavored with some vanilla extract. This is the easier method. Alternately mix egg yolks with sugar and beat until thick. Then beat the egg whites separately (with a clean whisk), and gently fold into the egg mixture and then combine with the mascarpone. Dust the top with powdered cocoa. Continue with other layers. Nothing is cooked, so the recipe is the epitome of simplicity. Top with some sweetened whipped cream and shaved curls of good dark chocolate.

Dairy

The following selection of dairy products ranges from the simplest of preservation methods to the more complex, though certainly not the most complex of cheesemaking. All these traditions ultimately hail from the same source, western Asia, whence they spread to Europe with cattle and thence to the Americas.

KUMIS (FERMENTED MILK)/CHINA

Originally made with mare's milk, common throughout western and central Asia, *kumis* is a naturally fermented product that can easily be made with cow's milk and often is today. Traditionally it was fermented in a hide container hung at the door of one's yurt, but buckets work just as well. If using pasteurized milk, a bacterial starter will be necessary, much like making yogurt. Unpasteurized milk, however, will be spontaneously colonized by lactobacilli in the air that lower the pH, turning it sour and preventing the growth of harmful bacteria, as well as by yeasts that convert the sugars in the milk to alcohol. *Kumis* has a very low alcohol content, but this also serves to preserve the milk. Mare's milk is very high in lactose, but this process transforms it into lactic acid, making it more digestible. Many people, in fact since ancient times, have extolled the health benefits of fermented dairy products such as this, and there is interest today. Several commercial brands or similar products, such as yakult, can be purchased in health food stores. To make *kumis*, start with raw cow's milk and add some sugar or glucose, as you want to encourage alcohol production. Leave open, covered with cheesecloth, in a room at about 70 to 80 degrees but not hotter, until it starts to bubble and smell sour, usually only a few days. Bottle this, refrigerate, and consume. It will be sour, refreshing, and slightly fizzy.

There are also several fermented cream-based products in Mexico, called *crema*. Some are thinner and more like crème fraîche and will not curdle when cooked; others are more sour, such as *crema agria*, much like sour cream.

QUESO FRESCO (FRESH CHEESE)/MEXICO

This is the simplest and easiest of cheeses to make. It is not fermented, so it is not meant to keep long. It is crumbled onto other foods, or baked, though note that it will not melt. Start with regular pasteurized whole milk—however much you want to make. A quart will give you a small disk. Heat the milk just below the boiling point, and add the juice of a large lemon. Some people use vinegar, but the lemon gives the cheese a much nicer flavor. Stir, with the heat on medium. You will see the milk coagulate and the curds begin to separate. Let them cook just a few minutes. Remove from the heat, and let sit for a few minutes to further encourage the separation of curds and whey. Then carefully pour this into a strainer lined with a double layer of cheesecloth. Let drain for about 30 minutes. Then lightly salt the curds, wrap them into a disc shape, and if you like either use within a day or two or place them, still wrapped, on a rack with a weight on top to press out more liquid. The longer you leave it, the more crumbly the final product. If left overnight, with increasing amounts of weight, you will have a disc of cheese that can be sliced.

MOZZARELLA AND RICOTTA/ITALY

Although also fresh cheeses, mozzarella and ricotta require the use of rennet. This can be ordered online from cheese supply shops, and it is very inexpensive. Traditionally it comes from the stomach lining of an unweaned calf, though it is now usually produced in a factory from genetically modified microbes. Whichever rennet you have, for the proper texture and consistency, raw milk is easier to work with, though pasteurized can also be used. First, heat a large stockpot to 90 degrees with 2 gallons whole milk, buffalo if you can get it, or cow's. Then, add ten drops of rennet diluted in cold water, stir, and leave the liquid, not to coagulate as with the lemon juice, but to set into a solid, jiggly mass, undisturbed. Maintain the heat at 90 degrees until it looks like a solid custard, and then cut the mass with a long knife into wedges and then crossways into batons. Wait about half an hour, still maintaining this temperature with a very low flame, and you will see more greenish whey separate from the curds. Turn the heat up to about 100 degrees for a few minutes to help the curds solidify. Next, gently lift the curds with your hands, place into a large bowl, and sprinkle with a little salt. Heat the remaining whey up to a temperature as hot as you can stand with your finger in it. Pour a few ladles full of hot whey over the cheese in the bowl—it should be hot enough to melt it. Using a stick, lift the cheese up and down and turn over until it begins to melt, and then start to knead it in the hot water. Fold it over, press down gently, and stretch like you would knead a dough. This is called *pasta filata*. You will definitely need to use the stick to help you, as the water will need to be very hot. Also, do not press down too hard or you will squeeze out the fat. After kneading, squeeze the end with your fist to create little knobs, which you can pull over and put into a container with some cooled whey. These are *bocconcini*. Or you can make larger, egg-shaped cheeses. These should rest in the refrigerator in their liquid until cold. Then serve with tomato and basil for a classic insalata Caprese, or slice and use on a pizza or anywhere you would mozzarella. The taste will be fresh and milky and very different from rubbery, store-bought, low-moisture mozzarella. If you like, you can also marinate these with olive oil, sprigs of thyme and rosemary, and perhaps some garlic, in small jars in the refrigerator.

Next, take the remaining whey and bring it up to a furious boil. You are now "recooking" the remaining cheese solids so they precipitate. As they rise to the surface, remove them with a skimmer and place in a plate. You should get about a cup or two at most. Salt these curds very lightly and use as you would any real ricotta, though on a slice of bread this is extraordinary.

Aged Italian cheeses naturally take much greater precision and expertise, either traditionally with direct hands-on practice with techniques for making and preserving the cheese, and today with scientific research. But aging cheeses is possible on a small scale. The only major difference in technique is that you will have to start with raw milk if you want naturally occurring cultures, or you will have to buy these cultures from a cheese supply shop. In either case, the curds are not pulled and stretched as with mozzarella, but pressed under successive weights, while still wrapped in cheesecloth. They are then unwrapped and salted well, and turned over for a few days in a space kept at about 55 degrees—a cave is ideal, especially with beneficial molds. Then this is sometimes rubbed with oil or fat, perhaps dusted with rice flour or some other material to encourage a rind to form. They can also be bound with cloth or dipped in paraffin wax to create a protective covering, but traditionally it is a naturally occurring rind. Sometimes they are smoked as well. These cheeses are then aged up to a year or more. The larger the cheese, the longer it will take to age, and the longer it will keep. The 2-gallon batch mentioned above will give you a manageable disc that should be aged and fairly hard and piquant within three months.

TRES LECHES CAKE/MEXICO

This is an oddly similar cake, custardlike in texture and soaked with, as the name indicates, three kinds of milk, namely condensed sweetened milk, evaporated milk, and cream. Start by making the cake. Cream together a stick of butter and a cup of sugar. Mix in five eggs one at a time and 1 tsp vanilla extract. Mix in 1½ cups flour and 1 tsp baking powder. Bake in a casserole or deep pie plate at 350 degrees for about 20 minutes. Test with a toothpick, which should come out dry. Then mix a can each of condensed sweetened and evaporated milk with a cup of cream in a bowl. You can also add in a little rum. Pour it over the cake and let soak in overnight. Then decorate with more cream, whipped stiffly with some sugar and a touch of vanilla.

EMPANADA DE CAJETA (PASTRY WITH MILK FUDGE)/MEXICO

Empanadas are often filled with savory ingredients, but there are also sweet pies. This one is filled with a Mexican version of *dulce de leche*, and fruit can also be added. To make the *cajeta*, take a quart of goat's milk and a split vanilla bean and heat very slowly with a cup of sugar; stir continuously. When it comes close to a boil, add in a pinch of baking soda dissolved in water. Stir frequently and maintain a low heat for a few hours until the mixture is reduced, thickened, and caramelized. You might need to add a little water from time to time, if it reduces too quickly. This can be used on ice cream, mixed with nuts to make a kind of fudge, or in a crepe. Here it is used to fill a sweet enclosed pie. The crust can be made simply with pastry flour, diced cold butter, and cold water, mixed until it comes into a ball. Let rest, then roll it out and cut it into 6-inch circles with a pastry cutter. Smear each circle with a dollop of *cajeta*, and fold over into a half circle. Seal with the tines of a fork or pastry wheel. You can also add a little cooked apple, pumpkin, raisins, or nuts. Bake these in a moderate oven until browned.

PANNA COTTA (COOKED CREAM)/ITALY

Among the simplest and purest tasting of desserts, tasting of fresh cream, panna cotta can be made simply with gelatin sheets or powder and just enough sugar to sweeten. Sometimes it is made with cream and milk. Most versions call for a little vanilla or lemon peel. Usually it is garnished with fresh fruit, a drizzle of chocolate or sauce of some kind, but it does taste best unadorned. Simply dissolve the gelatin in ½ cup milk. Heat 2 cups cream with ½ cup sugar and a little vanilla extract. Then pour in the milk, stirring until the gelatin is completely dissolved. Pour this mixture into little ramekins or fancy molds, and refrigerate until set. Loosen the edges with a knife, and turn out onto individual plates. Garnish as you wish.

{7}
Fish and Shellfish

Learning Objectives

- Consider the ways fish might be more constructively included in the Western diet using less-than-familiar techniques.

- Appreciate the wide range of species, especially small fish, once consumed everywhere and the very few fish now commonly eaten. Consider the impact these shifts have had on the world's oceans.

- Learn about how fish were and continue to be an important diet because of Christian Lenten restrictions.

- Think about the health benefits of a diet focusing on fish.

- Think about the impact of fish farming and why some species have been profitable.

FISH AND SHELLFISH have been a mainstay of coastal populations and those along rivers since prehistoric times. In Italy, Mexico, and China, fish and shellfish have usually been the most important source of protein for those in places with access to water. In Italy these areas include Liguria, the entire west coast facing the Tyrrhenian Sea, and the entire east coast facing the Adriatic, as well as the islands of Sicily and Sardinia, where fish is among the most important foods. The only river that qualifies as truly large in Italy is the Po River system, which supplies the north with seafood, but the Arno and Tiber also supply fish, and there are also large, fish-rich lakes such as Trasimeno and Lake Como in the north. Surprisingly, however, consumption of fish drops off precipitously as one moves inland, and especially up mountains, for the simple reason that, unless preserved, fish does not travel well. Only in modern times have freezing and canning technology and improved transport made fish a common food for noncoastal populations. But even then, apart from the coasts, fish is not a favorite food in the interior of Sardinia, simply out of long-standing habit.

The scenario of fish eaters is roughly the same for Mexico as in Italy. In places like Veracruz and the Yucatan on the Gulf of Mexico, throughout Baja California, and down the entire massive Pacific Coast, fish is a staple, cooked in ingenious ways and often preferred to meat. Lakes and freshwater species were also crucial in places like Tenochtitlan (now Mexico City), which was built on Lake Texcoco where fish were, in a sense, farmed.

China has an extremely long Pacific coastline where fish are the main source of protein. Fish are also abundant in massive river systems where civilization itself began. There are the Huang He or Yellow River in the north and the Yangzi (Chang Jiang) across the middle of China, which flows for 3,400 miles, supplying fish to millions of people, and last the Pearl River (Zhu Jiang) and its delta in the south (figure 7.1).

In China, Italy, and Mexico, the abundance of fish compounded with the inability to move it great distances inland stimulated methods of drying, smoking, and other forms

Figure 7.1. Market in Lantau, Hong Kong. © Huw Jones/the food passionates/CORBIS.

of preservation and fermentation. In the case of China, live fish were moved in tanks or kept in artificial ponds. There are also interesting reasons why certain fish came to be preferred in each of our cuisines, not necessarily simply dependent on those that could be caught. As we will see, some species were fished far from the homeland at great expense.

Other economic and cultural factors fostered the centrality of fish in the cuisine. In China the prevalence of fish has been partly the result of wet rice farming, which provides a habitat for freshwater fish such as carp and eels and also crabs and frogs. Fish ponds also supplied those without immediate access to fresh fish. In Italy since antiquity and in Mexico for the past five hundred years, fish was the only source of animal protein traditionally allowed by the Catholic Church during much of the year on fasting days, during Lent, the forty days from Ash Wednesday to Easter, on Fridays, and in the past on Saturdays and sometimes Wednesdays as well. Even inland populations would subsist on dried cod or sardines, and fish everywhere made up the traditional Christmas Eve supper. Although these rules have been changed since the Vatican II Council in the 1960s, people still traditionally eat fish on Fridays and always on Christmas Eve. In Mexico, the same applies for those along the Pacific coast and facing the Gulf. The healthy lifestyle of the so-called Mediterranean diet, the avowed longevity of the Chinese, and the healthy indigenous diet of coastal Native Americans can in large measure be attributed to the great proportion of fish in the diet, rich in omega-3 fatty acids, low in saturated fat, and full of essential vitamins and minerals.

Fish, of course, have mostly been caught wild, using nets or weirs that direct the fish into a kind of underwater basket. Line fishing with a rod and hook was practiced in prehistoric times, and large fish such as tuna are in a sense hunted, for example, in the traditional *matanza* slaughter of Sicily, in which the fish are herded and gaffed by men in small boats. Equally ingenious is the use of tamed and trained cormorants in China, which dive for fish but are prevented from swallowing them due to a ring or rope around their long necks. Even otters can be used for fishing, by chasing fish out of their hiding places into nets. Rather than an exhaustive list of fish and their uses, the following section groups basic fish and shellfish types together based on culinary usage within the three regions.

Large Fish

Large saltwater fish have usually been caught by commercial fishing operations rather than small-scale fishermen because of the size of the fish and the need for equipment, large boats, and many men to haul in the catch. Most of these fish are also caught many miles at sea in open water and require crews to be gone for days or weeks. Larger fish such as tuna, swordfish, shark, and sturgeon fall in this category. The main culinary distinction between these and smaller fish is that their flesh tends to be firmer and thus can be cut into steaks, which will hold together while being cooked. They are also deeper in flavor. Some can also be preserved—tuna belly can be salted and dried, as it is in Italy (*ventresca* is fresh, and when salted is called *mosciame*, which is related to the Spanish *mojame*, derived ultimately from Arabic)—or any part of the tuna can be canned. The popularity of canned tuna in the United States can be attributed, at least indirectly, to Italian American immigrants.

Of all large fish, sturgeon has traditionally been the most prestigious in Europe, partly for the size and delicate flesh, which was compared to veal, but even more importantly for caviar. This is the egg sack, or roe, whose eggs are salted and usually eaten raw. Although today nearly extinct in Italy, they used to be abundant in muddy riverbeds and lagoons. In the past caviar was also cooked, and recipes go back to the sixteenth century. Today they are mostly imported from the Black or Caspian seas. Sturgeon are not the only species that yield edible eggs, though; shad roe is another, and today lumpfish, salmon, and many other fish supply lesser forms of caviar. In Italy the roe is also processed into a product called *bottarga* (another word descended from Arabic) which is salted and dried. This can be made from tuna but is more often made from mullet.

Chinese cuisine makes use of a number of large saltwater fish, including many species of shark. Shark fin is one of the most prized and expensive of all Chinese ingredients. It is also rare but traditionally used at wedding banquets and other special occasions, though the practice of "finning," in which the fin is removed and the rest of the shark is left to die, has elicited great public outcry. In general, the Chinese make less use of tuna or mackerel and other oily fish, the preference being for whiter-fleshed species. There is a species of Chinese sturgeon that inhabits tributaries of the Yangzi River and other waterways, but it is today endangered and strictly controlled by the government.

Mexico is home to some of the finest sport fishing in the world, and fish such as swordfish, marlin, and sailfish are regularly captured. These are eaten, too, though are rare in traditional Mexican cuisine mostly because one needs complex rigging and motorboats to catch them. Increasingly, large game fish like tuna, dorado (mahi mahi), and wahoo are making their way into Mexican cookery, especially that which caters to tourists in places like Cabo San Lucas.

Although biologically not fish, we must include here sea mammals such as whale and porpoise. These have provided food not only for Asian cultures but were also considered delicacies in European courts, the marine equivalent of venison, up until the past few centuries. Why they went out of fashion in the West has probably to do with the prevalence of beef and the need to fish farther away from land. There were efforts to reintroduce them in times of war. Whales were, of course, hunted extensively for lamp oil well into the nineteenth century. This practice was only made obsolete with electricity and growing animal welfare concerns, as well as the recognition that these creatures' very existence was endangered by overhunting. But there is still a market for whale meat, primarily serving Japan, though interest has dramatically declined since moratoriums were placed on commercial whale catches. Japan, Iceland, and Norway remain the only countries that still commercially hunt whales, though Native Americans and Inuit often still do. The flesh of whale is actually similar to beef: dark and fibrous, with only the faintest hint of the sea.

Dried and Salted Fish

Codfish have been one of the most important species in Mediterranean cuisine, though ironically they are not caught there. For centuries they were fished in the Atlantic as far as the Grand Banks of Newfoundland, first by Nordic fishermen, then by Basques, Portuguese, and later the English. Exploration of North America in the late fifteenth and sixteenth centuries was largely in search of cod. Naturally they were hardly ever eaten fresh when caught so far out to sea, but rather dried in the cold, dry air as stockfish or salted as *baccalà*. This supplied both coastal and inland populations during Lent and other fish days. The principal virtue of preserved fish is that is it practically indestructible and will last unspoiled indefinitely. Stockfish must be pounded and then soaked, whereas *baccalà* can be merely soaked for a day or two (figure 7.2). In drying the fish takes on a toothsome texture and an unmistakable bouquet of the sea, which is undeniably alluring. In the past it was generally reviled on fine tables as food of the poor, though in typical fashion, the tables have turned and it is now appreciated as a simple rustic

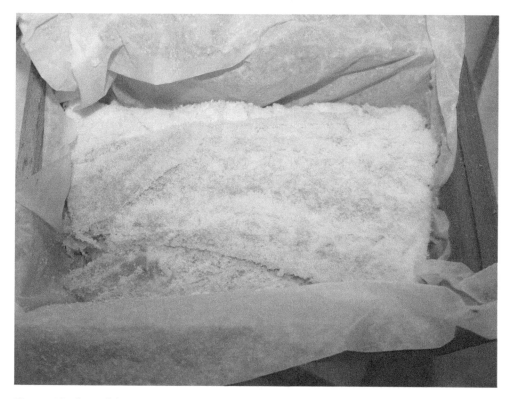

Figure 7.2. *Baccalà* is dried, salted cod, often sold in small wooden boxes from Nova Scotia.

and traditional food, reminding one of identity—and accordingly it is fairly expensive. Stockfish or *baccalà* are cooked in various ways, in a sweet and sour sauce in Rome, with pine nuts, raisins, sugar, and vinegar, which is a direct descendant of medieval dishes. In the classic dish of Vicenza, *baccalà* is gently stewed in milk with browned onions and garlic for several hours, garnished with parsley, and served with polenta. Equally delicious are fritters made of chunks of *baccalà*. In Genoa it is cooked with carrots, celery, and tomatoes, mushrooms and white wine, and finished off with potatoes and black olives, for a very hearty kind of stew. Finally, in Venice a fine purée is made with garlic and olive oil (*baccalà mantecata*).

Not surprisingly, codfish was also brought by the Spanish to the Caribbean and Gulf of Mexico, where here, too, it became a popular staple. In Mexico the classic dish is *bacalao a la Vizcaina*, which is eaten almost everywhere on Christmas. It is made with tomatoes, onions, garlic, peppers, olive oil, olives, and capers and finished off with potatoes.

Although not normally made with cod, dried, salted fish are also very important in Chinese cuisine, and in markets one sees specimens that look superficially scarcely different from *bacalao*. There are also dried and smoked eels, dried fish maw (air bladder), and similarly preserved fish of virtually every species available. Among the most expensive of all foods are dried scallops and abalone that one finds carefully graded in jars, though even these are cheap compared with shark's fin, which is the luxury food par excellence in this cuisine. The fins are dried, reconstituted, and shredded and cooked in a soup. Its appeal, apart from the cost, is all in the gelatinous texture as the fin is just fine, fibrous cartilage and has little flavor of its own. As mentioned, the practice of finning as well as needless killing of sharks has led to dwindling populations worldwide and a serious imbalance in marine ecosystems.

On the other end of the economic scale of dried seafood are tiny dried shrimp, which are used as a flavoring. They are used either pounded or whole in many Chinese dishes such as soups and braises or as an ingredient in dumplings. Dried shrimp are also popular in Mexican cookery, and when pounded and mixed with eggs they are used to make a dish for Lent called *romeritos*, a rosemarylike herb. Dried shrimp are also used to make tamale fillings and rich broths or are simply eaten as a snack. They are a cheap and convenient way to add a boost of fish flavor and texture. The connection between Mexico and Asia, in the case of dried shrimp, may not be coincidental. There was a long connection in trade between Acapulco and Manila and ultimately to China for trade, especially in silver. This brought certain ingredients (tamarind, for example) as well as cooking techniques, and the traffic in foodstuffs definitely went in both directions.

Almost all fish and shellfish that can be caught are also available in dried form, including the sea cucumber, an echinoderm with hundreds of little tentacles beneath that feeds on plankton in coral reefs. It is used in Chinese medicine and cooking and is widely reputed to be an aphrodisiac, most likely due to its resemblance to the cucumber. The sea cucumbers are also said to be good for the skin because of the high collagen content. They are used in Shandong cooking of the north as well as in Cantonese cuisine. The dark, dried creatures are soaked in water, chopped, and then usually braised with other flavorful ingredients like mushrooms, because they have little flavor of their own. Another sea creature, also always sold dried, is jellyfish. These are soaked overnight, sliced into strands or cut with scissors, and then usually seasoned with soy and oyster sauce, sesame seeds, scallions, and ginger and served as a cold salad (figure 7.3). The texture is pleasantly chewy. Last, squid are also sliced, seasoned, and dried and eaten as a popular snack, sort of like a jerky. Fermented fish products, sauces, and fish pastes will be discussed in the chapter on flavorings.

Medium-Sized Fish

There are a number of medium-sized fish taken from the Mediterranean that are not well known in the United States. These include the *orata* (gilthead bream) named for the golden spot on its brow, and the sea bream, which taste best simply baked. The red mullet and gray mullet are equally common in Italian fish stalls. The latter is also prized for its eggs, which are dried and salted as *bottarga*, which is shaved thinly and

Figure 7.3. Jellyfish are sliced and used in salads in China and are crunchy but relatively flavorless.

eaten much like prosciutto as an appetizer or ground into a sauce with olive oil and lemon juice and served on pasta. Widely accounted the most delicious of fish is the *branzino* (sea bass), which is sweet and succulent, especially when baked whole in a salt crust. This is done by making a paste of salt and egg white, covering the whole fish, and then baking it. Surprisingly, it keeps the fish moist without making it salty at all. Equally delicious is *pesce San Pietro* (St. Peter's fish), so called for the spots on its side acquired when the saint lifted it out of the water for coins to pay Caesar's taxes. This is also best baked or grilled. The salmon and trout are also appreciated in Italy, simply fried, or even in sauces over pasta.

In China fresh fish almost always means live fish, and not only are these found swimming in tanks in restaurants from which diners can choose their meal but also in markets there are large buckets containing fish brought home alive and killed only just before cooking or even in the process of cooking. Medium-sized fish such as red snapper or sea bass baked or fried whole in a wok are among the most popular dishes and are an essential part of the New Year's celebration, partly because the word for fish sounds like the word for abundance (*yu*). In contrast to most other sources of animal protein, fish is generally not cut into small pieces but merely slashed diagonally before cooking so the flesh absorbs seasonings and cooks evenly, and then it is served whole. This is because delicate fish would fall apart with most Chinese cooking procedures, and when cooked whole, most fish can be portioned by flaking with chopsticks, without the use of a knife. Whole fish are equally popular steamed, and in fact all parts, head and tail included, are considered delicacies. When fried even the fins take on a pleasant crunch. The very idea that one would discard such morsels, as is the case in the West, is unthinkable here.

Like Italy and China, Mexico has bountiful coasts. There are expectedly dozens of native fish species and even more ways to cook them. Fish such as red snapper are often simply marinated in chilies, orange juice, achiote paste, and herbs and then grilled whole in the Yucatan. The popular sierra (kingfish) is also cooked this way, or in a vinegar-based sauce (*escabeche*). The grouper (*mero*) is also very popular, with its firm, white flesh. So, too, are mackerel, mullet (*lisa*) from the Pacific, which are dried or smoked, tilapia (*mojarra*), which is appearing frequently in U.S. markets now, and *pampano* from the Gulf (figure 7.4). From the Gulf there is also tarpon (*sàbalo*), which is found in the United States as sable. It is often smoked and is among the most delectable fish anywhere. All these species can be split and grilled or used like meat to fill tacos. The fish taco, as found in California and increasingly elsewhere, made with fried breaded fish, cabbage, and a mayonnaise-based dressing, is a fairly recent invention, but simpler versions are indeed part of the indigenous cuisine and originally used grilled fish.

Salmon has become one of the most popular species worldwide and is increasingly farmed to meet the demand, though the flavor is decidedly inferior to wild. It is slowly gaining ground everywhere where it was not traditionally fished. We might say the same for the similar and smaller trout, though it is still mostly caught wild. These fish are cooked much like other medium-sized fish: grilled, fried, baked whole. The same is true

Figure 7.4. A small mackerel species common in Asian cuisine.

for bass, bream, tench, perch, and other strongly flavored freshwater species. These are also the fish most commonly cured, as in smoked salmon or trout, or salt-cured salmon, such as the Baltic gravlax, which is imported around the world as a luxury item. Increasingly, as globalization advances, such fish are intensively farmed in a few locales and shipped around the world, which of course includes our three regions. This is often due to degradation of spawning grounds, pollution, and in the case of salmon the proliferation of hydroelectric dams along rivers up which the fish used to migrate. Today and one can only expect in the future salmon will be largely farmed rather than caught wild. In Italy, China, and Mexico, increasing health consciousness also prompts consumers to turn to salmon and other omega-3–rich fish as a regular part of their diet, which will only exacerbate the problems.

In the list of medium-sized fish we must also include the flatfish, recognizable by two eyes on one side of their head, and normally a light underbelly and dark transverse side. These include small sand dabs and delicate sole, slightly larger flounder and fluke, all the way up to the gargantuan halibut. What distinguishes these in culinary terms is their extraordinarily light and white flesh, bordering sometimes on blandness. They also stand up to only the quickest and simplest of cooking methods, such as breading and frying, though one does see them stuffed. Sturdier halibut are sometimes sliced into steaks and grilled, or baked in a paper packet, comparable with the papillote method. They also hold up well to steaming, which is common in China and considered the best way not to ruin their delicate aroma. In Italy one sometimes find a whole platter-sized halibut roasted and served to feed an entire table at beachside restaurants. Flounder and its relatives are the fish most often served to those not very fond of fish.

At the other end of the fish-loving spectrum are eels and lampreys, traditionally eaten anywhere there are lagoons in China, Italy, and Mexico. Despite the tough skin and proliferation of bones, eels are actually one of the most versatile and tasty of fish. They hold together nicely when pickled or smoked and can be grilled easily or marinated or fried. In Italy they are braised or cooked in sauce, which is unusual for most fish. From the Comacchio lagoon on the Adriatic coast young baby eels or elvers (*cieche*) are a traditional dish, and full-grown eels are a Christmas specialty. Around Rome eels are quickly fried and then left to marinate in a sour, vinegar-based dressing for several weeks. In Chinese markets one will often find these live in barrels, and again practically any cooking method can be used.

Small Fry

Smaller fish are also important to these three cuisines. In Italian cuisine, there are sardines, which can be grilled fresh or marinated in olive oil or stuffed and baked in a casserole. Tiny anchovies are also salted dry or canned in olive oil, and these are not only a pizza topping but also used widely in cooked sauces. The tiny fish essentially disintegrate in cooking and lend a faintly briny flavor to tomato sauces and other dishes, or they are pounded with garlic and oil as a dip for vegetables, as in the Ligurian *bagna cauda*. The liquid of fermented anchovies is also poured off and used as a condiment (*colatura di alici*), a rare descendant of the ancient ubiquitous garum, which will be discussed with condiments under fish sauce in chapter 8. Small fish are also used fresh, battered, and eaten whole. In both China and Mexico, there are also small fish that can be fried whole, salted, and dried and used as a snack or cooking ingredient, or made into sauce. *Charales*, for example, in Mexico are dried and cooked with cactus in a green sauce or in *pozole* (a dish including whole hominy). Used fresh, they make a taco filling.

Freshwater Carp

In China, more than perhaps any other place on earth, freshwater fish are featured in the cuisine, and these are not primarily caught wild but rather farmed in ponds. This practice stretches back hundreds of years, and there was the added boon that these fish would eat malaria-carrying mosquitoes and other insects that spread disease. It is revealing that, historically malaria outbreaks would only occur when there was some disruption in the regular irrigation cycle due to war or other calamity, when the water was left stagnant. But in regular times the rich aquatic life formed a balanced aquatic ecosystem.

The main fish kept in irrigated paddies and ponds are carp, of several different species, because carp like muddy water, feed on waste, and will readily reproduce in ponds. The flesh of the carp, however, is white and delicate and for very good reason it is considered the best of fish. Like other fish, it is cooked whole or slashed and fried. But carp was also preserved in what many account to be the ancestor of sushi. At first this was not raw fish, but carp covered in salted-inoculated rice that was left to ferment, acting as a preservative. Before eating, the rice would be scraped away. Over time, and after this technique was introduced to Japan, the fish came to be consumed with the rice and also was eaten fresh. This newer type of sushi is normally found in Japanese-style restaurants in China.

Ceviche

Interestingly, Latin American culture has its own variation on the raw fish theme called **ceviche**, though in this case the fish is not exactly raw. Pieces of fish such as red snapper, tuna, or halibut are marinated in lime juice, and the acid actually changes the molecular structure, in a sense cooking it without heat and even inhibiting bacterial growth. The flesh of the fish also firms up, offering a pleasant texture and refreshing sourness. It has been suggested that the origin of this dish is in a Spanish ancestor called *escabeche*, which is fish fried then marinated in vinegar, onions, and other ingredients. This acted as a preservative, and the fish could be kept at room temperature for a week or so and eaten as is or reheated. How it came to be done with raw ingredients in Mexico is unknown, and though the dish appears to have originated in Peru, the use of citrus must postdate the arrival of the Spanish. In Mexico, ceviche is usually served as a salad with avocado or tomatoes, onions, and tortilla chips or crackers.

Similar preparations are also found in Italy, either introduced via Catalan cuisine, in which case *escabeche* becomes *scapece*, or possibly from Germanic cuisine, in which *sauer* becomes *saor* in the Veneto, referring to fish that have been fried then pickled in vinegar and sugar with onions and sometimes raisins and pine nuts. These are all derived ultimately from medieval cuisine, and relatives can be found throughout Mediterranean cuisine. Elsewhere in Italy it is called *in carpione,* perhaps referring to being originally done with carp, or from the fish *carpione* from Lake Garda, which is a salmon.

Shellfish

In culinary terms the word **shellfish** denotes both bivalves with hard shells as well as crustaceans. Both will be dealt with here as well as snails, polyps such as squid, though they have no shell, and other aquatic foods.

Crustaceans include large lobster, smaller crayfish and scampi, crabs, and shrimp. In flavor these are vaguely comparable as all have a very white, delicate, sweet flesh and in traditional cuisine were considered "bloodless" and easy to digest. In the case of lobster they are often deemed the most delicious of fish. The simplest way of cooking these is boiling or steaming, though this is more common in the United States than elsewhere. In China, for example, seasonings are often added right to the shell or to cut-up pieces, and then it is stir-fried or braised and each individual piece carefully picked and sucked out. Sometimes they are steamed or boiled as well. In Italy crabs (*granchi*) are often picked out of their shells and then stuffed back into them with breadcrumbs and seasonings, then baked. Soft-shelled crabs from the Adriatic are dipped in batter and fried whole. All these can also be incorporated into fish soups. The most famous of these is *cacciucco* from Livorno on the west coast. It contains crab, eels, cuttlefish, octopus, bream, monkfish, mullet, or basically anything caught that day cooked in red wine and tomatoes. Its descendant is the Italian American *cioppino* of San Francisco.

Shrimp are without doubt the one shellfish common and central to all three of these cuisines for the very simple reason that they are abundant. In China we find shrimp chopped finely and used as a filling for dim sum or poached gently as fish balls (figure 7.5). Or they may be simply stir-fried with vegetables or ground into a fermented shrimp paste, which is grayish and dried in the sun, and common in the south. Shrimp may also be simply boiled and served with a dipping sauce. In general shrimp are cooked as quickly as possible in a way that seals in their juices without making them tough. For example, they are butterflied (or in Chinese *fengwei*, meaning "phoenix-tailed"), which involves removing the dark digestive canal and then slicing open horizontally so they curl open when cooked. These can then be battered, quickly deep fried, and then seasoned with a sauce, perhaps sweet and sour. They are cooked with soy, sugar,

Figure 7.5. Dim sum at morning tea time, Hong Kong. © Imaginechina/CORBIS.

vinegar, and ketchup with chilies and peanuts—a classic *kung pao* (in Mandarin *gong bao*, meaning "palace guardian"). In addition, shrimp can be threaded on a skewer and quickly grilled until the color barely turns from brown to pinkish orange. Even live ones are cooked in a dish called drunken shrimp, in which shrimp are soaked in alcohol (in which they die) and then cooked by lighting the wine and pouring it over. Shrimp are enjoyed fresh throughout China, as there are both saltwater and river shrimp available. They are also now extensively farmed, and in recent years China has been the leader in world production. There are also mantis shrimp, which are not related to shrimp at all, which are dry roasted with salt and chilies.

In Italy there are various kinds of shrimp, from large *gambero rosso* and *cannocchia* or mantis shrimp, which are best grilled and simply seasoned, to smaller brown shrimp used to make sauces and served on pasta or in other mixed dishes such as risotto. Shrimp cooked *in umido* (gently braised in a little liquid such as wine or tomato) is also common. Scampi actually refers to a kind of small lobster rather than a shrimp. It is not a shrimp dish which is an American invention. An *insalata frutti di mare* is a common salad containing squid, octopus, shrimp, and sometimes mussels and firm-fleshed fish marinated in olive oil and lemon juice with parsley and served cold as an antipasto, especially in the summer. All these may also feature in the classic Adriatic *fritto misto*, in which the fish are coated in an ethereally light batter and briefly fried. This is nothing like the heavily breadcrumbed or even more oddly stuffed breaded shrimp found in the United States.

In Mexico, shrimp (*camarones*) are also popular, either cooked in a chipotle sauce or *pipián*, a creamy sauce based on ground pumpkin seeds and sour cream. They are also served cold in a shrimp cocktail or salad or grilled and served in a tortilla with avocado, which is a particularly appealing combination. *Arroz a la tumbada* from Veracruz, facing the Gulf, is a rice and fish broth dish that may be descended from similar dishes of African origin. Basically, shrimp can be cooked and used in practically any context in Mexican cuisine.

Bivalves

The list of popular **bivalves** would be endless but includes oysters, mussels, clams, cockles, razor clams, scallops, and abalone. Oysters have universally been considered the finest of these shellfish, partly because they can be eaten raw—as a kind of essence of the sea, or barbecued, cooked in soup or mixed into a stuffing. Huge shell middens (refuse piles) attest to the importance of oysters prehistorically throughout the world, but they have also been one of the earliest cultivated foods in coastal areas. Oysters have long been cultivated in China for food and pearls and normally are sold inland dried and salted and sometimes smoked as well, and there are special vendors who sell only dried fish products like these. They are also commonly made into oyster sauce, a ubiquitous flavoring. Italians have been enjoying oysters since antiquity, and since Roman times they were considered an aphrodisiac. They were farmed by seeding small oysters on long ropes that could later be hauled in and were kept fresh while transported throughout Italy, as oysters in a sense come with their own fresh seawater packaging and will survive several weeks out of water. Today oysters are cooked by sautéing gently with other ingredients, grilling directly in the shell, or baking with breadcrumbs, garlic, and herbs, then served with a slice of lemon on the side.

Other bivalves can be eaten raw, but they are normally cooked. Mussels and clams usually figure in soupy stews or soups or in a sauce served over spaghetti, with or without tomato. There are also classic dishes based on *vongole veraci*—a risotto in Venice made with clam juice, for example. Scallops, usually with the red coral and appurtenances still attached, are served in the beautiful shell, minimally adorned. However, in Italy these are also stewed in white wine or marinated in seafood salads, and the larger ones are grilled whole. The flesh of all these can also be lightly battered and fried. Smaller shellfish like cockles are normally served in broth or sauce. *Datteri di mare* ("dates of the sea"), which look remarkably like the fruit, are little dark mussels that can be eaten raw. In Mexico there are also numerous species of clam and mussel eaten, such as the pismo and *pata de mula* ("mules hoof"), so called because of its size and dark color. Mussels are often heated right on top of a *comal* until they open, or steamed.

The abalone is a similar shellfish, though unique in having only a single shell and large foot that it uses to attach to rocks. Abalones were once abundant up and down the American and Asian sides of the Pacific. Gathering them in the United States is now strictly limited, so one cannot purchase them fresh. But they are considered among the most delicious of shellfish. The flesh must first be sliced and pounded with a mallet. Thereafter it can be used in any recipe, sautéed simply, breaded and fried, or in a ceviche. In Chinese cuisine they are used like all other shellfish, as well as dried, and are even eaten raw and dipped in soy sauce, which is unusual in Chinese cooking. The canned versions pale in comparison to fresh.

Although unrelated to all these shellfish, sea snails, land snails, and other mollusks have also been common in these three cuisines. The ancient Romans considered them a restorative food, especially cooked in broth. Today they are cooked in tomato sauce. Garden snails, even those found in the United States, are easily processed. After a week or so in a barrel with cornmeal or other grain to purge them, they are boiled several times with several changes of water with a little vinegar to remove the slime and then removed from their shells and cooked and sometimes stuffed back into their shells. Snails also figure in Chinese soups, are eaten as snacks, and are used medicinally. In Mexico, snails were not only eaten but also their shells were burned to make lime for the nixtamalization processing of corn.

The polyp family, which includes squid, cuttlefish, octopus, and other relatives, is also important in all three of our cuisines. Apart from drying, polyps are normally cooked either very briefly so they do not become tough or recooked very long and slow so the flesh actually becomes tender again. In China, this process is often aided by turning the tubular body inside out and scoring the flesh in a diamond pattern, which ultimately makes it more tender and beautiful to look at. For example, in Cantonese cooking, little squares of scored squid are marinated in Shaoxing wine, soy, and sesame oil and then quickly stir-fried over searing heat with scallions and ginger and thickened with a little cornstarch dissolved in water. In Italy, calamari are cut into rings and lightly battered and fried quickly, or large squid may be stuffed and gently simmered in tomato sauce. The largest of squid can even be cut into steaks and grilled or battered and fried with a lemon-spiked sauce. The ink, especially of *seppia* (cuttlefish), can also be used to make a lurid black risotto or even a fresh black pasta when incorporated into the dough. Pounding serves to tenderize, and this is commonly done to whole octopus, which can then be marinated in olive oil, lemon, and herbs. Whole baby octopus is also fried as a common ingredient in the *fritto misto*.

Sea urchins are another strange sea creature that are eaten, sometimes even raw directly from the broken shell in Italy, where they are called *ricci di mare* ("sea hedgehog") on account of the spines. The best specimens taste sweet and are surprisingly subtle in flavor, sometimes even vaguely reminiscent of soft fruit or custard. They can also be gently cooked, either the whole animal or just the orangeish roe sacks, which are not actually eggs but the reproductive organs, which are considered, appropriately enough, an aphrodisiac.

Tortoises are a food uncommon today in the West but are popular in Asian cuisine and in the Caribbean. Sea tortoises, soft-shelled sea turtles, and freshwater snapping turtles and the like have been used in China since antiquity and are revered as a strength-giving food that aids in longevity, as turtles are thought to live many years. They are boiled to make soup or are stewed or roasted. The eggs are also eaten as a delicacy. Turtles were also eaten by indigenous Mexican peoples, as well as throughout the Caribbean. Turtle became a delicacy in the colonial era and afterward; the calipash (meat of the upper shell or carapace) and calipee (the gelatinous green substance lining the lower shell) featured on their own or in a turtle soup. The *caguama* is another Pacific turtle consumed until recently. The major concern over turtles is their endangered status, so as a food they are becoming rightfully rare.

Last, frogs must also be included in this roster of edible creatures. In Italy, unlike France, they were served whole, especially in Lombardy in rice-growing areas. Legs can also be used in a risotto or a soup. Frogs were prominent in pre-Columbian cooking of Mexico as well, and the tadpoles, according to early Spanish accounts, were considered among the best of all foods. Frogs are still commonly eaten in China, especially among rice paddies, and taste comparable, proverbially, to chicken. They are cooked in much the same way, in soup and stir-fries and quick braises. In Chinese markets one will often find live frogs as well, which are dispatched by cutting off the head, drawing out the entrails, and then dividing into four parts. These can be cooked any number of ways.

In Mexico the salamander (*ajolote*) is fried in a tomato sauce and eaten around the central lake regions. Although not a fish or amphibian at all, the iguana has been historically and remains a popular food in Mexico. Likewise, lizards and snakes are eaten in China, though they are more often found soaked in medicinal concoctions, which will be covered in the chapter on beverages.

Seafood in all three regions, as around the world, holds many promises and challenges for the future. We are advised to eat more fish in consideration of health, yet some species collect mercury and other toxins, especially the so-called top-feeders. Furthermore, it has become increasingly difficult to tell exactly which species are fished or farmed in an environmentally sound way, compounded by the fact that it is often difficult to tell where the fish comes from, and marketing very different species under similar names hardly helps. Chilean sea bass is only one example; it is actually a Patagonian toothfish. The challenge for our three cuisines will be to maintain indigenous traditions and recipes as local supplies are depleted, as techniques ideally adapted to local species are lost, and as fish is sourced globally. The homogenization of the supply of fish is already well apace; consider that China already gets much of its fish from Alaskan waters and no doubt Mexico and Italy will soon follow suit.

Study Questions

1. Describe fish farming both in the past and today and both why and how it has been undertaken. What benefits and costs are involved?
2. Why were preserved fish so important in Europe, and how were they indirectly responsible for Atlantic exploration?
3. How and why do some ingredients once considered fit for peasants later become esteemed and even expensive? What role does nostalgia play in the attempt to preserve dying traditions?
4. What role did the trade between Acapulco and Manila play in transferring ingredients and techniques?
5. Why are some seafoods considered aphrodisiacs?
6. Describe the difficult situation in fishing sustainably and the difficulty of negotiating health claims that encourage us to increase consumption.
7. What role has religion played in encouraging fish consumption in Catholic countries?

Recipes for Fish and Shellfish

Raw

CEVICHE/MEXICO

The most impeccably fresh fish is necessary here, preferably sushi grade if you can find it. The dish is actually native to South America but has been popular in Mexico for a long time (figure 7.6). The combination of fish is entirely up to you, but shrimp and scallops work very well, as does any firm-fleshed white fish like halibut or snapper, or even mackerel and tuna cut into small chunks. In Mexico ceviche is usually served with corn chips, avocado, a sprinkling of chopped cilantro, and chopped tomato. Or it can be served in a cocktail glass. The only essentials are plenty of lime juice; some chopped, mild, fresh chili such as jalapeno; salt; and chopped onions. You essentially just marinate the fish in the lime

Figure 7.6. A ceviche of tuna with green peppers and avocado served in a martini glass.

juice with the chili and onions, and the acid in a sense cooks the fish, or at least changes the protein structure so it becomes firm and sour. The marination should take at most 30 minutes—too long and the fish begins to fall apart and get mushy. Notice the balance of flavors here: salty, spicy, sour, pungent, and aromatic. And in the mouth there is the crunch of the vegetables and chips, the smoothness of the avocado, and the chewiness of the fish.

Cold

INSALATA DI MARE (SEAFOOD SALAD)/ITALY

Like much of the best of regional Italian food, this is an absolutely simple recipe and should use only the freshest fish available, in season and caught locally. The dish varies greatly from place to place, and thus making it exactly as it would be done there is not only impossible but also would necessarily vary from day to day. The predominant ingredient should be shellfish and mollusks. Shrimp are perfect, as would be larger scampi, if one could find them in the United States. Crawfish tails would not be amiss. Squid, cuttlefish, or small octopus are also requisite, though if using larger octopi, it is best to tenderize them by pounding on a rock. Small clams, *vongole veraci*, or in this country manila clams, can be used, as well as mussels.

The trick to this recipe is cooking each fish separately, because they have widely variable cooking times. Be sure to keep a bowl of ice water nearby to quickly chill the ingredients once they are cooked. The shrimp should be poached for about two minutes until just pink. The squid, cut into rings, should be cooked perhaps only 45 seconds to a minute. The shellfish should be cooked in a dry pan just until they open. Discard any that remain shut. If using a large octopus, simmer it whole until tender, about an hour, then peel the skin if it is tough and slice the tentacles and body. Baby octopi are much more interesting and can remain whole. They can be cooked much quicker as well. Exact timing for all these will depend on size, age, and many other factors, so do taste each as you proceed.

Once all the seafood is chilled, peel and devein the shrimp and remove clams and mussels from shells. Then toss with a light dressing of olive oil, lemon juice, salt and pepper, and a good sprinkling of chopped parsley. Some people also use some vegetables—julienned carrots or celery—and fennel goes perfectly, but it is not necessary. Never add garlic; it overwhelms the fresh taste of the seafood. Keep well chilled until ready to serve.

Poached

WEST LAKE SOUR FISH/CHINA

This dish comes from Hangzhou, the capital of the Southern Song dynasty in the Middle Ages, visited by Marco Polo and Muslim traveler Ibn Battuta, and probably the largest city in the world in the Middle Ages. It is made with freshwater carp from West Lake, traditionally kept alive in fresh water for two days to make the flesh white. Split the carp directly in half along the spine, but do not separate the two halves. Heat water in a wok and slide the split fish into the wok, boiling gently skin-side up for about 5 minutes or until the flesh is firm. If you need to, rearrange so both sides are evenly cooked, but do not overcook or it will fall apart. Remove the fish and lay it on a plate. Add to the water some black Chinese vinegar (Chingkang), soy sauce, slivers of ginger, a splash of rice wine, and sugar. Thicken the sauce with a cornstarch slurry and then serve the fish with the sauce poured over.

PESCE LESSO (POACHED FISH)/ITALY

Fish such as branzino or orata, or those more common in the United States like snapper or sea bass, are poached in a light court bouillon (broth made of vegetables) and often served with a sauce. Make the poaching liquid by boiling carrots, celery, onion, and parsley sprigs with some white vinegar and salt for 30 to 40 minutes. Strain and let cool a little. Poach the fish in this liquid at very low temperature, about 10 minutes per pound, or until the flesh feels firm. Remove from heat and let the fish cool in the bouillon. In a wooden mortar pound a clove of garlic with a handful of parsley and then slowly add olive oil until it forms a thick sauce. Serve the fish at room temperature, with the sauce spread on top. You can also simply drizzle with olive oil and lemon juice, or with mayonnaise. Save the poaching liquid for another dish as well.

Stewed

BACALÀ ALLA VICENTINA (COD VICENZA STYLE)/ITALY

Despite the name, this dish is actually made from stockfish (*stoccafisso*), which is merely dried but not salted cod, which comes from Norway. As an ingredient it has a long and important history, providing inland southern Europe with fish during Lent, and even coastal regions at the time when fishing was difficult in winter. Spelled *baccalà* with two c's, the term refers to salted, dried cod, here and elsewhere in Italy. Either salted *baccalà* or stockfish will work, but the latter will require longer soaking and some extra salt and perhaps even some pounding with a mallet to soften.

Start by soaking the fish, and changing the water a few times a day for at least 24 hours, and for true stockfish perhaps as long as 4 days until softened. Then make a *sofrito*, a chopped combination of sliced onions, a few anchovies and parsley, lightly fried in olive oil. Arrange half the *sofrito* in an earthenware casserole and lay the pieces of fish on top, then cover with the other half of the onion mixture. Then pour in milk, just barely covering the fish, and sprinkle on grated Parmigiano, despite what we are told about Italians never combining fish and cheese. Drizzle with some olive oil. Bake this in a very-low-heat oven for about 4 hours until all the liquid is absorbed. You may have to add a little liquid or white wine if it dries out too much. It should be served with polenta.

Fried

CALAMARI FRITTI (FRIED SQUID)/ITALY

We imagine fried fish as being heavy and sodden, but nothing could be farther from the results when done properly. Fried squid is usually featured in a *fritto misto*, which will include various types of seafood, but it can be done on its own as well. The key here is using olive oil—certainly not expensive, fruity, delicate extra-virgin, although there is no need to use the industrial-scale cans of chemically extracted regular oil, either. Just use a decent, inexpensive oil, in a deep pot heated to 350 degrees. It is not necessary to measure the temperature; it can be judged easily with a drop of batter. Herein lies the disagreement, though. Some will merely dust with flour and fry; others will coat heavily in a batter. Either is acceptable, but the coating must be both crispy and stay on the squid or it will be dreadful. Lighter is usually better.

If your squid is not cleaned when you buy it, remove the odd, clear, plasticlike strip inside and all the internal organs, including the eyes. Then cut up into rings; leave the tentacles whole and attached. You can just dust them with flour and fry in small batches until golden. The frying should take only a couple of minutes. Remove with a slotted spoon onto a rack to drain, sprinkle with salt, and serve.

Or, you can make a light batter. Some recipes call for egg, but this makes a soft, doughy crust. Preferably, just moisten some flour with white wine, or better, a vermouth or Cinzano. The bitterness and herbs work beautifully with the squid, and as alcohol evaporates quickly, the batter comes out very crunchy. Put all the squid into the batter and let sit a few minutes. Add a little more flour or wine if necessary. Drop pieces one by one into the hot oil and remove with a slotted spoon. Either way you cook these, serve with lemon wedges and nothing else. Marinara sauce makes no sense whatsoever.

HENG YANG SPICY SCALLOPS/CHINA

Henyang is in southern Hunan Province, so these include chili peppers but are otherwise uncomplicated. The key to the dish is to start with the best scallops that are sold dry, as opposed to floating in liquid. At a good Asian fish market you can find them in the shell, complete with the red roe, and it is a shame that this is rarely eaten in the United States. Divide each scallop horizontally and dry each piece thoroughly. This is crucial for browning, as a wet scallop will boil and become rubbery. In an extremely hot pan, simply sear the scallop halves in oil for about 30 seconds per side. Set aside. In the same pan fry some minced ginger, chopped red chili, and garlic. Add in some cloud ear mushrooms that have been soaked and chopped and some peeled and finely sliced water chestnuts, preferably fresh rather than canned. Add the scallops back in and simply moisten the dish with some soy sauce, a pinch of sugar, and a dash of vinegar. Serve at once.

COCONUT SHRIMP/MEXICO

Find the largest Gulf shrimp, remove the shell and devein, but keep the tail intact. Make a batter using a coarse-cooked grain like breadcrumbs or crushed corn flakes, or even pulverized corn chips, plus some dried, shredded coconut and a little chili powder. Dredge the shrimp in flour, then dip into beaten egg and then into the crumb mixture. Drop one by one into a deep fryer or pot with oil at about 350 degrees. Remove as they are done to drain on a rack. Serve a few per plate with some mango salsa. Simply chop up the mango flesh and mix with a finely chopped onion, some green chilies, lime juice, and salt. This recipe from Yucutan makes an excellent appetizer.

THE BEST STUFFED TENCH (CHRISTOFORO DI MESSISBUGO, 1549, P. 106)/ITALY

Take a good tench, and scale it, leaving the head attached to the skin, first opened down the middle. Then remove the good parts, cleaned well from the spine, and pound them finely with a knife, as you would do lardo to put in a soup, and pound dry oily herbs, then take pounded walnuts, a bit of hard cheese, and beaten egg yolks, and good spices, and whole well cleaned raisins, and mix everything well together, and with this mixture fill the skin of the tench from the side carefully, and if you add in a bit of finely chopped garlic, it won't be unbecoming. Then bind the tench to a thin board and fry it in good oil, and when it is fried, take away the board and garnish, placing on top oranges, or vinegar, or *cisame*, or *cameline* sauce, or *galantine*, or something else that pleases you for variety.

The tench is a freshwater fish somewhat like carp. Any similar fish with sturdy skin can be used. The directions require the deft removal of the entire skin, opened from the bottom, as fish are normally cleaned. Apart from returning to the table looking like a whole fish, there are no bones, and it can be easily portioned. The *cisame* is a sweet and sour sauce including honey and vinegar and often almonds, which is typical in northeastern Italy in this era. Cameline is a cinnamon-based sauce—Messisbugo's includes a pound of raisins, three crustless breads toasted, strong vinegar, a carafe of dark red wine, all sieved and added to a pound of honey with 1 oz cinnamon, ½ oz pepper, ½ oz ginger, ¼ oz cloves cooked gently with another pound of whole raisins. It is remarkably similar to Indian chutney. Galantine is similar sauce, though his version includes sugar and saffron. In final effect, note how similar this is to the Chinese West Lake fish discussed above.

Steamed

STEAMED FISH BALLS/CHINA

These are commonly found in dim sum restaurants, and unlike *quenelles* or dumplings made from fish, which are today ground in a food processor and were formerly pounded in a mortar, the proper consistency of the paste from which these are made is achieved by throwing the meat down rigorously into a bowl, creating a splat sound rather than a dull thud. As in much Chinese cooking, the correct technique is judged by the proper sound, which in this case renders the fish flesh completely tender. In this case you should use dace, which is a freshwater carp, or another light, white fish. Remove the head, tail, and skin and reserve for fish stock. Remove the fish flesh from the bones with the edge of a spoon, which will finely shred it. Throw vigorously with full force into a bowl repeatedly until the flesh disintegrates. Add to it egg white, salt, white pepper, a finely minced scallion, cornstarch, and a touch of sesame oil, and mix well with your hand. Roll into small balls. These can be placed on a lettuce or cabbage leaf and set in a bamboo steamer, for the lightest and most delicate results. Alternately they can be dropped into fish stock and poached, and they can even be deep fried. Fish balls such as these also figure in mixed dishes, with vegetables, as soups, or they can be served on a skewer as a street food.

HUACHINANGO A LA VERACRUZANA (VERACRUZAN RED SNAPPER)/MEXICO

Take one large red snapper about 3 pounds in weight, cleaned and scaled, but leave on the head and tail. Slash the skin on each side diagonally. Rub the fish with lime juice, salt, and oregano. Let this sit about half an hour. Then place in a large frying pan, with some heated olive oil. Add a finely sliced onion alongside the fish. Sear on both sides, and then add a good deal of chopped tomatoes, some pitted green olives and capers, and pickled jalapeños, sliced. Let the sauce come together as the tomatoes cook down and as the fish cooks through, on medium heat for 15 to 20 minutes. Baste with the sauce often. Serve.

⁂{ 8 }⁂
Fats and Flavorings

Learning Objectives

* Think about the indispensable role of fats as a carrier of flavors as well as a cooking medium.

* Appreciate the wide range of spices, their importance in history, and how they are used today and why.

* Consider the way spices and herbs are combined to create signature flavor profiles in each of the three cuisines.

* Speculate why some cuisines use spices liberally and in complex combinations while others relegate them mostly to desserts.

* Understand the humoral logic of adding spices and other flavorings to food.

* Appreciate the role of condiments not only in flavoring foods but also in stimulating the appetite, serving as digestive aids, and in improving the nutritional value of certain ingredients.

IN THIS CHAPTER we will explore fat-based cooking media, which also convey flavors as well as seasonings and ultimately how these work together to create flavor profiles in Italian, Chinese, and Mexican cuisines. As we shall see, although many of the building blocks of these cuisines differ, the ultimate effects are often comparable. The careful balancing of salt, sweet, sour, and bitter flavors as well as aromatic components such as herbal or spicy notes, and perhaps smoke, is combined with "mouthfeel" properties such as the acrid pungency of garlic or mustard or the heat of chilies or black pepper. These similarities reveal certain basic underlying rules of logic that inform cooking in these cuisines. These principles also explain the logic of sauces, whether the piquant *pico de gallo*, the soy and vinegar dipping sauce for dim sum, or the Genoese *bagna cauda*.

Fats

After water, **fats** are the most important cooking medium in all world cuisines. They provide necessary calories and nutrients as well as distinctive flavors characteristic of each cuisine. Fats also capture seasonings and transmit them to the tongue. Fats may be derived from either animal or vegetable sources. They may be simply excised directly from an animal and used for cooking or cured for longer storage. In the case of plants they can be pressed from seeds or fruits such as olives or highly processed with industrial methods of extraction. Fats have variable smoking and burning points, so some are appropriate for sautéing and frying, while others are merely applied as a condiment or added to cooked dishes for flavor. For each cuisine the flavor and texture of the preferred fat are more than incidental; they lie at the base of the basic flavor profiles, and although soy and other bland vegetable-based oils have frequently replaced more traditional fats in the interest of health, in terms of gastronomy this is only to be lamented. Another consideration, the issue of saturated fats aside, is that all fats have the same calories per volume. Furthermore, some fats that were originally marketed as healthy are now considered exactly the opposite—hydrolized vegetable shortening and transfats in particular. With this in mind, we discuss the positive attributes of traditional cooking fats.

Lard

Practically any animal fat can be used for cooking, and in the process of roasting the external and internal fat melts and in a sense bastes and moistens the flesh. This is also the case with boiled dishes and soups that contain fatty flesh. Here, however, we are concerned with fat that is intentionally removed from the animal to be used in a different context. Of all animal fats, pork fat has been the most significant historically. This is partly due to the large deposits of fat that accumulate in the belly of the pig and elsewhere, which are easily removed intact and either salted or cured. Ruminants do not carry such large fat deposits—with the exception of fat sheep's tails. The popularity of pork fat can simply be attributed to its abundance wherever pork is raised, its ability to be stored at room temperature, and its flavor and texture. In the wake of years of nutritional admonitions against consuming saturated fats such as **lard**, which largely served the interests of margarine producers, lard is enjoying a minor resurgence, not only because of its unique flavoring capabilities but also as a natural and traditional product, which when used in moderation need not be banished entirely from a healthy diet. Given the sometimes erratic swings in opinion, future nutritional findings may very well promote lard as the ideal fat.

Lardo in Italy refers to slabs of fatback cured in marble tubs and then thinly sliced and served on bread or melted in a pan for frying. The most famous of these is *lardo di Colonnata* from Tuscany, usually eaten raw as an appetizer. Pig fat can also be rendered fresh and cooked down as *strutto* (similar to what in the United States is labeled lard, though not usually hydrogenated) or beaten, called *lardo battuto*, into a spreadable white form and sometimes flavored with salt and herbs. The fat may also have some meat attached, as with *pancetta*, a rolled, cured, and usually unsmoked bacon, or *guanciale*, made from pork jowls. These are chopped and melted in a pan, which provides a base in which to fry onions, garlic, carrots, and celery in a classic *battuto*, sometimes with peppers and tomatoes. This forms the base for various fricasséelike dishes using meat or vegetables, as well as soups. The remaining bits of meat flavor the dish after the fat has melted. Also worth mentioning is caul fat, the lacy membrane lining the viscera of the pig. This has been used traditionally to wrap around meats while roasting to provide a continuous baste as the fat globules melt. It can also be wrapped around meatballs with the same effect. In areas of Italy where pig rearing has been practiced, primarily in the center of Italy but also in the north and south, lard in its many forms was the traditional fat of choice.

Lard has also been important in Chinese cookery and is used to make pastry, though for cooking, oil has gradually been replacing lard for supposed health reasons. A relative of bacon called *lop yuk* (*la rou* in Mandarin) is cured in soy and Shaoxing wine and used much the same way as in Italian cooking, not eaten in crispy, thin slices as we do bacon, but as a flavoring agent. It may be used to flavor rice or green vegetables such as mustard greens, either added to fried dishes, or even steamed in others. Sometimes this bacon is also smoked, which further dries the flesh and helps prevent bacterial growth. Fat is also something appreciated for its own sake in Chinese cuisine, not merely an adjunct to meat, but glistening cubes of unctuous adipose tissue, crispy on the outside and soft and yielding within. One must keep in mind that in all cultures with a relatively low caloric intake and great energy output in daily labor, fat is considered a food not to be wasted or squandered, but reveled in and appreciated for its own sake.

Since the sixteenth century lard has also been a key ingredient in Mexican cooking, both for frying and for making *masa* dough, the base for tamales and tortillas, as well as making molés. The lard preferred in Mexico is made by slowly rendering bits of fatty pork in a big pot over low heat, lending it a brownish color and deep flavor. The crunchy bits left over (*chicharrones*) are eaten as a snack, similar to the *ciccioli* of Italy, but can also be incorporated directly into cooked dishes with the fat. The white, clarified, and practically flavorless lard sold in U.S. markets is a poor substitute. The Mexican use of lard is taken directly from Spanish cookery, but the context is often quite different. Many indigenous dishes would have been virtually fat-free originally, and there was even widespread reported aversion to fat among the Aztecs, who only gradually began incorporating lard as a cooking medium or flavoring. The many recipes that call for lard might rightfully be considered of mixed heritage.

Suet and Butter

Suet is the fat of bovines, normally shredded and used to make pie crusts and puddings. The finest is said to come from around the kidneys. In flavor it is not entirely dissimilar to butter but does take on a beefy flavor and deep caramelized color in cooking. It does not play a major role in our three world cuisines, only insofar as beef itself is used, though there are exceptions, as in the northern Mexican cattle ranges. The most likely reason is that beef fat, along with mutton, was rendered for tallow, used for making candles, and mixed with lye for soap. Interestingly, the fat was reserved as a perquisite to professional chefs in Early Modern Italian kitchens expressly for this purpose.

The more common fat derived from cattle is, of course, **butter**, and it can be made from cow's milk, which is most common, or sheep's and goat's milk. Butter was probably independently invented around the world at various times in history, though it was known to Indo-European herdsmen who brought it to the Middle East, Europe, and India in ancient times. Likewise, the herdsmen of western Asia and Tibetans with their yaks made butter. The process is utterly simple. The milk is left to sit for a while until the cream naturally separates on top. This is skimmed off and whipped or churned until the fat molecules break and realign in solid clumps that separate from the liquid or buttermilk. Buttermilk, despite the name and association, is a low-fat food and is sour and thick because of bacterial cultures. The butter clumps are then drained, kneaded, and washed to remove any remaining water. At this point it may be salted, which extends shelf life. The butter most familiar in the United States, called sweet cream butter, is made from fresh pasteurized milk and thus is relatively neutral in flavor. In Europe butter is often cultured; that is, made from slightly fermented milk, giving it a certain tang, which is probably closer to the original farmhouse butter that was unpreserved and unrefrigerated. Good artisanal butter tastes of whatever the cows have been eating seasonally, even picking up a yellow tinge from certain plants. Sometimes there are even noticeable haylike barnyard flavors, just as with cheese. Butter can also be clarified, which involves slowly heating to evaporate excess moisture, skimming off the foam that forms on top, and then straining out the browned milk solids. The result is that it lasts much longer and has a higher smoking point, so it can be used for frying. Though this technique is usually associated with French cuisine, it is also used in northern Italy and sometimes in Mexico.

In cooking, butter is foreign to East Asia, and this again is the result of a lack of dairying tradition, though in the western provinces of China and especially among

Muslims, one finds many dishes cooked with butter, as well as flat breads brushed with butter and herbs. Salty butter is also used by Tibetans in their tea. In Mexico its use is largely due to European influence—it is spread on bread and cooked in dishes, mostly of French ancestry. Culinary historians often trace the prevalence of butter in cooking to early classical French haute cuisine, particularly to La Varenne, Massialot, and the cookbook author known only as L. S. R., who in the seventeenth century introduced butter-based sauces and the classic roux—a combination of butter and flour cooked and used as a base for other sauces. But it is important to remember that a full century earlier butter had been at the center of Italian cooking. Two historic cookbooks, Messisbugo's *Banchetti* of 1549 and Scappi's *Opera* of 1570, use butter with a heavy hand that we do not normally associate with Italian cookery. This was not only in baked goods but also as a cooking medium and a flavoring for cooked dishes like pasta and chicken along with sugar and spices like cinnamon. Italian cuisine has obviously evolved and veered away from an obsession with butter, but it remains the case that Italian cookery, especially in the north, does use butter in sauces and for cooking in ways that are uniquely Italian and not merely borrowed from the French. A good risotto would shrink without the good hit of butter at the end, and one can find pasta dressed with a simple butter and herb mix as often as with garlic and oil. The classic fettuccine alfredo is merely egg noodles with a lot of butter and Parmesan cheese—though in the United States cooks usually add cream as well. We must also consider the role of butter in classic pastries and pie crusts.

Other animal fats such as duck and goose fat, although common in regions such as the southwest of France, are not very important in our three cuisines. Whereas France can be easily divided into regions that prefer butter, oil, or animal fat, and even this is a stereotype, it is important to note that today most people have access to a variety of fats in all locales. In Italy certain dishes are always prepared with pork fat, others with olive oil, others with butter regardless of region. While oil predominates in China, some recipes call for animal fat, and the same holds true for Mexico. So a simple formula for preferred fat for each region is not possible.

Olive Oil

Western culture is currently in the throes of an ardent love affair with **olive oil**. This is partly due to the fascination with Mediterranean cuisine in general, but also the scientific recognition of olive oil as one of the healthiest sources of fat in a diet, as it is mostly monounsaturated; that is, liquid at room temperature and with antioxidant polyphenols, it is less likely to oxidize or clog up arteries. Olives have been pressed in Italy since antiquity, either broken with a hand mill and weighted with heavy stones or crushed with sophisticated millstones and pressed with complex levering systems through various filters. The process of making what is cold-pressed extra-virgin olive oil is unchanged, though now machines do most of the labor. Traditionally it is pressed in a tall, cylindrical tower of round fiber mats, lined with olive pomace, which is pressed hydraulically (figure 8.1). The crushed juice of the olive must then be decanted or separated from the bitter water

Figure 8.1. Men operate an olive press to extract the oil at a factory in Italy. © Vittoriano Rastelli/CORBIS.

and residue in a centrifuge, and after a few weeks to clarify, it is ready for consumption. The centrifuge prevents oxidation of the oil that may occur in slow decanting.

Oils differ significantly depending on the variety of olive used and the state of maturity of the fruit, which should be just ripe or nearly so. Some oils, such as that of Lucca in Tuscany, have a peppery afterbite in the back of the throat. Others, as from Liguria, are sweeter and more mellow. Some have a grassy, herbaceous flavor, while others are fruity, like that of Apulia. The flavor depends partly on the level of acidity, which legally must be below .8 percent to be labeled extra-virgin. These oils are best used as a condiment, drizzled over a salad with vinegar, droplets anointing a crust of bread, or even added to a plate of food at the very end to add a touch of unctuous piquancy. American diners are sometimes shocked by the ways Italians use olive oil, as a table condiment more than a cooking oil, for example, drizzled on pizza, into a soup, or over a piece of meat. Olive oil does figure in cooking, too, and is used for most recipes involving sautéing or frying and is even used in baked goods such as cake. But the finest olive oils, in fact, should not be cooked. Their subtleties of flavor and aroma are destroyed by heat.

Beneath these are lesser oils processed with heat or chemical solvents to extract every last drop. These are usually used only for cooking, especially when frying in oil, and using the best would not only be a waste of money but also would ruin the flavor. These oils have a slightly higher smoking point as well. Contrary to popular belief, one can deep-fry foods in regular olive oil, and this is common in Italy, especially for battered fish and vegetables or artichokes. What is labeled as merely olive oil has been processed with heat and solvents, while virgin olive oil may be heated but not refined with chemicals. It can also have a higher acidity than extra-virgin, which some people enjoy. The term 100 percent pure olive oil is meaningless, as this usually contains refined oil and is scarcely different from plainly labeled olive oil. The term *lite* or *light oil* is essentially a travesty, being light only in flavor and color but not in calories. All olive oil is fat and has the same caloric value by volume—120 calories per tablespoon.

After Spain, Italy produces the most olive oil in the world, but again, it is important to read labels carefully. Much oil labeled Italian has merely been processed there and may have been grown in northern Africa or elsewhere. This does not necessarily mean it is inferior, but when looking for oil with a distinctive taste of place, one can be easily misled. Olive oil is a very recent and not much appreciated addition to Asian cuisines, though one does find health-conscious cooks using it. It has been an import to Mexico for hundreds of years. Though olives are grown for oil today, most Mexicans have always bought their oil from Spain, and it was an intentional imperial policy to discourage production in the colonial world so as not to damage Spanish exports. This changed after independence, and Mexicans began to grow olives extensively in the nineteenth century and introduced the olive to California as well.

Sesame Oil

Sesame oil occupies a comparable culinary niche in Chinese cooking as olive oil does in Italian, though it is by no means as ubiquitous in usage. There are relatively light and refined sesame oils that are suitable for frying, but more important is toasted sesame oil, which is darker and intensely fragrant. This is used as a condiment, to flavor fillings, or drizzled onto a vegetable dish at the last moment. Ironically, sesame is not native to Asia, but rather Africa, or, according to some authorities, India. In all these places sesame seeds are pressed into oil, ground into a paste (similar to the Middle Eastern tahini), and made into sauces. Sesame seeds can be used on their own too, the various colors (red, white, or black) lending both flavor and visual appeal to any number of dishes. Sesame oil is uncommon outside Asia, but the seeds on their own are used extensively in Italian and Mexican cuisine, normally atop baked goods like cookies and pastries or bread. In Mexico they are also ground with other seeds and nuts in sauces.

Peanut and Nut Oils

Although native to South America, the peanut was introduced to China soon after direct contact with Europe in the sixteenth century. Peanuts are as elsewhere a common snack food and are incorporated into many savory cooked dishes to lend a contrasting crunchy texture, or ground into a paste used in sauces. **Peanut oil** is very important in Chinese cuisine, especially in Cantonese cooking. Its nutty flavor and high smoking point provide an ideal medium for stir-frying. Keeping in mind that peanut oil is expensive, usually more so than other oils, and that even though boiling and steaming are far more common cooking methods, peanut oil is nonetheless ideal for quick, intensely hot cooking in a wok. Olive oil and many other oils with less flavor would degrade before the ingredients cooked, and this probably explains why olive oil has not been important in China.

True **nut oils** have a low smoking point and are almost never used in cooking. Walnut oil, almond, hazelnut, and more exotic types such as argan oil from North Africa are used mostly in dressings or as a condiment. Their intense aromatic flavor is best used sparingly, on salads, or to add depth to a cooked dish or vegetable. Some of these do make their way into pastries, though, particularly almond oil.

Soy, Corn, and Other Vegetable Oils

Among the most common fats on earth today, normally labeled merely "vegetable," is the oil processed from soybeans. Its prevalence is not the result of traditional Asian cookery, as other soy products are. It is a modern industrial invention. Oil cannot be simply pressed from soybeans like oil from olives or sesame. To extract oil from soy, the beans must be dried, cracked, rolled flat, and then mixed with the solvent hexane. The resulting oil, however, has a high proportion of linoleic acid, which makes it prone to oxidation; that is, going rancid. Thus, much of the oil is hydrogenated, making it into "shortening." This basically involves forcing hydrogen molecules into the oil so it becomes semisolid at room temperature. The structure of the fat in the process changes into what is now called a transfat, which cannot be processed by the human body and poses a serious danger to health. Nevertheless, because of the shelf stability of products made with shortening, it remains a major ingredient in processed foods around the world, especially mass-produced pastries, snack foods, and crackers, though this is beginning to change. Fluid soybean oil, if perhaps less dangerous, is used just as often in fried and baked goods, as well as in ordinary households. It has practically no flavor whatsoever, but being cheap has become common, in China as well as among Mexicans and Italians, who were not very long ago sold soy oil as a heart-healthy, unsaturated fat. The majority of the world's soybeans are grown in the United States, and much of these end up as soybean oil.

The use of corn oil is somewhat similar. It is processed using highly industrialized methods from field to plate, from the fertilizers and pesticides used in growing corn, sometimes genetically modified corn, to the harvesting, shipping, and extraction of oil. Through expelling and solvent extraction plus heavy refining, the modest percentage of oil in the corn kernel yields a golden oil with a high smoking point but relatively little flavor. The reason both soy and corn have become prevalent around the world is due to the massive volumes grown in the United States, subsidized by taxpayers.

The range of other vegetable oils are also today processed with similar methods, and on a huge scale. There is rapeseed oil, which has a longer historical pedigree but today is mostly grown in North America and is familiar by the Canadian name canola. It is oil pressed from seeds of the rape plant in the Brassica family, a cousin of turnips. The value of these plants is partly in crop rotation systems, but also the use of oilseed cakes as fodder and fertilizer. Thus, they are very valuable to farmers but perhaps less so in the kitchen. There are also oils pressed from cotton seeds, safflower, and the North American native sunflower. These all share the benefit of being mostly unsaturated fats, which are judged to be healthier than saturated fats from animal sources or from tropical plants like coconut and palm. We have yet to learn the long-term health effects of these highly processed and often genetically modified plants. In culinary terms, these oils are increasingly used around the world, including our three regions, again, because they are cheap, have a long shelf life, and remain stable at high heat. These advantages may well be offset, however, by their meager gastronomic value, and the replacement of traditional fats with vegetable oils is often to the great detriment of taste.

The exception in this regard, perhaps the only one, is avocado oil, which is a green-tinged and flavorful oil used normally for dressings and condiments, though surprisingly it can be used for deep frying as well. Pumpkin seed oil is a dark, intensely flavored seasoning and, like avocado, is another American native.

Herbs and Spices

The distinction between **herbs** and **spices** is often arbitrary and sometimes confusing. In general culinary parlance, herbs refer to the leafy green parts of plants used fresh or dried that are grown in temperate climates, historically obtained locally and at little cost. In other words, they are plants that grow wild or in gardens and are snipped as needed by the cook, in wealthy and more modest households. Today they are increasingly imported from around the world and sold dry in jars. Spices, in contrast, are typically dried bark, roots, and seeds of plants that are for the most part grown in tropical climates and thus have always been exotic imports to Europe, as well as to China and Mexico. The confusion arises because some plants are an herb when the leaves are used fresh, as with cilantro, but a spice when its seeds are used dried, called coriander. Furthermore, some spices are grown in temperate regions, but are expensive, as in the case of saffron, or inexpensive, as with mustard. Ginger, grown in China and used fresh, is considered a spice, mostly because in Western culture it was known in dried form as an import.

It is also important to remember that herbs and spices have in all three cultures served the dual purpose of culinary embellishment and medicinal corrective. In the Western system of humoral physiology, plants were categorized by their ability to heat and dry or cool and moisten the body. This category was usually determined by flavor rather than actual temperature in degrees. That is, a spicy "hot" flavor—and we still use this term in a fundamentally humoral way—would burn the tongue and heat and dry the body. Bitter flavors were considered "dry" in the same way we call a martini dry. Sour flavors were thought to be cooling, as we know well from a mouth-puckering glass of lemonade to counter the summer's heat. Medicines were categorized with the same logic, and thus spices might be considered quasi-medicinal foods, which counteract the body's own imbalances and also, fascinatingly, correct recipes that would themselves be considered humorally imbalanced. That is, a hot and dry spice like pepper might be used to balance a cold and moist ingredient like fish, and a sour or cold and dry food might be used to counteract a hot and moist one. Many historians have argued that this humoral logic underlies many culinary traditions and even indirectly influences our own taste preferences today.

China has its own ongoing tradition of medicinal use of herbs and spices, both for illness and to gently rebalance the forces of yin and yang in particular dishes. Yin and yang refer to the two primordial forces that govern the universe and the human body. They are polar opposites but depend on each other, just as hot and cold, dry and moist, soft and hard are relative terms rather than absolute qualities. A patient diagnosed with an excess of yang might be prescribed a soft and cooling food considered yin to bring the body into homeostasis. Thus, as in the West, Chinese cuisine has a complex theoretical underpinning that still influences basic food combinations. Foods are also categorized by their ability to increase *qi*, which is the basic force that strengthens the body and enables it to fend off illness, and many medicinal soups are eaten expressly for bolstering *qi*.

The Aztecs also routinely categorized foods as hot and cold, and this survives in Mexican folk medicine. The difficulty in reconstructing the original Aztec system lies in the fact that the Spanish introduced the superficially similar humoral system that was subsequently laid over the indigenous system, creating a hybrid, much like the cuisine itself. In any case, many Mexicans still routinely think of foods as hot or cold and herbs and spices are still commonly used in cooking to balance dishes, even unwittingly, according to a medical system of native and European ancestry.

The discussion that follows covers the major plant-derived seasonings in each of the three cuisines and suggests how they have been used traditionally and in the present. Readers who would like an encyclopedic coverage of world herbs and spices are advised to consult one of the reference works listed in the bibliography.

Mediterranean Herbs

The range of **herbs** grown throughout the Mediterranean is staggering, and all are used extensively to add flavor, aroma, and color to cooked dishes, in pickling brines, used fresh in salads or with cheese, sprinkled on a pizza—their uses are absolutely ubiquitous. Historically they have been used simply because they grow everywhere wild, and for those with limited budgets, herbs are a source of free flavoring, livening up practically any savory dish. Nowadays, they are, of course, cultivated and purchased, but many Italians still keep their own kitchen garden, which may be no more than a few pots of herbs on a balcony. One can scarcely even imagine Italian cuisine without oregano in the tomato sauce; basil pounded into a pesto; parsley chopped with a *mezzaluna*, liberally adorning the plate; or rosemary flavoring a roast. It is no doubt this traditional dependence on fresh herbs that in part makes Mediterranean cooking fundamentally healthy. It depends not on fat-based sauces or stock reductions but on the vibrant, fresh flavor of herbs. Many herbs also dry well, and this naturally was a way to store them out of season. Their flavor in some cases even intensifies, though it is decidedly different from when fresh. Oregano, for example, holds up very well when dried, but parsley loses all its flavor, unless freeze dried, which is a poor substitute for fresh. Basil also lacks luster dried but is still used in a pinch for sauces. The best way to use herbs is, of course, to pick them and put them immediately into the pot, something quite feasible in a Mediterranean climate, practically year-round.

The principal herbs used in Italian cooking are oregano and its cousin marjoram, basil, parsley, and rosemary, to which should be added sage—especially in pork dishes—and thyme, and to a lesser extent mint, tarragon, savory, and chervil. In former days one might also have found artemisia species (both wormwood or absinthe and southernwood) and rue (a bitter but alluring herb) as well as tansy, though these more frequently made their way into medicines and liqueurs (figure 8.2). So, too, were fennel fronds used as well as the seeds, the swollen base as a vegetable, and even the pollen from the yellow flowering heads, called *fior di finocchio*. Along with these are a welter of wild and local herbs picked fresh and used both in traditional folk medicines and in cookery, such as pimpernel, lovage (similar to celery leaves), borage, hyssop,

Figure 8.2. Rue is a bitter herb once widely used in Italy for flavoring pestolike sauces.

lavender, and violet, plus many other flowers once used extensively in cookery, such as rose. Today rosewater is one of the few remaining ingredients of what was once a long list of floral flavorings.

Mexican Herbs

Mexicans today use the full range of native species and herbs brought from Europe since the sixteenth century. The appreciation for the traditional species above all is still strong, and substitutes from a jar do not always achieve the desired effect.

Epazote is perhaps the most distinctive native Mexican herb, used extensively simply because it grows everywhere (figure 8.3). Its botanical name *Chenopodium ambrosioides* means "ambrosia-like goosefoot," which it somewhat resembles, as does its alternate genus name *Teloxys*, which means "pointy tip," again close but not exact. Imagine a long, serrated, spearhead-shaped leaf with a pointy tip. Its flavor cannot be adequately compared with that of any other herb, either. The translation of the original Nahuatl name gives some suggestion: "sweaty skunk." Its most common use is with beans, and it is said to prevent flatulence. *Epazote* is an integral part of Mexican cooking and well worth trying, especially fresh.

Figure 8.3. *Epazote* is widely used in Mexican cooking and for an herbal tea, shown here.

Cilantro or coriander leaf is one of the herbs frequently associated with Mexican cooking, even though it is native to the Mediterranean. Interestingly, it is even more popular in Asian cookery, hence known sometimes as Chinese parsley, though used infrequently in its homeland. The fresh leaves have an appealingly soapy aroma that melds nicely with lime juice, which probably accounts for its use in salsas such as guacamole. Usually underappreciated, much of the flavor is in the stems and roots, and these should not be discarded. Why exactly this plant became so popular in all Latin American cuisines must be a fortuitous accident, as it vaguely resembles several native species. It probably stood at first as a substitute, and since it grows so prolifically and easily, it gradually pushed out the native plants to some extent.

Not to be confused with this is the similar sounding culantro (*Eryngium foetidum*), a long, thin, serrated leaf, grown mostly in southern Mexico and native to the Caribbean. It is not eaten raw but rather typically cooked in soups and stews and tastes similar to cilantro. It is also used throughout Southeast Asia.

Hoja santa (holy leaf, *Piper auritum*) is a large, spade-shaped leaf grown in Veracruz and the south used both to season sauces such as an uncooked *molé verde* and meat or rice dishes or even to wrap tamales. Some compare the flavor with root beer or aniseed, but it is in fact a cousin of black pepper. The comparison to root beer is actually not far-fetched, as it is safrole that gives both this plant and sassafras, from which root beer was once made, their distinctive aroma. Its flavor goes especially nicely with fish wrapped in the leaves and baked and served with tomato sauce.

There are also a number of thin, leafy herbs reminiscent of tarragon, such as *hierba de conejo* (*Tridax coronopifolia*), *chepiche* (*Porophyllum tagetoides*), and the round-leafed *papaloquelite* (*Porophyllum ruderale*). All these pungent herbs are used fresh and generally are unavailable outside Mexico. Another distinctive herb is the avocado leaf, which is toasted and ground and added to bean dishes or to season meat. It has an aniselike flavor. So-called Mexican oregano may also be a number of different herbs and is generally different from Mediterranean oregano. It is usually *Lippia berlandieri*, a relative of verbena, but in local usage many relatives as well as completely different plants are also called oregano.

Chinese Herbs

The Chinese have used herbs in medicine for millenia, stretching back to the mythic celestial emperor Shennong, who tested every available species on himself. Today the Chinese enjoy what is no doubt the most extensive and complex pharmacopoeia on earth. It is difficult to distinguish between herbs used solely for therapy and those used in cookery because the two overlap so thoroughly. Some familiar herbs are used principally in cooking, though. Basil, both the species familiar in the Mediterranean and holy basil, probably came to China via India in Tang times or earlier. Both are used in soups and go especially well with noodles. Unlike in the West, though, they are rarely chopped, but merely added whole to a dish at the last minute of cooking. Cilantro is

Figure 8.4. Lily buds, the unopened flower of the day lily, dried, are a common ingredient in hot and sour soup.

another now-ubiquitous herb, introduced from the West, whose leaves and roots are used in soups and stir-fries. It is often paired with scallions and ginger in a dipping sauce or strewn onto steamed fish to improve its aroma. The easy adoption of these plants reflects the general attitude of Chinese culture toward all living creatures, every single one of which serves some culinary or medicinal function. Unlike Western civilization, which at first looked askance at strange, exotic plants, the Chinese welcomed everything.

Apart from these imports, China also abounds in native species. Lily buds ("golden needles") are a native herb common especially in the north (figure 8.4). The dried buds are soaked and then added to dishes like hot and sour soup or mu shu pork (*muxi rou* in Mandarin). For medicinal soups, a variety of herbs including these is often purchased prepackaged, though traditionally it would be supplied by an apothecary. One such mixture is called *qing bu liang* ("to clear, restore, and cool") and contains seven herbs: Job's tears, lotus nuts, dried longan, lily buds, Solomon's seal, and Chinese yam. These are herbs in a medicinal rather than culinary sense. Other Chinese culinary flavorings will be discussed under spices, because they are usually dried.

Spices

Although **spices** offer practically no nutritional value, they have always been highly sought after. They have always been expensive, and among all foods are a marker of status and distinction. For most of recorded history they have been imported at great expense over thousands of miles, stimulated voyages of exploration, and incited colonial wars of conquest. Columbus's accidental "discovery" of the New World was initially motivated by the quest for Asian spices; he was searching for a westward sea route to China, the supposed source of spices. The Portuguese had at about the same time found a direct sea route around the Cape of Good Hope on the southernmost tip of Africa, across the Indian Ocean and then directly to India and Indonesia. The question remains: Why would Europeans go to such lengths to obtain a seasoning ingredient?

Spices, pepper, in particular from India, but also cinnamon from Sri Lanka, nutmeg and cloves from the Moluccas, ginger and galangal root from Southeast Asia, as well as a handful of other spices less known today were all at the center of medieval cookery. Many of these had been known since antiquity, but with the revival of Mediterranean trade after 1000, aided in part by the brief conquest of the Holy Land, Europeans developed a seemingly insatiable hunger for the exotic luxuries of the East. Spices were among the very few luxury items available for those with disposable income. They were lauded by physicians as powerfully hot and dry medicinal drugs, suitable as an aid to digestion and as a culinary corrective. In fact, Europeans had little clear idea exactly where these spices came from. They were carried either overland along the various roads, which have come to be known as the Silk Road, or shipped around the coasts of the Indian Ocean in small dhows by Arab merchants. By the time they arrived in trading depots like Alexandria, Beirut, and elsewhere in the Middle East, their cost had become astronomical. They were then picked up mostly by Venetian merchants who further inflated the price before redistributing through the rest of Europe. In many ways, Venetian cuisine retains the medieval penchant for spices used in surprising contexts.

It is by sheer chance that of all spices pepper is the only one remaining as a universal flavoring on savory dishes. At one time one might have found cinnamon sprinkled liberally on food throughout Italy on aristocratic tables, along with sugar, which was no less valued as an exotic import. There were sauces based on dense combinations of spices such as ginger, nutmeg, cassia, grains of paradise, or melegueta pepper from the west coast of Africa, as well as long pepper (a little catkin-shaped pepper relative) and tailed pepper or cubebs, another relative, spiked with vinegar and thickened with breadcrumbs. These combinations conjure up modern Indian cuisine, which was actually influenced by the same source as medieval European cuisine—medieval Baghdad and Middle Eastern cuisine in general—though many of these spices are originally from India, too. These very same spice combinations were brought to Mexico with the conquistadors. Thus, we find cinnamon in chocolate or Asian spices in molés. Mexico also established its own direct trade with Asia with galleons crossing the Pacific from

Acapulco to Manila. There they brought silver and traded it for spices, among other luxuries.

The Chinese were equally interested in spices imported from their neighbors to the south. Apart from indigenous or proximate spices such as ginger, cassia (a relative of cinnamon), and star anise, all of which continue to be used extensively in China, they also traded extensively in Southeast Asia for pepper, cinnamon, cardamom, and turmeric, and to the West for saffron and coriander. Thus, it is safe to say that thanks to their rarity and relative imperishability, spices were the first food item of consequence in global trade, which in many ways presaged the traffic in foods predominant in the modern era. But it remains to be seen how these spices are used today.

Spices in Italy

In Italian cooking, spices have retreated into the background since the eighteenth century. This was in part due to the influence of French cuisine, which had already begun to marginalize spices to desserts and sweets, main courses having depended on herbs alone and used merely to flavor intense, meat-based reductions, with the *bouquet garni* (a tied bundle of herbs such as parsley, thyme, and bay leaves) removed. It is also true that the greater volume of spices arriving in Europe in the sixteenth century, and eventually their lowered cost, meant that they no longer served as markers of status. But there remain in Italian cookery several rudiments of the earlier obsession with spices. These appear, not surprisingly, in the descendants of medieval sweet and sour sauces, in *mostarda di frutta*, and with a habitual grating of nutmeg in gnocchi dough or cinnamon in a Venetian risotto. Hints of spices still occasionally appear in cured salami and meat products or a game stew. Saffron is still widely used in Italian cooking as well, the tiny stigmas of crocus lending a bright-yellow hue and elusive aroma to dishes like risotto or rich breads like *pandoro* and even some pasta dishes and fish soups. Clearly as elsewhere in Europe, spices are mostly used in confections—either "confetti," which are either candied spices or today a generic name for pills, or baked into cookies, cakes, or in the direct descendant of medieval cuisine, the Sienese *panforte* made with dried fruits and a bevy of spices, or *pan pepato*, a dense cake spiked with pepper.

Italy also has its own indigenous spices of great importance. Fennel, perhaps the most important, is the seeds of an umbellifer, used ground as a seasoning, especially for salami (*finocchiona*), as well as aniseed. Coriander and cumin are also sometimes used in Italian recipes and can be grown there. All these have also been used to make confetti. Mustard has also been used widely in Italian cooking, though not as extensively as in France or northern Europe. Interestingly, the Italian word *mostarda* derives from the condiment made with grape must and fruits. Other native spices find their way into cooking, often gathered wild, such as juniper berries in a game stew or myrtle berries and bay berries in a sauce. Bay leaves dried or fresh are also a common addition to a simmering tomato sauce or stew. They are removed before serving but lend a unique aromatic, resiny flavor. They can also be interlaced with meat on a spit for roasting.

Mexican Spices

We have Mexico to thank for what is one of the most ubiquitous spices globally—chili peppers. When dried and ground or crushed into flakes, these form the basis of seasoning mixes and condiments used on every continent. As has been discussed, chili peppers spread almost instantaneously to China and Southeast Asia with the Portuguese, as well as to Africa and the Middle East and somewhat more slowly to Europe. Chili peppers are culturally distinct from the aforementioned Asian exotics for one simple reason: they can be grown in nearly any warm climate. Thus, they were easily transplanted and never really needed to be imported. This meant that chili has since the sixteenth century been the primary source of heat for those who could not afford black pepper, as important in Sichuan as in Hungary, in northern Africa as in Indonesia. They became an integral part of so many cuisines around the world simply because they are easy to grow from seed and were never protected the way cloves and nutmeg were, for example, by the Dutch in the Moluccas. The nearly universal appeal of chili peppers may serve a psychological function as well. Apart from stimulating the metabolism, causing sweat that may cool the body, the heat of chilies also stimulates a rush of adrenaline and speeds the heartbeat, much as a ride on a roller coaster would. It may be precisely this cheap and ultimately safe thrill that accounts for the popularity of chilies.

Mexico does have deep traditions of using other spices as well. Vanilla is among the most important, though we rarely think of it as a spice but as a flavoring or extract. It is the seed pod of a particular orchid of southern Mexico that must be fastidiously pollinated by hand. The pods are fermented and turn black but should remain supple and moist. These can be split and scraped directly into mixes for baked goods, sauces, or even beverages. The original chocolate, for example, was prepared with chilies, vanilla, achiote paste, and other various flowers.

Achiote is the seed of *Bixa orellana*, crushed into a paste or heated in oil and used to color dishes a bright orangeish-yellow (figure 8.5). It is used in much the same way saffron would be in Europe or turmeric in Asia. Coloring food is actually highly appreciated in these three cuisines, though it is rarely accounted for in gastronomy. As an ingredient elsewhere it is called annatto and appears in butter, confections, and many other industrial products. In Mexico the crushed seeds are usually mixed with oil to be used in marinades and sautéed dishes, and the appreciation for brightly colored dishes made with it stems both from Aztec usage and the coincidentally similar Spanish medieval appreciation for bright yellow foods.

Another spice originating in Mexico and the Caribbean is the curiously named allspice, apparently suggesting that its flavor resembles a spice mix, but this is hardly the case. Its Latin name *Pimenta dioica* derives from the Spanish name, a diminutive feminine form of the word *pimento*, meaning "black pepper," which was probably wishful thinking. It is an ingredient in molés and also used to flavor meat as well as pastries.

Apart from these natives, Mexican cuisine also uses many Asian and Middle Eastern spices, in the same way they were used in medieval Spanish cookery. Cinnamon is commonly used in pastries, as is cumin, which is a standard in savory dishes. What

Figure 8.5. *Annatto* (*achiote*) is used to make dishes a bright yellow.

is known in the United States as chili powder or Mexican seasoning is often nothing more than chili, oregano, and cumin. An example of the way sweet spices are sometimes incorporated into a sauce would be an *adobo*, a kind of barbecue sauce used on chicken or pork that is made from dried red chilies, spiked with *piloncillo* (a type of sugar) and a touch of nutmeg, cinnamon, ginger, and garlic. All this is ground together or, more commonly today, whizzed in a blender. These types of flavor combinations, as mentioned, ultimately go back to Persian cookery, making several Mexican dishes distant cousins to Indian ones, also influenced by Mughal invaders from Persia.

Chinese Spices

Ginger is the single most distinctive flavoring in Chinese cuisine. As an item of export it was normally dried as a whole "race" or hand of ginger and then pounded into a powder as needed. Today it is usually sold preground. Most ginger in China is consumed fresh, though, and may be either the relatively mature and fibrous form familiar in the West, which must be finely chopped or grated, or young, green ginger, which is merely sliced and can be eaten more like a vegetable. These are well worth seeking out in Chinese markets. Ginger can also be pickled, preserved in syrup, or candied. It is believed to be native either to southern China or northern India but is unknown in wild form. It has been used in China as far back as there are written records. It is not only used in practically every conceivable cooked dish but also as a flavoring in tea, wine, and sweets. The Chinese consider it a hot and strengthening food and an aid to fertility. Ginger, while aromatic, also supplies some spicy heat to Chinese cuisine, although it is quite a different heat from that of capsicums.

A similar rhizome (a mass of roots) is galangal root (*Alpinia officinarum*), once well known in Western cookery but now exclusively used in Asia. It is similar to ginger, but spicier and more mustardlike in flavor. It is also tougher and is usually merely added to food in slices while cooking but not consumed. Mention should also be made of ginseng, which, although typically associated with medicine, is added to soups and of course tea in ways that are somewhere between daily cooking and therapeutics. It is considered the most potent plant for increasing orthogenic *qi*, or strengthening the body.

Sichuan peppercorns (*fagara, Zanthoxylum simulans*) are one of the Chinese spices that only made it into world trade fairly recently. They have an initial heat that is somewhat similar to that of black pepper and are also highly aromatic, leaving a numbing tingling on the tongue and lips, an experience to which the Sichuanese are said to be addicted. There is a particular appreciation here for the combination of chilies and Sichuan peppercorns called *ma la* ("numbing spiciness") (figure 8.6). The tiny, reddish seed husks are added to soups and marinades, roasted and ground in stir-fries and braises, and even used as a dip or sprinkled on dishes at the end. These should not be confused with pink peppercorns (*Schinus terebinthifolius* and its relative *S. molle*), which

Figure 8.6. A market in Guizhou Province, China. ©Lawrence Manning/CORBIS.

FATS AND FLAVORINGS

are native to Brazil and Peru, respectively, and have a relatively mild flavor. Sichuan peppercorns are also included in five spice powder along with star anise, cassia, cloves, and fennel. Other versions use ginger and cassia buds in place of Sichuan pepper, or even licorice and other spices. Cassia buds, incidentally, were once a common spice in medieval European cookery, but today are unknown. As the name suggests, they are merely the dried buds of cassia, a relative of cinnamon, and look rather like a fat clove but taste intensely of cinnamon. Whatever the mixture, five spice powder is used in braised, "red-cooked" dishes as well as on roast pork, and it goes perfectly with beef as well. It lends a deeply aromatic nutty flavor that combines at once sweetness, heat, sourness, and a hint of bitterness.

Other spices one may encounter in Chinese cuisine include several forms of cardamom, black, white, and green, all of which come from India. The larger and intensely smoky and aromatic black cardamom is most commonly added to soups or braised dishes and not consumed. Other types have thinner skins (the green being fresher, the white is the same variety that has been blanched), and normally the seeds are removed from the pod and crushed. It has a highly aromatic and resiny flavor, somewhat akin to camphor, and has also been used as a spice in cooking. Star anise (*Illicium verum*) is a small, brown, star-shaped seed pod with eight radial points. It goes perfectly with pork and duck dishes and can be used ground or whole, although it is not eaten whole as it remains woody. Although anethole is the familiar licorice or aniselike scent, it is totally unrelated to these plants. Chinese cookery also makes use of true licorice root in confections and dried fruits and even to flavor savory dishes.

In the south of China, especially among ethnic minorities, one also finds spices more typically associated with Thai or Indonesian cookery, such as lemongrass, kaffir lime leaves, and tamarind. Tamarind is a sweet/sour legume, the dark red pulp of which is loosened in water and strained before used in cooking (figure 8.7). Interestingly, it originates in Africa, made its way both eastward to Southeast Asia and thence to Mexico, rather than across the Atlantic. In western China, one also finds cumin, fennel, and cloves, especially in Uighur cooking.

Flavorings

While we have discussed many sauces and condiments in the context of their principal ingredient—for example, soy with beans, oyster sauce with shellfish, tomato sauces with the plants—there remain many essential flavoring agents that properly belong here as building blocks of the three world cuisines.

Sugar

Sugar is the most universally used flavoring on the planet, and there is strong evidence that we are biologically programmed from birth to enjoy sweetness, first in the form of mother's milk, and later in other foods such as fruits, which contain fructose. Our

Figure 8.7. The pods of the tamarind contain a sticky, sour pulp used in many dishes around the world.

bodies, of course, use energy in the form of the simple sugar glucose. As a refined sweetener or sucrose, sugar is a relative newcomer in the human diet. Sugar cane is native to New Guinea, and from there was brought to the mainland and domesticated in ancient times, becoming important especially in India where the juice of the cane was pressed, boiled, and dried into crystalline form. This crude form of sugar as well as jaggery (made from palm sugar) reached China sometime between the fifth and seventh century AD, and perhaps a little earlier in ancient Rome, though there it was extremely rare and used primarily as medicine. It was only in the Middle Ages that sugar came to be used in cookery, interestingly not exclusively in sweets and desserts but as a universal flavor enhancer in practically any dish, especially on elite tables, as it was still a rare and expensive luxury item imported along with spices from Southeast Asia. The sweet-and-sour flavor combination of sugar with vinegar or verjuice (unripe grape juice) was especially popular in medieval cookery, as it still is in Chinese cuisine. It was only after the production of sugar in the Caribbean and Brazil, using African slave labor, that the supply increased, the cost eventually decreased, and industrial manufacture made possible the consumption of sugar by the masses. At the same time it was gradually marginalized in culinary terms, relegated to the final dessert course of a meal or used to sweeten beverages.

Nonetheless, there remain culinary rudiments of sugar's former popularity in our three cuisines. Sugar in the form of rock sugar is often added to savory dishes in China, to round out the flavor of a soup, or in a glaze for a barbecued dish. In Italy, a pinch of sugar is often added to a tomato sauce to balance its acidity, or even in dishes directly descended from medieval cookery, such as fish preserved *in saor*. In Mexico disks or truncated cones of unrefined sugar (*piloncillo*) are used in dishes such as *capirotada*, which is a savory, sweet bread pudding with cheese, or *arroz con leche* (rice with milk), as well as in beverages like *café de olla* and *atole* (figure 8.8). The most common use of sugar, however, is in confections, which are countless. Curiously there are some regions that seem addicted to sweets, such as Sicily and Guangdong Province, while others scarcely like them at all, such as Tuscany or Hunan. This may be a matter of heat, though why that would induce a sweet tooth is unclear. It may be merely that proximity to sugar-producing regions and a long history of commerce in sugar gives certain people a predilection toward it. This certainly makes sense in the case of Sicily with the culinary influence of Arab cuisine and its own sugar cane production, and southern China with its proximity to Southeast Asia and local production. Mexicans

Figure 8.8. *Piloncillo* is a brown sugar sold in little cone shapes.

seem to be addicted to sugar, and once again, facing the Caribbean centers of world production made it accessible and affordable. Sugar from sugar beets is identical to that of sugar cane after processing.

Honey

Honey was the original source of sweetness in earlier cookery. The Romans used it in all manner of savory dishes with vinegar and garum, as did the ancient Chinese at the same time. It is odd to think that these once-similar cuisines took such divergent paths down the road. Today honey is used mainly as a sweetener for teas or in some confections. Honey is collected from the wild, as well as from domesticated honeybees of the *Apis mellifera* species, but there are other honey-producing species. Honey also ranges in flavor and color depending on the flowers from which the bees have been collecting nectar, and they do generally make a beeline for a single flower while in bloom. These range from dark chestnut or buckwheat honey found in the Alps as well as Northern China to delicate orange blossom and now-ubiquitous clover. There are also other sources of sweetness; for example, sapa, a common condiment in ancient Rome, made from boiled-down grape juice and now known as saba, which is also the base for balsamic vinegar.

Other Sweeteners

There are also other derivatives of sugar worth mentioning: first molasses, which is a by-product of the sugar-refining process, and the more commonly caramelized sugar, which is used to give color and depth of flavor to many sweet and savory dishes. Last, in the past century various artificial sweeteners have been introduced in the interest of weight loss or for diabetics. Saccharine was the first of these, which several decades ago bore a health warning label, being suspected as a carcinogen, though this warning has since been revoked. Although it is stable in cooking, it lends food an unpleasant bitter aftertaste and makes a poor substitute for natural sweeteners. The same must be said for aspartame, which breaks down in cooking but is used in diet beverages and chewing gum, and sucralose (first marketed as Splenda), which, although it can be used in cooking, is primarily used in industrially manufactured products.

Vinegar

Just as sugar provides a universal form of sweetness, **vinegar** is the prime source of sourness in cooking. Technically vinegar should only refer to wine soured through bacterial fermentation of ethanol, whereby it turns to acetic and other forms of acid. But the term is also used for malt vinegar derived from beer (once known as alegar), apple cider vinegar, as well as Asian vinegars derived from rice wine. Any alcohol can be made into vinegar through natural acetobacter in the atmosphere. The bacteria create a slimy raft or "mother" on the top of an open container and convert the alcohol into acetic acid. This mother can in turn be used to convert other batches of wine. This process happened all too often accidentally before modern winemaking equipment and extensive use of sulfur.

Chinese black vinegar is made from glutinous black rice and sometimes millet or sorghum (figure 8.9). The best is said to be Chinkiang (from the city on the east coast now spelled Zhenjiang) and is aged until it has a deep, smoky flavor. It is sometimes compared with balsamic vinegar, though the similarity is superficial. There is also red rice vinegar as well as milder whiter versions. Vinegar is used in Chinese cooking mostly to perk up cooked dishes or in contexts like sweet and sour soup, rather than in salad dressings and the like. It is also used medicinally and for pickling.

The now famous balsamic vinegar of Modena in Italy is made with boiled-down grape must, usually of white Trebbiano grapes. This is put into barrels washed down with vinegar, which begins fermentation. It is aged by successively decanting into smaller barrels of different woods, which might include oak, chestnut, cherry, mulberry,

Figure 8.9. Bottles of black vinegar and rice wine, essential flavorings in Chinese cuisine.

ash, pear, and juniper, in which the vinegar is exposed to the air through a square opening in the barrel that is covered with cloth. Each wood is said to contribute its own unique flavor in the course of years of aging, from twelve up to one hundred or more. Balsa wood is incidentally not among these woods; the name *balsamico* derives from its medicinal qualities. The aging takes place in the attic or a loft rather than in a cellar because the vinegar is supposed to evaporate and concentrate very slowly. A set of barrels with vinegar is considered a family heirloom and is handed down from generation to generation. At the end of the process a tiny bottle sells usually for at least a hundred dollars, and often much more. This process refers only to genuine, registered *aceto balsamico tradizionale di Modena*. There are also much quicker and less expensive versions that are made from ordinary red wine vinegar colored with caramel and sweetened with sugar. While they are fine for an everyday salad, the true balsamic is a thick condiment, best savored in droplets on Parmigiano cheese, fruit, or even steak. It is also drunk after a meal as a digestive, in a tiny, thimble-sized shot.

Italy also produces ordinary white and red wine vinegar, which are used in salad dressings with olive oil; to pickle vegetables like peppers, giardinera, or mushrooms *sottaceti*; and as an ingredient in cooking where a sour note is desired. Vinegar itself may also be flavored with herbs suspended whole in the bottle and are preserved by the acid.

The use of vinegar in Mexico is much the same. It is used on salads and occasionally in cooked dishes, though there are indigenous varieties made from fruits like pineapple, which are fermented and left to sour. Vinegar also perks up sausages like chorizo, which in Mexico is cooked, unlike its Spanish ancestor.

Salt

The use of **salt** is nearly universal on earth. It can be evaporated from seawater or mined from the earth directly. The distinctive flavors of various salts depend on the organic material and minerals remaining if the salt is left relatively unrefined. Trapani sea salt is thus distinguished from *fleur de sel* from Brittany, from Maldon sea salt, from pink Tibetan salt, black salt from the Ganges, and so on. Since practically all world cookery depends on salt, places where salt could be processed or mined thus became important economic centers and quite wealthy. It is not surprising that it has always been among the most important sources of tax revenue for governments around the world. It is also of prime importance as a preservative for meats, vegetables, and for making condiments. Why some cultures seem to have a predilection for salt, just as there are some with a sweet tooth, is probably the result of long ingrained usage and available sources. Long, habitual use changes the palate, so that unsalted foods seem bland and flat. Saltiness in cooking may also derive from condiments such as soy sauce rather than pure salt.

Salt is normally merely added to dishes when cooking or presented on the table, once quite formally in an elegant salt cellar or mounded in a little pile on the edge of a plate, today in more hygienic but quotidian salt shakers. Salt can be used as a cooking medium, too. In Chinese cuisine, a whole chicken may be buried in salt set in a casserole or wrapped in leaves and baked. Intriguingly, the salt keeps in the moisture but does not flavor the dish excessively. The effect is much the same as "beggar's chicken," in which the bird is enclosed in fresh clay and baked, to be cracked open upon service. In Italy, a salt crust is made for fish, covered in a mixture of salt and egg whites, sometimes with herbs, and then baked.

As mentioned, in Chinese cuisine, soy sauce is the typical source of saltiness. It is basically the descendant of an earlier condiment called *jiang* ("sauce") made from fermented, ground soybeans and used two millennia ago. It is similar to, and the ancestor of, Japanese miso paste. Soy sauce was merely a liquid drawn from the *jiang*, called *jiangyou*. Our word *soy* is a corruption of the Japanese name for this condiment, *shoyu*. Chinese soy tends to be darker and stronger than the Japanese one, but there are both light and dark versions, the latter sometimes colored with molasses. Light soy is meant to give food flavor, while dark is more commonly used for color, often in braised meat dishes. The light versions are used in dips and added to virtually any recipe using any cooking technique. Indeed, soy sauce has come to be the single flavor most commonly associated with Chinese cuisine.

Saltiness can come in other forms, such as oyster sauce, the finest of which is made with oysters, but just as often with other fish. It is always thick and pungent. Fish sauce itself, *nam pla* or *nuoc mam*, which is normally associated with Southeast Asian cooking, is also used in southern China, especially Fujian, where it is called *yulu*. It is a thin, brownish liquid made from whole fermented fish and has an extraordinary pungency that mellows with cooking. Whole, fermented black beans are also a source of salt, made by soaking and steaming small soybeans and then letting them ferment for several months. They are sold either dried, which are preferred, or in brine, and are in flavor similar to other fermented soy products. Before using they must be soaked to remove some of the salt, and then they are used in sauces or merely added as an ingredient in cooked dishes.

Monosodium Glutamate

There has been much controversy in the past few decades over the use of **monosodium glutamate** (MSG) in cooking. Its use as a flavor enhancer is relatively recent, being first processed in Japan only a century ago, and it is mostly still made there by the Ajinomoto Company. It is also used in China (and called *wei jing* or "flavor essence"), though perhaps less frequently than in Chinese restaurants in the United States in the past fifty years. After reports of headaches and studies that suggested that it is a neurotoxin, these effects came to be known as the "Chinese Restaurant Syndrome." Subsequent research has been unable confirm these effects, though anecdotally it is still widely perceived as dangerous, such that Chinese restaurants often advertise that they use no MSG. Strangely, it has also been found that glutamates serve as a flavor enhancer in many foods naturally, such as beef, mushrooms, fermented soy products, and seaweed. These have recently been identified in gastronomic circles as the fifth flavor (umami), ironically the subject of fervent culinary interest. These natural forms, however, may differ from the processed industrial form of MSG, and in consequence there is no real reason to use powdered MSG or seasonings made from it. Other than in Chinese restaurants, most of this industrially processed MSG appears in snack foods such as nacho-flavored corn chips, crackers, and instant soups.

Study Questions

1. Why does nutritional science seem to change its position about certain fats? Is this due to the advance of scientific knowledge, or are there deeper cultural and economic reasons for these swings in official orthodoxy?
2. Why are some fats preferred in some places but not in others?
3. How do our modern gastronomic concepts of hot for spice and dry for drinks ultimately hail from humoral physiology?
4. Why were exotic spices in such demand in the past, and why did this change?
5. How are herbs and spices used in medicine in these three cultures?
6. What influence did spices have on trade and the economy globally?
7. Physiologically and psychologically, what effect do hot spices have on the body?
8. Why has sugar been so popular historically but now is relegated mostly to desserts? What surprising role does it play in flavor combinations?
9. How can we reconcile the current interest in umami with the general dislike of MSG in cooking generally?

Recipes

Sauces

PESTO GENOVESE/ITALY

This is the classic green sauce used with pasta in Genoa. The word *pesto* derives from *pestle*, with which it is pounded. It has many ancestors stretching back to the Middle Ages. These were basically herbs (usually parsley) and garlic pounded with olive oil and nuts and thickened with breadcrumbs. Today, pesto is normally made with a fragrant, tiny-leafed basil variety, but any type will serve. The key to the sauce is a capacious wooden mortar and pestle. Begin by pounding a bunch of basil to a fine pulp with a small clove of garlic and a handful of pine nuts. Gradually add in a handful of grated Parmigiano and continue pounding. Gradually drizzle in the fruitiest extra-virgin olive oil you can find until it becomes a sauce consistency. It will be thick. The key to using the sauce is to add cooked pasta in a big bowl with the sauce and stir vigorously with a ladleful of the cooking water. This will make the sauce smooth and heat it up, but it will remain still raw.

PICO DE GALLO (TOMATO SALSA)/Mexico

Pico de gallo ("peck of the rooster") is really not a salsa but more of a salad made with the same ingredients. Use only fresh tomatoes of the best quality, chopped into pieces. Add salt, chopped cilantro, chopped jalapeños, and chopped onion. Squeeze some lime juice on top, and serve at once. This mixture, more finely chopped and left to marinate or even cook together briefly, is what is known as salsa in the United States. A pico de gallo can also be made with fruit, such as mangoes, pineapples, and papaya, and dusted with chili powder.

BAGNA CAUDA (HOT SAUCE FOR VEGETABLES)/ITALY

This is a simple sauce used for dipping cardoons (a stalk-shaped relative of artichokes) and other raw vegetables like fennel bulbs, carrots, bell peppers—anything crunchy. It is kept warm at the table in a small terra-cotta dish with a candle burning below, hence the name "hot bath." The ingredients vary, but typically it contains garlic pounded with anchovies, olive oil, and butter. Sometimes cream is used, sometimes truffles or parsley. Start by gently browning the chopped garlic in the olive oil with the anchovies. Add the butter, whisk vigorously until frothy and amalgamated, and then transfer to the warm serving vessel. Place at the center of the table, and everyone dips as they please.

MOLÉ VERDE WITH TOMATILLOS/MEXICO

Confusingly, tomatillos in Mexico are referred to as *tomate*, while red tomatoes are called *jitomate*, which comes right from the Nahuatl *xictomatl*. They are actually unrelated, the former being *Physalis vulgaris*. The green molé is made by first removing the papery husks from the tomatillos and washing off any sticky residue. Next, chop the tomatillos and cook them in a pan, barely covered in water. When softened, put them into the bowl of a mortar or in the blender. Next, toast a handful of pumpkin seeds and sesame seeds in a dry skillet and add them to the tomatillos. Do likewise with a few slices of onion, garlic, and some seeded serrano chilies, until toasted with darkened spots. Add them as well. Season with cumin, salt and pepper, and a touch of cinnamon and fresh herbs such as cilantro, *epazote*, and *hoja santa* (a peppery herb with a licoricelike flavor). Pound or blend everything into a smooth liquid. Finally, fry this mixture in the pan with a little lard or olive oil. Thin with more water or chicken stock. The sauce can be poured on top of chicken or into a pot with simmering pork. It can be folded into tortillas or served with virtually anything.

GUACAMOLE/MEXICO

This is among the few condiments that remain almost completely unchanged since Aztec times—even the Nahuatl name *ahuaca-mulli* is similar. Traditionally it is made and served in a stone *molcajete*. Cut the avocados in half without cutting through the seed. Twist the two halves until they come apart. Firmly plant the knife's edge into the seed and twist to remove. Scoop out the avocado flesh with a spoon. Immediately squeeze lime juice over the avocado flesh and pound. Add chopped tomatoes and onions, plus salt and pepper to taste. Add chopped cilantro. It should be a little chunky, not smooth.

If you like, serve this with *totopos*, or as they are called in the United States, tortilla chips, which are simply cut-up leftover corn tortillas, quickly fried until crisp.

XO SAUCE/CHINA

This is a new sauce that is all the rage in Hong Kong. It appears to get its name from well-loved cognac, the finest varieties of which are labeled XO. The ingredients are expensive, so perhaps it deserves the title. Begin with a handful of dried scallops, soaked until they fall apart. Pour off the soaking liquid. Then put the scallops in a small bowl and set the bowl in a bamboo steamer and steam for about 15 minutes. Shred finely. Chop a handful of peeled shallots, a garlic clove, a handful of small red chilies, and the same amount of dried shrimp. Add a slice of *jin hua* ham or some prosciutto. Salt-cured fish, usually mackerel, is also common; use some as well if you can find it. You can pound these finely or puree in a food processor. Then, heat up a few cups of oil in a small pot on medium-low heat, add all the ingredients, and cook gently, until fragrant. Be sure not to let the chilies burn. Store in jars in the refrigerator. Use as a dipping sauce or to flavor stir-fries.

TOMATO SAUCE/ITALY

There are as many versions of tomato sauce as there are Italians in Italy and abroad. Some contain meat; others none at all. Some are cooked quickly; others long and slow. Some use fresh tomatoes; others canned. Some are from the north; others are southern. Here are two: the simplest, and the most renowned and luxuriant—the ragù Bolognese.

While any good, ripe tomato will work, some are too sweet or watery or flavorless to make a decent sauce. Most round supermarket tomatoes fall into the latter category. The best bet is what is labeled in the United States a Roma tomato, which is pear-shaped and very meaty with few seeds and little juice. Simply heat olive oil in a pan as hot as it will go and throw in a clove of garlic, unpeeled, and a sprig of oregano or basil. After a minute throw in the tomatoes, cut into large chunks, enough to fill the pan but not so they are piled on top of each other. Ten tomatoes is probably enough. Do not stir. You want a nice char on them. After a few minutes, stir and add a good glug of red wine and a pinch or two of salt. Continue to cook until the tomatoes have fallen apart completely, which should be another few minutes. Pass this mixture through a food mill into a bowl, and discard skins and seeds. Return the sauce to the pan, reducing a little further if you like. This can be used in other recipes, or pasta can go directly from the pot into this pan of sauce. This is the simplest of cooked tomato sauces.

The ragù Bolognese is properly a meat and tomato sauce, with meat definitely dominating. Start with a pound each of ground beef and pork. If you like, a good flavorful cut works best, especially skirt steak, which you can grind yourself or chop very well. Finely dice a thick slice of pancetta (unsmoked bacon) and put it at the bottom of the pot to melt slowly. Add a few finely chopped garlic cloves, more or less to suit your taste, and finely diced carrot, celery, and onion. One of each works fine. Cook these until just tender and flavorful. Push these vegetables to the edge of the pot and add the meat, letting it brown at first, then stirring to mix with the vegetables and adding about ½ tsp salt. Let this continue to cook until all the liquid has evaporated. Then, add either a small can of tomato paste or 2 cups tomato purée. Some people like the more concentrated tomato flavor, but it is also nice as just a high note. Then add a cup of hot milk and some freshly grated nutmeg. Let this simmer 3 or 4 hours, adding white wine to top off if necessary and making sure nothing sticks to the bottom of the pot.

MOLÉ POBLANO/MEXICO

As the most quintessential of all Mexican foods, there is bound to be disagreement about the ingredients, origin, and how best to prepare it. Most Mexicans buy the sauce ready-made in the market, but it is, like all good things, worth making yourself and adjusting to your preferences. What cannot be denied is that this is a combination of a basic native Mexican sauce with European, African, and Asian ingredients, which could only have been possible in the colonial era. Most important, the spice of chilies should balance with the sweetness of just a hint of chocolate and fruit, the creaminess of nuts, everything harmonizing into a deep and well-rounded sauce. Traditionally it is served on turkey but is equally good with chicken.

Start with a handful each of dried chilies: mulattos, anchos, and pasilla. Remove the stems and seeds. Heat these on a dry skillet or griddle (*comal*) until dark and fragrant, and put into a large bowl of water to soak. Add a handful of raisins as well. Most recipes will tell you to use a blender, and that is how most people would do it today, but try a *metate* (large, stone mortar). It will take a lot of patience, but it does work. First, toast your spices in a dry pan over high heat: pepper, coriander, whole cinnamon stick, whole cloves, and aniseed. Put these into the mortar and pound to a fine powder. Next, gently toast almonds, pumpkin seeds, and sesame seeds, and add these to the mortar and pound. Peanuts are nice, too. Incidentally, you can also fry these ingredients in lard. Add the soaked raisins to the mix and also work into the paste until smooth. Then, finally add the chilies and continue pounding or grinding with your *metate*, gradually thinning with the soaking liquid. Or you can blend the chilies and then add all the other ingredients. Put the mixture into a large pan with grease and fry. Add a wedge of Mexican drinking chocolate. Add toasted breadcrumbs or a stale tortilla. Add salt to taste. Again, there should be a blend of all the flavors. Then cut up your turkey, separating the legs and thighs and cutting the breast with the bones into several large hunks. Put it right into the sauce to cook, gently, for about an hour. Keep an eye on it and add water if necessary.

CHILI SAUCE/CHINA

Most sauces (soy sauce, oyster sauce, plum sauce) in China are bought premade, but chili sauce is one that can easily be made at home and is better tasting than what you buy in a jar. The simplest of these is just chili oil, used extensively in Sichuan cooking and as a condiment. Crush a handful of dried small, red chilies, discarding stems but not seeds. Heat this in a pot containing a couple of cups of oil. Let cool and infuse, then put into jars. You can also make chili sauce from fresh peppers, either red "bird" chilies or a milder variety, such as jalapeños. Cut these up with the seeds and boil in just enough water to cover until barely softened, maybe 10 minutes. Add some salt and sugar and a good dose of rice wine vinegar. Then pound into a smooth paste or blend thoroughly.

BLACK VINEGAR SAUCE/CHINA

This is really just a dipping sauce composed of other ingredients in bottles, but it is quite addictive. Pour some black vinegar in a small bowl, then add some finely diced ginger and a little soy sauce. If you like, add other ingredients such as a little sugar, some ground white pepper, or, even more interestingly, some ground Sichuan peppercorns and perhaps some sliced scallion. This goes wonderfully with dumplings.

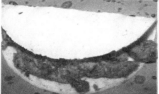

⁂{ 9 }⁂
Beverages

Learning Objectives

+ Learn why some cultures depend largely on water as a beverage while others do not, and see how waterborne pathogens may have played a part in shaping these preferences.

+ Appreciate how food and beverages are matched in a complete gastronomic experience.

+ Consider why many beverages are mass produced and the impact this has had on human health.

+ Appreciate the importance of wine culturally and gastronomically and how fermented beverages have come to play a symbolic role in world religions.

+ Understand how many beverages were at first used medicinally and how and why this has shifted to more widespread consumption.

BEVERAGES ARE an essential component in every meal. They hydrate the human body and facilitate digestion and serve as a complement to the solid foods being eaten. Careful matching of food and beverages, such as wine, is also a major part of gastronomy. Moreover, alcohol as a product of nature and culture has become an integral part of human civilization.

Milk

Human **milk** is, of course, the first liquid we taste as infants, but drinking a cow's or other ruminant's milk in adulthood is an exception for humans. Historically, northern Europeans and a few other peoples on earth, such as the Maasai in Africa, drank milk. The ability to digest lactose is an unusual and developed trait. Most Asians and Native Americans, as well as southern Europeans did not drink nonhuman milk as children and are thus lactose intolerant. Fermentation breaks down the lactose, and *kumis*, a fermented mare's milk, is found in western Asia and Tibet, and likewise yogurt is an altered and more digestible product. Today, however, milk drinking is on the rise in Asia, especially for young adults because it is perceived to stimulate growth, and indeed average heights have been increasing in places like China. Dairy cattle were introduced to Mexico in the sixteenth century, but drinking milk is much less common than cheesemaking, and the same is true in Italy.

Water

The prototype of all beverages for children and adults is naturally **water**, consumed as the sole refreshing liquid for the greater part of our history as a species. With the advent of urban living, and probably even before, a realization that water could be tainted led humans to experiment with juices of fruit and steeped grains, which in turn led to the discovery of alcoholic beverages. These not only lasted longer without spoiling but also were considered healthier, and they often were because harmful bacteria are killed by the process of fermentation. It is ironic in a sense that we have now returned to the idea of water as the purest drink, at least when it issues pure from mountain springs with beneficial minerals. It is also ironic that what was once the cheapest and most ubiquitous drink has now become a commodity, sold everywhere in plastic bottles, which often cost more to make than the water inside.

Italians have always appreciated the therapeutic benefits of mineral water, both to bathe in and to drink. The ancient Romans were avid connoisseurs of water, and many water sources they first identified throughout Europe are still in operation today. Famous springs bottle mineral water, which may be still or naturally bubbly, or as is said in Italian, *frizzante*. Although at first the benefits of different types of water for various ailments was empirically discovered, now bottles always bear a label listing specific quantities of minerals as certified by a scientific laboratory. Italians generally keep bottles at home and order them for the table in restaurants. The most ubiquitous of brands is San Pellegrino, but Galvanina, Panna, Ferrarelle, and Fiuggi are also popular brands, and there are hundreds of others as well. In some places people still fill their own bottles at natural springs, and preferences for the specific flavors of water are fiercely defended. The ubiquity of bottled water stems from a time when municipal water supplies were considered dangerous, but the habit has lingered, even in places where the water is excellent. Take, for example, Rome, where icy cold water issues from public fountains and is perfectly safe and delicious.

This has not been the case in Mexico, however, and while local people always consumed the water, tourists were always warned to assiduously avoid it. Today even Mexicans themselves have taken to bottled water almost exclusively, and it is one of the fastest-growing sectors of the food and beverage industries. Mexicans have the second-highest consumption of bottled water by volume in the world, following only the United States. Italy has the highest rate in Europe, trailing only slightly behind China, though Italians have the highest per capita consumption in the world. The major concern over this dramatic increase in bottled water consumption is environmental. Even taking account of recycling efforts, the energy required to make and transport bottles, as well as the ultimate waste in plastic, is creating serious and unprecedented problems. One can certainly appreciate the gastronomic appeal of unique mineral waters, though sometimes people purchase for ten times the price what is essentially tap water. The Chinese never drink water straight from the tap; water is always boiled and drunk hot or at room temperature. Not drinking tap water may stem from concerns over safety, whereas the temperature preference probably derives from Chinese medicine.

Soft Drinks

The traditional solution in Mexico to purifying water, and for making it tasty, has always been to flavor it with acidic fruit juices, creating a light and refreshing set of drinks called *aguas frescas*, which are still sold from big glass jars on the streets. Many of these are made with native fruits and have counterparts in pre-Columbian times. *Aguas frescas* can be flavored with citrus juice, tamarind, or jamaica (dried hibiscus blossoms), or milky *horchata* made from rice. Actually any fruit—mango, guava, melon, usually the more sour the better—will do. These drinks in some cases evolved into carbonated soft drinks, mass manufactured and sold in cans. These, of course, compete with imported sodas such as Coke and Pepsi.

Soft drinks in China also range from the very modern supersweet sodas, which are rapidly gaining in popularity, to more traditional cooling beverages based on fruits or herbs. A brief stroll in a Chinese grocery will reveal an array of these, including grass jelly drinks, basil seed, and exotic fruits, most common in the south. Iced tea, often lemon flavored, is most popular in Hong Kong. Bottled green tea is also popular and is even mixed with whiskey as a bar drink. The beverage consumed with meals would have been the water in which rice or vegetables were boiled, or more often soup, which should not be considered a first course as it is in the west but rather a drink, taken throughout the meal and also at the end. There are even cold, sweet soups to finish a meal, which we would probably classify as beverages. But these habits are changing, and the soft drink industry in China as well as exports are a truly booming market.

Italians also have their own unique soft drinks. Bitters based on orange peel are popular, as well as bottled carbonated drinks like Chinotto and Limonata. There is also a huge variety of syrups that are added to carbonated water, so-called Italian sodas manufactured by the Torani company, which come in raspberry, strawberry, orgeat (almond), and exotic flavors like chocolate macadamia, crème de banana, mandarin orange, and even peanut butter.

All three countries import Coca-Cola, Pepsi, and other soft drinks, and indeed often bottle them domestically, sometimes with their own formulas. The Coke in Mexico is often preferred, even abroad, because it is made with sugar rather than high-fructose corn syrup. In China, the Coca-Cola Company, under its own name and under the brand name Smart, sells coconut-, peach-, and watermelon-flavored sodas. These countries all have their own local and rival brands of course, too. In Mexico there is Jarritos, which comes in flavors like tamarind, jamaica (hibiscus), and guava. In China, the medicinal Wanglaoji is sold in little red cans and is drunk specifically to counteract spicy heating foods. Although there is not much to say about these from a perspective of gastronomy, sweet sodas have been identified as the major culprit in rising obesity and late-onset diabetes rates. High-fructose corn syrup is the main reason: a 12-ounce can of Coke contains about the equivalent of 9½ teaspoons of sugar, a 32-ounce Big Gulp over 23 teaspoons. Ironically most sodas began as quasi-medicinal drinks, tonics, nervines, or stomach relief medicines.

Chocolate

Although we do not immediately think of **chocolate** as a beverage, this was its original form, and it is indeed native to Central and South America. Cacao is the plant from which chocolate is made, the football-shaped fruit of which grows obliquely from the trunk of the tree (figure 9.1). Inside is a creamy white mass, which can be eaten, as well as seeds that are removed, dried, and fermented. The seeds are then crushed and mixed with other flavorings and repeatedly rolled on a heated *metate* to create a smooth paste, which is formed into disks. To these water is added to make chocolate. Originally it was a drink very different from what we think of as "hot chocolate" or cocoa, which is usually made with milk and sweetened with sugar. The Maya, from whom we get the name *ka-ka-w*, and perhaps even the Olmec before them, drank a spicier concoction that was poured from one vessel to another from a great height to achieve a froth. The earliest recipes come from the Aztecs, who took cacao from the south as a tribute, and it was even used as money. They added vanilla, chili peppers, ear flower (*Cymbopetalum penduliflorum*), and *achiote*, which gave it a deep blood-red color, and sometimes honey. They also drank it cold. The emperor Montezuma was said to drink it all day from a golden cup, as it was apparently considered an aphrodisiac.

With the arrival of the Spanish, Old World flavorings were added to chocolate, such as sugar, cinnamon, pepper, anise, and ground almonds. To achieve the froth on top, they used a *molinillo* (little wooden mill), basically a rod with wooden rings on the end. Eventually the hotter spices were removed, but chocolate as it is drunk in Mexico today is directly descended from the colonial beverage. Of course, chocolate can also

Figure 9.1. Freshly harvested and opened cacao pods waiting to have their seeds removed, in preparation for making chocolate. The Mexican state of Tabasco produces much of the world's cacao. © Keith Dannemiller/CORBIS.

be used as a cooking ingredient, though it is not as ubiquitous in molé sauce as one is often led to believe.

As a confection, early forms of edible chocolate were invented in Spain and elsewhere in Europe, and chocolate was even promoted for a while as a medicine. But it was not until Dutchman Coenraad van Houten in the nineteenth century invented a process whereby the cocoa butter could be removed using an alkali (called Dutch processing) that cocoa powder appeared. The cocoa butter in turn was later used to create solid, edible chocolates with a smooth texture using a process called conching. Today, perhaps coincidentally, chocolate with a high percentage of cacao is becoming more common, sometimes even flavored with what may seem incongruous chili peppers and other hot spices, but in fact this is much closer to its original form than the familiar bland milk chocolate.

In Mexico another popular hot drink is *atole*, which is based on cornmeal, sometimes toasted, or corn starch. It is usually flavored with *piloncillo* sugar, vanilla, and cinnamon and can range in consistency from a thin drink to a thick porridge. When flavored with chocolate it is called *champurrado*. It can also be flavored with fruit or nuts. It is the sort of warming drink appreciated at Christmas time or on chilly mornings for breakfast.

Tea

The **tea** plant (*Camellia sinensis*) grows wild in a region covering northern India and southwest China in the province of Yunnan. According to legend, tea was discovered accidentally by the Emperor Shennong (c. 3000 BC), who was intrigued by the aroma of leaves that had accidentally fallen into his cup of hot water. After tasting the tannic beverage, he pronounced it good and proceeded to add it to his extensive list of medicinal herbs. This is said to have happened nearly five thousand years ago. In fact, tea was well known by the first century BC and became a truly popular drink by the Tang dynasty when it was also extensively cultivated. There were even poems extolling its virtues as well as *The Classic of Tea* by Lu Yu, written in 780. It describes all the necessary utensils, methods of service, and ways to appreciate tea, and also scholarly opinion. For example, a quotation from *The Dissertation on Food* by Hua Tuo is "If bitter tea be taken over an extended period, it will quicken one's power of thought."

At this time tea was generally steamed and then pressed into hard cakes for transport, and the cakes were later broken off into hot water. It was bitter and astringent and generally thought of as a medicinal drink (figure 9.2). Lu Yu appreciated tea in its unsullied form without any additions and derided the common custom of adding onion, ginger, orange peel, peppermint, and the like. Further refinement came in the Song dynasty (960–1279), when powdered tea was developed and was appreciated, whipped into a green froth, in fine porcelain vessels. Interestingly, it is the froth that is often considered the finest part of hot beverages—not just chocolate and tea, but coffee as well. The Song also flavored tea with flowers such as jasmine and rose. Not until the Ming dynasty (1368–1644) was black tea, in which the leaves are oxidized, invented. This was mostly for export, as it could withstand extremes of temperature and humidity, but it was not preferred by the Chinese.

To this day most tea in China is quite unlike what is found in teabags in the West, which is a black oxidized (not technically fermented) leaf tea. In China, tea is usually green or even white. It is a much more delicate, aromatic, and subtle drink, never taken with milk, and usually unsweetened. To prepare tea, the server pours hot water below the boiling point first into the teapot to warm it up (figure 9.3). This is discarded, the leaves are added, and more water is gently poured over the leaves into the pot or cup. For the finest teas, this is also poured out immediately. This is done to awaken the leaves and remove some of the tannins. More hot water is added, and the leaves are left to steep for a few minutes, then the tea is carefully poured into a series of tiny cups, ensuring that

Figure 9.2. Ingredients for a medicinal Chinese tea.

the same strength is in each cup. The tea leaves are also often reinfused several times, and the different flavors and aromas in each successive steeping are appreciated. It is also important to note that, except for banquets and dim sum, tea is not usually consumed with meals but rather before or after. At the table one also normally serves others from the teapot, but not oneself.

There are also many different types of tea, which were historically categorized by the color of the finished brew—white or colorless, pale yellow, green, blue—which one strains to imagine—and red and black. A precise system of classification is perhaps less important than identifying the different types of Chinese tea one may encounter. First, there is white tea, probably the earliest form of tea, made only from unopened buds that are dried in the shade. The best is said to come from Fujian Province. Green tea is dried after picking, either in the sun, in a basket over a charcoal brazier or even cooked in a wok, and then rolled into various shapes. The leaves dramatically unfurl in the hot water. There are also more modern industrial processes of heating and drying the leaves. The most famous of these is Dragon Well from Hangzhou, though the easiest to find in the United States is gunpowder green tea, which resembles little pellets of gunpowder. Oolong (*Wulong*) is another type of tea made from whole leaves and is semioxidized and

Figure 9.3. The traditional clay pot for making hot medicinal drinks and soups in China.

processed in a long and complicated procedure. The leaves are often rolled into little balls or in long quills. One popular type of oolong is named for the Bodhisattva Kuan Yin (Iron Goddess of Mercy). Last, there is black tea, which is first "withered," then rolled and twisted and then thoroughly oxidized. It has the deepest, richest flavor. There are also other types of tea. *Pu er* is actually fermented and can range in color from green to black and is normally sold as a Frisbee-shaped disk wrapped in paper (figure 9.4). The flavor is deep and earthy, and the finest are even aged for many years. The most famous comes from Yunnan's Six Famous Mountains. There are also teas flavored with jasmine or chrysanthemum or dried lychee. Some are smoked over pine, like *lapsang souchong* (*lapushan xiaozhong*).

Coffee

The places that first come to mind in connection with **coffee** might be Colombia, Brazil, Java, or even today Kenya. But all these locations only started coffee plantations in the colonial era. Coffee is native to Ethiopia. Said to have been discovered by a young goatherd watching his goats become especially frisky after nibbling the red berries of a

Figure 9.4. *Pu er* is a dark, deep tea that is actually fermented.

wild bush, the first use of coffee as a drink is unknown, but it was in general use by the Middle Ages.

Each individual "cherry" of the plant contains two beans, and these are picked only as they ripen over the course of several weeks. Originally these were merely dried and ground and then steeped in water, and coffee is still made this way in Yemen. More commonly they are roasted, which releases aromatic compounds, and then ground and steeped in hot water. The stimulating effect of the drink was noted as early as the tenth century by the Persian physician al-Razi, and coffee eventually became the preferred beverage of devout Muslims, to whom alcohol is forbidden. Clearing the mind and keeping one alert, it was said to be ideal for those engaged in prayer and meditation.

It was not until the seventeenth century that coffee made its way to Europe, and specifically to Venice, where coffee houses opened, imitating those of the Middle East. Soon it spread to France and England and became the most popular drink for any time of day, with the coffee house a prime spot for socializing and discussing politics and business. It has been said that coffee's popularity, especially in northern Europe, can be attributed to the Protestant work ethic and the desire to stay awake for long hours working. Chocolate, with less caffeine and generally soothing and relaxing, was thus more appropriate for aristocrats of southern Europe, for whom indulgence and luxury were the cultural ideals. Italy proves the exception to this rule, for Italians have been avid coffee drinkers longer and more persistently than any other European nation.

The mode of taking coffee in Italy is quite different from elsewhere, though. At least in modern Italy, coffee is taken on the go, and as quickly as possible. Espresso is drunk while standing at a bar, in barely a few sips. In fact, if one asks for *caffè*, it is always a minuscule shot, according to one authority a mere 25 to 30 milliliters of coffee in a 170- to 190-milliliter cup. It is intensely strong, deeply roasted, and bitter, and thus the flavor lingers, and the jolt of caffeine hits almost immediately. Surprisingly, Italian coffee is not as high in caffeine as one might expect; it is always made of arabica rather than robusta, which goes into most American supermarket blends, and has a higher caffeine content.

Some people in Italy do order coffee with milk (*caffè latte*) or *macchiato* with a dab of milk, but cappuccino with frothed, hot milk is only drunk in the morning, sometimes with a roll or pastry. Preparing coffee has also become an art, sometimes successfully imitated outside Italy, but just as often not. The barista takes the ground coffee and tamps it into a mesh-lined container with a handle that fits into the espresso machine. The best machines can cost thousands of dollars. The machine shoots hot water through the grounds at a precise temperature and pounds per square inch of pressure, and the espresso issues from beneath within thirty seconds, with its signature top of creamy froth. At home, most Italians use a stovetop aluminum coffee maker called a *moka* or *caffettiera*, with the familiar double-cone shape, pinched in the middle. The bottom chamber holds the heated water and a container with the grounds. The hot water bubbles up and pushes the liquid coffee into the top chamber, from which it is poured.

Fermented Drinks

The discovery and use of fermented alcoholic beverages is nearly universal among complex ancient civilizations. Whether fruit or grain based, the alcohol content serves to kill harmful bacteria in the water supply and gives the drink a longer shelf life and, of course, confers a pleasant buzz. There is good reason to believe that the neurological and physiological effect of alcohol replicates in more dramatic form what human hormones do naturally. That is, under stress or in pain, the brain can release endorphins and natural painkillers, which ease the burdens of daily toil. Our brains also have a built-in system of editing memories, not only to erase the unpleasant ones but also to save those of importance amid the wash of external stimuli. Again, alcohol and perhaps other intoxicants behave much the same way in helping us forget. It is not merely that they help us relax, but perhaps even make the brain function more efficiently.

Alcohol obviously also serves a social function, lubricating the wheels of interaction in group settings, helping people lose inhibitions that would otherwise cut them off from their community. But alcohol also comes with its own dangers, foremost when consumed in excess. Many civilizations have limited the consumption of alcohol, as with other drugs, to ritual use. The intention was to mark off the state of inebriation from ordinary daily life, when one must be sober and obedient. Only at certain times would ordinary people be allowed to experience this state of ex-stasis, literally being beside one's self, in communion with the gods. At least this was the hope of priestly authorities, whether we are referring to the fermented corn beverages among the Aztecs, wine among the Romans in their Bacchic orgies, or rice and sorghum drinks among the Chinese. Despite these efforts, alcoholic beverages were consumed regularly and came to form an indispensable part of the gastronomic experience.

Beer

Beer is, according to the archaeological record, the first fermented beverage regularly consumed by the earliest civilizations such as the ancient Sumerians and Egyptians, who were avid swillers of beer. This was no doubt the result of grain being their main source of calories, and because crushed grain mixed with water so easily ferments, the discovery of beer, as with bread, was probably inevitable and perhaps accidental. We know that Sumerians merely added water to crushed grains, and their beer was so thick it had to be drunk through a perforated straw to strain out the solids. In a certain sense beer should be considered a food rather than a simple beverage, especially in those places where it provided a good proportion of daily calories, as in northern Europe.

Beer can actually be made of any cereal grain, but typically it is barley, which has been allowed to sprout and is then heated and raked in a process called malting. This is then ground and heated in water and then fermented with wild yeast into beer. The process is, of course, much more complicated today, with specialized brewer's yeast and both bottom and top fermentation. Historically there were many different flavorings added to beer around the world, from herbs such the medieval concoction called *gruit*, to spices, to the now most common hops. These additives both flavor and help preserve beer. Most of these developments occurred in Europe among Germanic peoples, and in fact the ancient Romans considered beer a drink only for their barbaric northern neighbors.

As for our three cuisines, beer such as lager and ale—northern European–style beers—is a relatively recent and foreign addition. In Italy it is mostly in the Alpine north that one finds beers such as Peroni, Moretti, and Forst, though today they are increasingly consumed by younger Italians in bars. Despite the popular success of Mexican beers such as Corona, Dos Equis, and similar brews, this, too, is a fairly recent introduction from Europe, but as around the world, young people prefer beer over traditional drinks. Even China was introduced to beer by the Germans in Tsingtao, one among many now-popular brands. In a certain sense, Mexico and China have always been brewing their own types of beer, the former from corn, the latter from rice.

Rice Beer (Wine)

Although we generally call beverages brewed from **rice "wine"** because of their relatively high alcohol content, Japanese sake, for example, is about 15 or 16 percent, they are really a kind of beer. Actually many American brands of beer, such as Budweiser, use rice in the brewing process, and we might arbitrarily call these sake. In any case, the general word for alcoholic beverages in China is *jiu*, which is sometimes transliterated as *chiu* or *chiew* on a bottle. *Jiu* is usually qualified by type, thus *huang jiu* (yellow alcohol) for rice, *shao jiu* for distilled alcohol (or *bai jiu*, meaning white alcohol), or *pi jiu* for barley-based and other kinds of beer.

Historically, the Chinese have brewed many grains, including millet and sorghum, not merely rice. The most famous of rice wines is from Shaoxing in the central coastal Zhejiang Province, made with glutinous rice and water from Lake Jiang. Although called yellow, its color is more amber to light brown due to oxidation. Thus it resembles in color and vaguely in flavor a Spanish sherry, which is sometimes called for in recipes as a substitute. That is to say, Shaoxing is both consumed as a beverage and used for cooking. Chinese grocery stores often sell the lowest grade of "rice wine" for cooking, but it is worthwhile using better quality brands, which can only be determined by tasting, and to some extent by price. The cheapest rice wine should not be consumed as a beverage. The best are aged for nearly ten years and are often sold not in a bottle but rather a green ceramic gourd-shaped bottle or a brown crock.

There are also lighter rice wines more closely resembling sake, and they, too, can be drunk heated in winter or chilled in summer. These and other varieties are consumed with meals as a regular beverage and are commonly drunk at festivals and weddings for toasts. They can also be used for offerings to the ancestors. These drinks vary by level of sweetness, some being extremely dry such as *yuan hong*. *Hua diao* are a little sweeter, followed by *shan niang*, and then the extremely sweet *xiang xue*.

Corn Beer

Posolli is the original fermented drink of Native Americans of Mexico. It is basically made of the same nixtamalized corn dough used to make tortillas. Balls of *masa* dough are left to ferment under leaves or in a container for several days. A complex interaction of yeasts, bacteria, and molds makes the dough not only last longer but also free nutrients. These lumps are then thinned with water, flavored with salt and chili, or sweetened with honey or sugar and then drunk. It has a relatively low alcohol content. *Posolli* was of great ritual importance to the ancient Maya, remains important among their descendants, and is generally considered an Indian beverage today. *Atole* can also be soured and fermented, and in this form can be served flavored with ingredients like chilies, beans, squash seeds, or herbs.

Pulque

Pulque is a drink made from the fermented sap of the maguey or century plant and was the most widely consumed alcoholic beverage in preconquest Mexico. Among early

Mexican peoples it was a sacred drink, only to be taken on solemn occasions, and offered ritually to the gods. The Aztecs, who called this *uctli*, had strict rules about consumption, and intoxication was severely prohibited. Its production is unique: the plant sends out a stalk after many years of growth (hence the name century plant), the bud of which is cut out, and the sweet sap or *aguamiel* (honey water) is collected and fermented with naturally occurring *Zymomonas* bacteria, rather than yeast. The harvesting of the sap can continue for several months without killing the plant. *Pulque*, once fermented, is a white, thick liquid and is sometimes flavored with fruit juice. It turns rancid very quickly and thus must be consumed fresh. Today it is still made and sold in *pulquerías*, and there is now even a way to can and bottle *pulque* (figure 9.5). However, it has also long been tainted with a social stigma, as a drink of the poor and *campesinos* (peasants), and as cheap beer proliferates, its continued existence may be threatened until perhaps it is rediscovered as a novelty.

Figure 9.5. Mexican young people drink *pulque* in Mexico City. In Aztec times, *pulque* was the highly esteemed drink of the elders, priests, and warriors, a nectar that according to myth oozed from the four hundred breasts of the goddess Mayahuel. © Daniel Aguilar/Reuters/CORBIS.

Wine

Wine is without doubt the most ancient and revered of all beverages in Western civilization. The *Vitis vinifera* grape grows wild throughout the Mediterranean, and its transformation into wine was almost assuredly an accident; crushed grapes in contact with yeast-laden skins ferment all on their own. Wine is in a sense a product of nature rather than human invention. Of course, ancient civilizations nonetheless hoping to curb intemperance cast its invention in mythological terms, either given by the god of disorder Dionysus, or Noah, the first to plant vines and get drunk. Human intervention does also play a major role, in training vines on trellises; culling the grapes; tending to their harvest and methodically crushing, fermenting, and aging of wines. All these were thoroughly understood by ancient Roman times, and there were even detailed treatises on how to invest in viticulture, such as that by Cato the Elder. The Romans also stored their wine in amphorae dated with vintage and origin, and apparently there was a highly developed and sophisticated appreciation of fine wines, stretching from then right to the present (figure 9.6). We must also note that most of the great wine-growing regions of Europe, including Bordeaux and Burgundy, were first planted by the Romans. Wine was also considered nutritious, as the food that most easily converts into blood. Thus, the miracle of the Eucharist was not very difficult to imagine; since 1215 it was decreed that the wine celebrated in the Catholic Mass literally transforms into the blood of Christ, even though the "accidents" or external appearance and taste still appear to be wine. In a nutshell, wine is among the most important foods in Italian culture and religion.

We tend to assume that the attention to soil type, exposure, traditions of oenology—all those factors now expressed as "terroir"—are somehow a recent discovery. In fact, hundreds of years ago, these very same considerations went into wine making. For example, the Veronese physician Giovanni Battista Confaloneri, writing in the 1530s, discussed the way exposure to the sun and soil type determine the flavor of wine, as well as its effect on the body. He described some wines as bitter, some as salty, others as sulfurous, or with a taste of asphalt. Today wine critics use language as poetic.

Methods of wine making, of course, have as significant an impact on the final flavor of wine, and there has been a dramatic improvement in recent decades, which has both brought Italian oenology to the same level as the French and elsewhere and has also encouraged a great deal of innovation in both equipment and wine-making technique. It is also true that the same basic procedures for making wine are used today as thousands of years ago. Grapes are picked when ripe and crushed in a press. Yeast is added, and the mash is left to ferment with the skins, which gives the juice color, tannins, and flavor. The yeast, which forms a cap above the must, is pushed down regularly, and then after a few weeks the solids are strained. The wine is then placed in a wooden barrel or, today, a cement or stainless steel container, where it is free of oxygen and continues to ferment. Sulfates are also usually added at some point to stop the fermentation process. The wine can be consumed anywhere from a month or two to many years after aging in bottles.

Figure 9.6. The funerary monument for Sentia Amarantis, a Roman tavern keeper, shown pouring wine from a barrel into a jug. © Alfredo Dagli Orti/The Art Archive/CORBIS.

Italian wines today fall into several groups based on region and grape varietal. The production of many wines is strictly controlled by law, especially those designated as *denominazione di origine controllata* (DOC), which has been in place since the mid-1960s and now includes more than three hundred types. DOC wines are not only controlled but also guaranteed as well, and these are the most elite of wines. Regular wine is usually labeled *vino di tavola* (table wine), though some more experimental wines may be labeled as such because they do not abide by the legal definitions. Thus, labels are no absolute assurance of quality, but they are a decent indication. For example, Chianti, the familiar Sangiovese–based wine from Tuscany once found everywhere in raffia-covered bottles, may be very expensive, especially in the case of the so-called Super Tuscans. The finest have an unmistakable hint of black cherry. However, those that are made in new ways or with a different proportion of Cabernet cannot be labeled Chianti at all. Furthermore, much of the appeal of certain types comes from astute marketing. The name *Chianti* is everywhere familiar, but today less so the historically more refined Vino Nobile di Montepulciano and Brunello di Montalcino, also made in Tuscany. These are among the best wines made in Italy. Ranking with these are the great wines of Piedmont, based on the Nebbiolo grape variety, called Barolo and Barbaresco, and others made from Barbera. These deep-red wines are extraordinarily complex with hints of truffles and spices, and as they have strong tannins, they age very well.

Apart from these top-end wines, most of the wine produced in Italy and most of which is consumed is both inexpensive and eminently quaffable. Whether the popular light Valpolicella from the area around Verona, or deeper, fruitier reds from the south like Primitivo from Puglia, most Italian wine is meant to be a simple accompaniment to a meal rather than an investment. Most Italian wine is also consumed locally, without fuss or pretention. In any supermarket in Italy one will find a wide selection of local wines for a few dollars per bottle. There are also fine white wines based on grapes such as Pinot Grigio from the Alto Adige or Vernaccia from San Gemignano, as well as excellent sparkling Prosecco. Perhaps only to be found in Italy is the simple rustic and fizzy red wine called Lambrusco. We must also mention sweet wines such as Marsala and the lovely brownish Vin Santo.

Wine also plays a major role in Italian cooking, either a splash in a pan to create a sauce, a glug or two into a tomato sauce, or a whole bottle in which to braise meats or as a marinade. Again, wine is absolutely at the center of Italian culture and its cuisine.

Surprisingly, wine has also been made in China for many centuries. Grapes were introduced there in Han times from the west, though archaeological evidence suggests that it was being made even before then. In Tang times wine drinking became elegant and fashionable. Today the Chinese wine industry is growing, and there is enormous potential for exporters of wine as the Chinese economy flourishes and appreciation for European wines grows. There is also, incidentally, wine made now in Mexico, though interestingly, there was not a major industry in colonial times, and most wine was imported from Spain or later produced in Chile and California, where the climate is more hospitable to viticulture. Today there are major wineries in Baja California and south of the border with Texas.

Distilled Drinks

Europeans learned to distill wine using an alembic from the Arabs in the Middle Ages. Who originally invented the process has been the subject of great controversy, and there were types of distillation in ancient times, but it is generally agreed that it was not invented in Western Europe. Distillation of alcohol was first done by alchemists and physicians seeking a restorative beverage that would prolong life, and such spirits or *aqua vita* (water of life) continued to carry medicinal associations nearly up to the present. For example, a cordial was originally meant to soothe the heart (*cor* in Latin). The idea was that as wine was thought to nourish the body by being converted into blood, a concentrated form of wine would do so more directly without the effort of digestion.

The aperitif, likewise, was thought to open the body's passages by means of a bitter scouring property, as a prelude to a meal, and a digestif would help the "concoction" of food in the stomach as it was being broken down by heat.

The first popularly consumed forms of distilled spirits in Italy was *aqua vita*, which was merely wine heated in a ceramic or glass alembic until the steam rose and passed through a series of cooler coils, wherein the alcohol would condense and drip into another vessel. In northern Europe this would be called brandy, a corruption of the Dutch phrase for "burnt wine," which after being aged in wooden casks took on a darker color and more complex aroma. In Italy the spirits would have been consumed unaged and clear or would be flavored with any number of medicinal herbs and spices such as cinnamon or juniper or green herbs such as wormwood or anise. These are all direct ancestors of drinks that still survive as schnapps, gin, absinthe, and sambuca. Complex combinations of herbs were also used in early distillates and still are in liqueurs such as the lurid yellow Galliano and Strega (meaning the "witch"), or amaretto, which is flavored with apricot kernels, not almonds, as is commonly thought. Italians also make their own liqueurs, such as *limoncello* with lemons or *nocino* with walnuts soaked in alcohol, though these liqueurs can be purchased too, as can Frangelico made with hazelnuts. Among the digestives one also finds surprises, such as Cynar made with artichokes, or Fernet, a dark, bitter after-dinner drink with an alluringly complex flavor. There are also many Italian aperitifs, which are essentially wine flavored with herbs and fortified with alcohol, such as vermouth made by Martini and Rossi and Cinzano. There is also the much stronger and luscious Campari, flavored with orange peel, herbs, and spices.

The spirit most often associated with Italy, however, is grappa, which can be made either directly from wine or from the leftover grape skins after pressing, fermentation, and then distillation. The latter process tends to be harsh and powerful, while the former, though clear, retains the bouquet of the wine varietal from which it was made. Grappa is rarely used to mix cocktails, but rather it is taken straight, after dinner as a digestive. In Italian shops one always sees elaborate blown-glass bottles containing the most expensive grappa. Today, of course, Italians consume spirits from around the world as well.

China has a long history of distillation, and many historians contend that it was Chinese alchemists who invented the process, and it was transmitted from there to the Mediterranean. Many Chinese distillates are made from sorghum and have long histories, such as *megui lu jiu* (rose dew alcohol), which resembles Italian rosolio. *Fen jiu*, in contrast, is a potent, clear drink, more like gin, while *wu jia pi* has been compared with whiskey. There is also *gaoliang*, a potent sorghum distillate from Taiwan. The Yushan brand from Jade Mountain, the highest peak in Taiwan, is widely available in the United States in Chinese groceries and comes in a brown crock. It has a delicate, floral bouquet that is reminiscent of melons and jasmine. One has to wonder why such exquisite products have not made their way into mainstream U.S. culture, as have spirits from so many other parts of the world. There are also dozens of medicinal alcohols flavored with therapeutic herbs like ginseng and exotic ingredients such as snakeskin. The most famous Chinese spirit is called *Moutai* from Guizhou, which is clear and potent and similar to grappa or vodka and is served on state occasions. Every locality has its own distilled beverage, and the variety is as wide as anywhere else on earth.

In Mexico, the great celebrated spirit is tequila, which is a colonial invention, though based on the native blue agave plant, which looks like a huge pineapple. In pre-Columbian times the related *maguey* (century plant) was brewed into the aforementioned alcoholic beverage *pulque*, and it has been speculated that the first tequilas were merely *pulque* distilled. Today, however, tequila is made from the entire plant, with the outer leaves removed, then the core cooked and distilled, not merely the sap of the agave fermented (figure 9.7). Tequila originates from a city of the same name in the state of Jalisco in central western Mexico, and today there are hundreds of brands that vary greatly in quality. Some are made from 100 percent blue agave; others are distilled with sugar and are thus a kind of tequila and rum mixture. White tequila is the most familiar variety, which is relatively bland and is usually mixed into cocktails such as the margarita. Nothing like the drink

Figure 9.7. Piñas are the inner core of blue agave, used to make tequila.

made with horrid, syrupy mixes one finds in the United States, a properly made margarita consists only of lime juice, tequila, and a dash of orange liqueur such as triple sec or, more extravagantly, Grand Marnier. The rim is usually salted as well, but the drink is not sweet or frozen. In an entirely different class from *blanco* are the premium-aged tequilas, which are sipped, more like a brandy, and generally are not used in mixed drinks. Classifications include *oro* (gold), which is usually just colored with caramel to resemble aged tequila, *reposado*, which has spent some time in an oak barrel gaining color and some complex aroma of vanillin and other compounds, and *añejo*, which has been aged several years, and lastly the more recent *maduro*, which has been aged at least three years. Mezcal is a similar drink, made in Oaxaca and other regions outside the official denomination of tequila, using a different process and not necessarily with blue agave. The worm at the bottom of the bottle is basically just a gimmick conjured up in the last century.

When we consider spirits around the world, the general trend has been for huge, multinational corporations to buy up smaller manufacturers. In the past decade consolidation has left the majority of distilling operations in the hands of a few companies such as Diageo, which owns brands as disparate as Cuervo tequila, Tanqueray gin, Smirnoff vodka, and Captain Morgan rum. Whether this trend will push out the smallest producers of fine tequila, traditional grappa, and still relatively unknown Chinese spirits is yet to be determined. However, the fact that the Chinese have become avid consumers of cognac and Italians and Mexicans are still as fond of cocktails that are now found everywhere does not bode very well for the small operations.

Study Questions

1. Explain how attitudes toward water have changed through history and why.
2. How did soda pop originate, and what is its current use? How do many beverages follow this same trajectory?
3. Speculate why caffeinated drinks have become so popular in capitalist countries.
4. Why are alcoholic beverages so important to the rise of civilizations?
5. How is beer made, and what forms does it take around the world?
6. Why is wine invested with mythical significance?
7. Consider why various places make spirits from different ingredients and flavorings. How does their usage also differ in these places?

Recipes for Drinks

Nonalcoholic Drinks

HORCHATA (RICE DRINK)/MEXICO

Horchata is descended from a Spanish drink of the same name, but it is made with *chufa* (tiger nuts) pounded with water. The word *horchata* suggests that it was first made with barley (*hordeum* in Latin). The Mexican version usually uses rice. Originally this would have been made with pounded rice soaked in water. You can grind the dry rice first in a blender dry, or with water. Either way works fine. It should be about a cup of rice to 4 cups of water. Let this sit, with a cinnamon stick, sugar, and vanilla, in the refrigerator for several hours so the rice is completely hydrated and the flavors meld. Drink cold with a dusting of cinnamon.

AGUAS FRESCAS (FRUIT WATERS)/MEXICO

Along with *horchata*, in Mexico one also finds a whole range of refreshing fruity drinks, which are simple to make by soaking fruit in sweetened water. The most popular are tamarindo—made with the pulp of fresh, sour tamarind pod—and jamaica, which is made by infusing the deep-red, dried hibiscus flowers into tea and then chilling. Mangoes, papayas, prickly pear cactus (*tunas*), lime, watermelon, and strawberry are all popular. The fruit is simply pureed coarsely with water, sugar, and perhaps lime juice and left to infuse, and then can be strained or not. Even more simply, the fruit is coarsely mashed with a fork, and sweetened water is added.

TEA/CHINA

It is customary in the West to simply steep tea leaves in a big pot and pour off as desired. Traditionally and to this day it is done quite differently in China. The leaves can either be placed in a small ceramic pot or directly in the cup. The water is brought up to just below the boiling point, so as not to shock the leaves, drawing out bitterness. The hot water is poured over the leaves and immediately poured out, to awaken the flavors. Then more hot water is added as the first infusion. The leaves are used again, several times. Each successive infusion is said to yield a slightly different flavor and aroma, not merely being weaker. The tea is not usually sweetened, and milk is not added. Tea is drunk mostly before a meal and after. When a guest arrives, it is always customary to offer tea, handing and receiving the cup with both hands.

BUBBLE TEA/CHINA

Originating in Taiwan, these recently very popular drinks are somewhere between a starch meal and a beverage, if one thinks of the contents. They can be fruit flavored or milk based. This is just one version. First boil tapioca "pearls" until soft, let soak according to package directions, and then cool under running cold water. Brew green or black tea and sweeten with white and brown sugar. Let cool. When ready to drink, mix the tapioca with the tea and milk. It is strangely refreshing and chewy at the same time. Instead of milk you can add fruit syrups or even pureed fruit, which closely approaches a smoothie, but it is still very good.

GINSENG TEA/CHINA

Ginseng is a root said to be a universal therapeutic in Chinese medicine, hence the Latin genus name *Panax*, meaning "panacea." The root is shaped like a man, and this may account for the medicinal associations, though modern science does record many benefits. The best ginseng is said to come from Korea, though it is also grown in China and North America. There are relatively inexpensive instant versions and little pieces of root that can be bought and steeped for tea. A soup is also made by boiling, chopping, and steaming the root for hours, which brings out its therapeutic strength. Whatever form you find or can afford, the tea does have a stimulating and invigorating aroma.

ITALIAN SODAS

In Italy, there are many intriguing sodas premixed and tiny bottles of bitters, such as Chinotto and Limonata. There are also premixed syrups. The Torani brand is the most familiar in the United States and Italy. It comes in dozens of flavors. The syrups are simply mixed with soda water and served on ice. Despite these modern manifestations such syrups ultimately descend from medicinal syrups popular since the Late Middle Ages. These could be based on dark cherries (on which cough syrups are still based) or sour and acerbic fruits to combat the summer's heat. They are made simply by boiling down the fruit with sugar, straining, and reducing. Or in the case of fruits that can be juiced, such as pomegranates, the liquid was simply cooked down, sometimes with the addition of warming spices. These were diluted with cool water and drunk and are without doubt the ancestor of modern commercially branded sodas, which also had medicinal associations at first. Although today carbonated water is most frequently used, in the past natural bacterial fermentation would have given any of these fruit- and root-based syrups some fizz in sealed bottles. (This is also the origin of root beer.) Making your own homemade soda is as simple as letting any fruit- and sugar-based liquid ferment for a few days open and then capping it, with a little extra sugar, which will convert into just a little alcohol and create CO_2 bubbles.

Alcoholic Drinks

MARGARITA/MEXICO

Despite the frozen, fruit-flavored concoctions one commonly sees in the United States, the margarita (meaning "pearly") is one of the simplest and most elegant of drinks. A mix is out of the question. Start with a small plate of coarse sea salt. Rub the rim of the glass with the cut edge of a lime and dip the entire rim in salt. In a cocktail shaker mix two parts clear tequila, one part lime juice (small Key limes, if you have them), and one part orange liqueur such as triple sec, Cointreau, or Grand Marnier with some ice cubes. Shake gently and strain into the salt-rimmed glass. Garnish with a slice of lime perched on the rim.

NEGRONI/ITALY

Evocative of the 1950's era *Dolce Vita* obsession with Italy, this drink is slightly bitter and perfectly refreshing on a hot summer's day as an aperitif. It is a variation of a simpler cocktail made simply with Campari, sweet red vermouth, and soda water in equal proportions. This version adds one part gin to the Campari and vermouth, garnished with a slice of lemon.

BELLINI/ITALY

This is the simplest of drinks, invented at Harry's Bar in Venice, and apparently named for a certain pale yellow shade of cloth in a fifteenth-century painting by Giovanni Bellini. Take a white peach; peel, pit, and puree it in a blender. Put a few tablespoons in the bottom of a champagne flute and slowly pour Prosecco over. This is the local sparkling white wine. It can be made with other fruit purees as well and goes under various other names. The basic flavor combination is not something new, though. Peaches or melons cut into cubes and soaked in white wine was a popular drink since the Renaissance, so perhaps the Bellini is unwittingly well named.

LIMONCELLO (LEMON LIQUEUR)/ITALY

Although you can purchase Limoncello in a bottle, it is much better to make at home, where you can sweeten to taste. The commercial versions tend to be sickly sweet and lower in alcohol content. You must be patient though, as it takes a long time for the flavors to develop. Remove the peel, but not the white pith, from six clean, organic lemons. If you use regular store-bought ones, be sure to scrub well. Put the peels in a bottle of the strongest grain alcohol you can find. The alcohol content differs from place to place, but be sure to use neutral spirits rather than 40-proof vodka. Let this steep for about a month, or until it is golden colored and fragrant. Then pour off about ¼ to ½ cup, and add a little cooled sugar syrup to replace what you have poured off. (This is made simply by melting ½ cup sugar and ½ cup water and reducing.) Let this sit another month or up to a year. If you like, you can strain out the lemon peels and decant into a fancy bottle. This is most commonly drunk after a meal, though it is very nice to start as well.

NOCINO (WALNUT LIQUEUR)/ITALY

Also homemade, this is a digestive liqueur made from green, immature walnuts. It was very popular around Modena, where a festival is held in its honor. The walnuts are traditionally picked on June 24, on the feast of St. John the Baptist, when the shells are not yet formed inside. You simply cut about thirty walnuts into quarters. Be sure to wear gloves. Cover these with 2 cups sugar and let them sit for about two days. The liquid in the walnuts will dissolve the sugar, and they will begin to become fragrant. Then add grain alcohol in a big jar to cover and wait as long as you can. Add spices like whole cinnamon and cloves to the jar as well. After at least several months, strain the dark liqueur into another bottle. It continues to improve with age.

CHINESE COCKTAILS

China is not generally known for mixed drinks, although Western-style ones can of course be ordered in bars. Usually spirits are drunk straight. Nonetheless, a clear Maotai made of sorghum or other liquor makes an interesting neutral base for cocktails. The juice squeezed from grated ginger, lychee, or any fruit all go very nicely in a cocktail. Try making melon juice or watermelon juice by squeezing the flesh through a cloth, or making an infusion of kumquats. This is still a relatively unexplored realm.

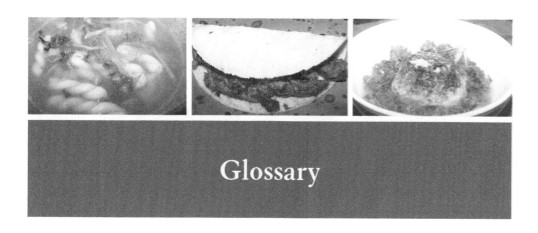

Glossary

acetobacter Naturally occurring bacteria that convert alcohol into acetic acid or vinegar. They require oxygen for this process, which is why traditional vinegar barrels or crocks are usually unsealed.

alembic A word derived from Arabic referring to a still used for the distillation of alcohol from grape wine, fermented grain, or other fermented liquid. It is usually made of copper, though a glass flask is structurally similar. It normally includes a pot, which is placed over a flame; a cooler spiral coil in which the alcohol condenses; and a receptacle to catch the alcohol.

annatto/achiote The seed of a tropical plant, *Bixa orellana*, either crushed into a paste or used to color oil and subsequently used in cooking to impart a bright, yellowish-orange color and subtle, peppery aromatic flavor.

aspic Any meat or vegetable set into a flavored, usually savory, gelatin base and served cold. It may be served in thin slices or with contents in a decorative arrangement within a relatively clear, molded shape of gelatin.

Aztecs The imperial power that dominated central Mexico from the fourteenth to early sixteenth centuries, with whom the Spanish conquistadors fought under Hernan Cortés.

bagna cauda A "hot bath" or sauce based on olive oil, anchovies, and garlic, sometimes with other ingredients, kept warm in a ceramic pot into which raw vegetables are dipped.

bagnet vert A Piedmontese green sauce based on parsley, garlic, anchovies, oil, and vinegar, usually served with boiled meats.

bain marie A water bath used to gently cook food, usually sealed in a vessel to prevent scorching.

battuta A mix of finely chopped onion, celery, and carrots.

black kale (Italian, *cavolo nero*). A thin, dark variety of kale used in Italian dishes such as *ribollita*, a thick bean and bread soup.

braise/braising A cooking technique wherein food is gently simmered in a small amount of liquid, covered, for a long period to render meats or other foods tender while remaining whole.

cabrito A young kid goat or dish made from goat in Mexico. It is also served in Italy for Easter, where it is called *capretto*.

chitarra A box string with metal strings over which a sheet of pasta is laid and rolled with a rolling pin to create thin, flat noodles. The word is cognate with the term *guitar*.

choux paste A term in French cuisine, derived from the word for "a little cabbage." It is a type of dough made with boiling water and butter into which flour is added and then eggs. Baked, it is used in cream puffs and fried for dishes like fritters and churros.

cochinita A suckling pig that can be cooked in various ways: roasted, or in a steamer box called a Caja China in Cuba, or in the classic *cochinita pibil*, cooked in an earth oven in the Yucatan.

comal A small, flat, circular cooking griddle used in Mexico for tortillas and for charring vegetables. Originally made from clay, these griddles are now almost always made of steel or iron, seasoned much like cast iron with fat.

congee Chinese porridge.

contorno A side dish, typically of vegetables or potatoes, that accompanies a main dish in Italian cuisine.

court bouillon A simple broth made of vegetables, sometimes flavored with lemon juice or vinegar. Used to poach other ingredients such as fish.

dim sum A meal typically eaten in a Chinese restaurant specializing in this type of cooking, in which a variety of small dishes, including dumplings and other foods that can be eaten in one or two bites, are brought to the table from which diners choose. Tea is the traditional accompaniment.

dulce de leche A confection made from milk and sugar slowly cooked down until thick, spread on bread or used as a filling for pastries.

extrusion The process of squeezing a fine paste of ingredients through a tube with the use of a syringe, cookie press, or pastry bag.

five spice powder A dark, aromatic mix of spices used in Chinese cuisine consisting of ground cassia, star anise, cloves, Sichuan pepper, and fennel seeds. The ingredients sometimes vary.

fond The mass of browned particles that accumulates at the bottom of a pan when sautéing ingredients, especially meats. It forms the base for a sauce when liquid is added and the bits are scraped up.

fritelle A kind of fried doughnut popular in Venice at Carnival time—not to be confused with *pizzelle*, which is a sweet and crisp flat, wafflelike confection cooked with a long-handled iron and often flavored with aniseed.

garum An ancient Roman sauce based on fish such as anchovies or mackerel. Modern *colatura alici* is similar, though made from juice poured from salted anchovy filets. The original was made with whole ungutted fish, as are Asian *nuoc mam* and *nam pla* fish sauces today.

gordita Meaning "little fat one" in Spanish, it is a corn-based cake stuffed with other ingredients.

Han dynasty A formative period in Chinese history, 206 BC to AD 220, roughly coinciding with ancient Rome.

hominy Whole corn kernels that have been soaked in lye or calcium hydroxide, a process called **nixtamalization**, and from which the seed coat has been removed. Eaten in stews in Mexico. The same corn is also ground to make tortillas.

horno A clay oven set on the ground, introduced to Mexico by the Spanish, used to bake bread.

Instacure The proprietary name for salt and sodium nitrite (6.25 percent) used to cure meats. Instacure #1 is used mainly for fresh cooked sausages, and Instacure #2 is used for dried, fermented products and also contains sodium nitrate, which slowly converts into nitrite though bacterial fermentation. Potassium nitrate or saltpeter was the traditional cure but is less dependable.

jiaozi A family of steamed dumplings in China containing ground meat and sometimes vegetables, which can be steamed, boiled, or fried.

kumis Fermented mare's milk from central Asia, now sold as a health food and often made from cow's milk.

lardo Cured pork fat, the classic version of which is *Lardo di Colonnata* from Tuscany, which is cured in marble coffers and traditionally served raw and sliced thinly.

locavore A person who only eats food drawn from a limited geographical radius to conserve fuel used in shipping, to minimize pollution, and in order to eat seasonal and fresh ingredients.

lovage An herb similar to celery leaves, usually cooked in soups or stews.

mandoline A rectangular box with an inset *v*-shaped or diagonal blade, over which foods are passed to create extra-thin slices or decorative shapes, such as the waffle-cut french fry style.

masa Dough made of nixtamalized ground cornmeal, made into tortillas or tamales, though the proportion of ingredients differs with each.

Mestisaje The official policy of Mexico recognizing the mixed ethnicity of the people descended from Native American and Spanish ancestors.

metate A saddle-shaped stone slab used for grinding corn into *masa*, the dough used for tortillas and tamales.

mezzaluna A curved, half-moon-shaped blade with handles on either end used to chop herbs by rolling the blade back and forth on a board or in a special shallow wooden bowl.

milpa A small backyard garden in which corn and vegetables are grown for household use in Mexico. The varieties of plants tend to be traditional and suited to the soil, and seeds are saved from season to season, unlike in industrial agriculture.

Ming dynasty The ruling dynasty of China from 1368 to 1644.

molcajete A bowl-shaped mortar with three legs made of volcanic stone used to pound relatively soft ingredients such as tomatoes or avocados.

molé Any sauce in Mexico, the most famous of which is the molé poblano of Puebla made with dozens of ingredients including chilies, spices, nuts, and a small amount of chocolate.

molinillo A wooden rod with a carved finial and free-floating disks that is used to froth hot chocolate when inserted into a mug and rolled back and forth between the palms of both hands.

mouthfeel The sensory reception of foods when entering the mouth and being chewed, which includes textures such as slimy, crunchy, chewy, sticky, and so forth.

must Pressed grape juice with the skins and seeds before fermentation into wine.

Nahuatl The language of the **Aztecs**. Versions of the language are still spoken by approximately 1.5 million indigenous peoples in Mexico. From this language are derived our words for tomato, avocado, chili, and chocolate.

Napa cabbage A long, thick variety of cabbage with tender leaves that is used widely in Asian cooking.

nixtamalization The processing of field corn with lye or calcium hydroxide so the kernel swells and the outer seed coat can be removed. The basis for hominy/posole and corn masa. The process also makes available nutrients otherwise indigestible in regular corn.

olla A traditional rounded, earthenware clay pot in Spain and Mexico.

Olmec The first major civilization in Mesoamerica, thriving roughly from 1500 to 400 BC around Veracruz and from which many cultural traits, including culinary, derive.

pancetta A rolled and cured but typically unsmoked type of bacon that is browned and melted as the base of other dishes in Italy.

panadería A Mexican bakery.

panettone A yeasted, slightly sweet bread enriched with eggs and studded with candied fruits. Eaten as a dessert in Italy. Many similar types of bread abound, such as *pandoro* and the dove-shaped Columba Pasquale for Easter.

parchment paper Paper used in baking that will not burn even in very hot ovens. Used underneath breads and pastries, it obviates the need for grease, which can burn at high temperatures. It can also be folded into little packages for cooking fish and other delicate ingredients.

pastería A shop that makes fresh pasta in Italy.

peel A wedge of wood or metal affixed to a long handle used for baking bread, pizza, and other foods that are cooked directly on the floor of an oven.

piloncillo A small, cylindrical cone of brown, unrefined sugar used in Mexican cooking.

polenta Any grain cooked with water or other liquid to create a porridge, most typically made with coarsely ground, but not nixtamalized, corn.

posole Another term for **hominy**, as well as a dish made from whole nixtamalized corn kernels.

prosecco A light, effervescent white wine from Italy that is sometimes compared with champagne.

pulque Fermented milky juice from the maguey (agave or century plant), the ancestor of the base for what is now distilled tequila. Considered sacred among the Aztecs.

qi In traditional Chinese medicine the energy force that prevents illness, sustains the body, and flows throughout the universe. It can strengthened by eating certain foods, and the proper distribution of *qi* through the body can be aided with acupuncture.

quern A rotary, hand-grain mill on which a stone top turns on a convex base incised so as to finely crush the grain. The earliest milling technology after a simple mortar.

queso fresco A number of fresh cheeses used in Mexican cookery.

ramekin A small, cylindrical ceramic vessel, usually made of red earthenware or porcelain, in which foods are baked or poached in a *bain marie*.

ricer A cylindrical press that passes the cooked potato through small holes.

risotto A typical rice dish of northern Italy made with a starchy, short-grained variety such as arborio, vialone nano, or carnaroli.

roux A cooked mixture of flour and fat used to thicken and flavor sauces. Originating in seventeenth-century classical French haute cuisine but used widely around the world today. It is normally based on butter, but in Louisiana it is oil based.

salame The Italian term for pork meat ground, cured in a casing, and eaten thinly sliced and raw. Typically spelled *salami* in the United States. *Salumi* refers to all cured pork products. Some varieties are based on or include beef or other meats such as horse or donkey. There is also a *cotto* or cooked variety, which is the one familiar in U.S. groceries.

salumeria An Italian shop that specializes in cured meats.

Shaoxing wine A brownish, oxidized wine or, technically, beer, based on rice used for drinking and in cooking. This is only the finest variety of many rice wine types.

Slow Food Movement An organization founded in Italy whose goals are diametrically opposed to fast food and devoted to saving food traditions, threatened species, and dying culinary arts.

slurry Any mixture of ingredients usually used to thicken soups or stews. It can be made from flour and water or cornstarch or other starches as well.

soffrito An Italian technique similar to the Spanish *sofrito* in which finely diced aromatic vegetables such as onions, celery, and carrots are gently sautéed as the base for complex dishes. Around the globe, versions may include peppers or tomatoes as well.

Song dynasty The ruling dynasty of China between 960 and 1279, notable for the development of a navy and the invention of gunpowder, the compass, and other devices. From 1127 to 1279 it was referred to as the Southern Song after northern China was defeated by Mongols.

sous-vide A modern cooking technique in which food is cooked very slowly under water sealed in a plastic bag at precise temperatures. Chicken à la King boiled in a bag is an early industrial version of the technique.

sweetbreads The thymus gland of animals, usually calves. It is a very light, soft meat, cooked gently without assertive sauces.

Tang dynasty The ruling dynasty of China from 618 to 907, a period of greatness when China was connected to the West via the Silk Road.

tannin An acid found in tea, tree bark, and other sources that has an astringent puckering effect in the mouth, which gives the sensation of being cleansing.

terroir The idea that food should reflect the soil, climate, and processing traditions of a particular region. It was first applied exclusively to wine and the properties imparted by particular locales and is now used for many other foods as well.

Three Kingdoms A period of disunity and turmoil in China when the Wei, Shu, and Wu kingdoms succeeded the Han dynasty in the 200s.

toothsome A term used to describe foods that are chewy and delicious.

torta In Mexico, a kind of sandwich on a roll; in Italy, a flat, baked tart.

totopos Mexican term for fried tortilla chips, served with dip.

umami Refers to the fifth flavor, occurring naturally in foods like mushrooms and Parmesan cheese but also found in any product high in glutamates, including monosodium glutamate. It is said to intensify other flavors.

verjuice The unripe juice of grapes or sometimes other fruits, used widely in medieval cooking as a souring agent and still used sometimes in place of lemon juice or vinegar.

yin/yang The two primordial opposite forces in traditional Chinese medicine and cosmography that govern the universe and the human body. They are maintained in balance through diet, exercise, and other factors such as weather. Foods that are hot or cold, soft or hard, dark or light are also described as yin or yang for the effects they have on the body.

Bibliography

General Reference Works

Albala, Ken. *Food Cultures of the World Encyclopedia*. Santa Barbara, CA: Greenwood Press, 2011.
CIA (1997). See the Perry-Castañeda Library Map Collection at the University of Texas. Mexico: www.lib.utexas.edu/maps/americas/mexico_pol97.jpg
CIA (2001). Perry-Castañeda Library Map Collection at the University of Texas. China: www.lib.utexas.edu/maps/middle_east_and_asia/china_rel96.jpg
CIA (2006). Perry-Castañeda Library Map Collection at the University of Texas. Italy: www.lib.utexas.edu/maps/europe/italy_admin_06.pdf
Davidson, Alan. *The Oxford Companion to Food*. Oxford: Oxford University Press, 1999.
Fernández-Armesto, Felipe. *Food: A History*. Oxford: Macmillan, 2001.
Hu, Shiu-ying. *Food Plants of China*. Hong Kong: Chinese University Press, 2005.
Katz, Solomon H., ed. *Encyclopedia of Food and Culture*. New York: Charles Scribner's Sons, 2003.
Kiple, Kenneth. *A Moveable Feast: Ten Millennia of Food Globalization*. Cambridge: Cambridge University Press, 2007.
Kiple, Kenneth, ed. *The Cambridge World History of Food*. Cambridge: Cambridge University Press, 2000.
Sauer, Jonathan D. *Historical Geography of Crop Plants*. Boca Raton, FL: CRC Press, 2004.
Strong, Roy. *Feast: A History of Grand Eating*. London: Jonathan Cape, 2002.
Tannahill, Reay. *Food in History*. New York: Crown Publishers, 1998.
Toussaint-Samat, Maguelonne. *History of Food*. Oxford: Basil Blackwell, 1992.
van Wyk, Ben-Erik. *Food Plants of the World*. Portland, OR: Timber Press, 2005.
Zohary, Daniel, and Maria Hopf. *Domestication of Plants in the Old World*. Oxford: Clarendon Press, 1993.

Scholarly Sources

Anderson, E. N. *The Food of China*. New Haven, CT: Yale University Press, 1988.

Cappatti, Alberto, and Massimo Montanari. *Italian Cuisine*. New York: Columbia University Press, 2003.

Chang, K. C., ed. *Food in Chinese Culture*. New Haven, CT: Yale University Press, 1977.

Cheng, F. T. *Musing of a Chinese Gourmet*. Hong Kong: Earnshaw Books, 2011.

Coe, Sophie. *America's First Cuisines*. Austin: University of Texas Press, 1994.

Coe, Sophie, and Michael D. Coe. *The True History of Chocolate*. London: Thames and Hudson, 1996.

Diamond, Jared. *Guns, Germs, and Steel*. New York: W. W. Norton, 1997.

Dunlop, Fuchsia. *Shark's Fin and Sichuan Pepper*. New York: W. W. Norton, 2008.

Foster, Nelson, and Linda S. Cordell, eds. *Chilies to Chocolate: Food the Americas Gave the World*. Tucson: University of Arizona Press, 1996.

Fraser, Evan D. G., and Andrew Rimas. *Empires of Food: Feast, Famine and the Rise and Fall of Civilizations*. New York: Free Press, 2010.

Fussell, Betty. *The Story of Corn*. New York: North Point Press, 1992.

Gentilcore, David. *Pomodoro: A History of the Tomato in Italy*. New York: Columbia University Press, 2010.

Gollner, Adam Leith. *The Fruit Hunters*. New York: Scribner, 2008.

Helstosky, Carol. *Garlic and Oil: Food and Politics in Italy*. Oxford: Berg, 2004.

Jones, Martin. *Feast: Why Humans Share Food*. London: Oxford University Press, 2007.

Lee, Jennifer. *The Fortune Cookie Chronicles: Adventures in the World of Chinese Food*. New York: Twelve, 2008.

Lee, Seung-Joon. *Gourmets in the Land of Famine: The Culture and Politics of Rice in Modern Canton*. Palo Alto, CA: Stanford University Press, 2011.

Lin, Hsiang Ju, and Tsuifeng Lin. *The Art of Chinese Cuisine*. Clarendon, VT: Tuttle, 1996.

Liu, Xinru, and Lynda Norene Shaffer. *Connections across Asia: Transportation, Communication and Cultural Exchange on the Silk Roads*. New York: McGraw-Hill, 2007.

Long-Solis, Janet, and Luis Alberto Vargas. *Food Culture in Mexico*. Westport, CT: Greenwood Press, 2005.

Mair, Victor H., and Erling Hoh. *The True Story of Tea*. London: Thames and Hudson, 2009.

Mazoyer, Marcel, and Laurence Roudart. *A History of World Agriculture*. New York: Monthly Review Press, 2006.

McNeil, Cameron L., ed. *Chocolate in Mesoamerica*. Gainesville: University of Florida Press, 2006.

Newman, Jacqueline. *Food Culture in China*. Westport, CT: Greenwood Press, 2004.

Norton, Marcy. *Sacred Gifts, Profane Pleasures: A History of Tobacco and Chocolate in the Atlantic World*. Ithaca, NY: Cornell University Press, 2008.

Parasecoli, Fabio. *Food Culture in Italy*. Westport, CT: Greenwood Press, 2004.

Pilcher, Jeffrey M. *¡Que Vivan Los Tamales!* Albuquerque: University of New Mexico Press, 1998.

Pilcher, Jeffrey M. *The Sausage Rebellion*. Albuquerque: University of New Mexico Press, 2006.

Schafer, Edward H. *The Golden Peaches of Samarkand: A Study of T'ang Exotics*. Berkeley: University of California Press, 1963.

Schafer, Edward H., and Benjamin E. Wallacker. "Local Tribute Products of the T'ang Dynasty." *Journal of Oriental Studies* 4:213–48 (1957–1958).

Schivelbusch, Wolfgang. *Tastes of Paradise: A Social History of Spices, Stimulants and Intoxicants*. New York: Vintage, 1992.

Serventi, Silvano, and Françoise Sabban. *Pasta: The Story of a Universal Food*. New York: Columbia University Press, 2000.

Simoons, Frederick. *Food in China: A Cultural and Historical Inquiry*. Boca Raton, FL: CRC Press, 1991.

Sinoda, Osamu. "The History of Chinese Food and Diet." *Progress in Food and Nutritional Science* 2:483–97 (1977).

Smith, Andrew. *Peanuts*. Urbana: University of Illinois Press, 2002.

Swislocki, Mark. *Culinary Nostalgia: Regional Food Culture and the Urban Experience in Shanghai*. Palo Alto, CA: Stanford University Press, 2009.

Symonds, Michael. *A History of Cooks and Cooking*. Urbana: University of Illinois Press, 1998.

Van Bath, Slicher. *The Agrarian History of Western Europe*. London: Edward Arnold, 1963.

Visser, Margaret. *The Rituals of Dinner*. New York: Grove Weidenfeld, 1991.

Wrangham, Richard. *Catching Fire: How Cooking Made Us Human*. New York: Basic Books, 2009.

Wright, Clifford. *A Mediterranean Feast*. New York: William Morrow, 1999.

Zhang Wengao, Jia Wencheng, Li Shupei, Zhang Jing, Ou Yangbing, and Xu Xuelan. *Chinese Medicated Diet*. Shanghai: Publishing House of the Shanghai College of Traditional Chinese Medicine, 1990.

Cookbooks

Bastianich, Lidia. *Lidia's Family Table*. New York: Knopf, 2004.

Bastianich, Lidia. *Lidia Cooks from the Heart of Italy*. New York: Knopf, 2009.

Bayless, Rick. *Authentic Mexican*. New York: William Morrow, 2007.

Chen, Teresa. *A Tradition of Soup*. Berkeley, CA: North Atlantic, 2009.

De Medici, Lorenza. *The Renaissance of Italian Cooking*. New York: Fawcett, 1989.

Dunlop, Fuchsia. *Land of Plenty: Authentic Sichuan Recipes*. New York: W. W. Norton, 2001.

Dunlop, Fuchsia. *Revolutionary Chinese Cookbook: Recipes from Hunan Province*. New York: W. W. Norton, 2006.

Hazan, Marcella. *Essentials of Classic Italian Cooking*. New York: Knopf, 1992.

Hom, Ken. *Complete Chinese Cookbook*. Richmond Hill, Ontario: Firefly Books, 2011.

Hsiung, Deh Ta, and Nina Simonds. *The Food of China*. Sydney: Murdoch, 2001.

Kennedy, Diana. *The Cuisines of Mexico*. New York: Harper and Row, 1986.

Kennedy, Diana. *The Essential Cuisines of Mexico*. New York: Clarkson Potter, 2000.

Kennedy, Diana. *From My Mexican Kitchen*. New York: Clarkson Potter, 2003.

Lo, Eileen Yin-Fei. *Chinese Kitchen*. New York: William Morrow, 1999.

Santibanez, Roberto, Romulo Yanes, and J. J. Goode. *Truly Mexican: Essential Recipes and Techniques for Authentic Mexican Cooking*. Hoboken, NJ: Wiley, 2011.

Tom, Irene Kwok. *The Mystique of Chinese Culinary Creations*. Orinda, CA: Irene Kwok Tom, 2006.

Yan, Martin. *Martin Yan's China*. San Francisco: Chronicle, 2008.

Yan, Martin. *Martin Yan's Feast*. San Francisco: Bay Books, 2003.

Young, Grace. *The Breath of a Wok*. New York: Simon and Schuster, 2004.

Young, Grace. *The Wisdom of the Chinese Kitchen*. New York: Simon and Schuster, 1999.

Index

abalone, 4, 266, 273–74
Acapulco, 266, 276, 298
acetobacter. *See* bacteria *and* vinegar
achiote, 230, 297, 300, 318, 339; paste, 140, 218, 223, 268, 300
ackee, 188
acorns, 18, 33, 49, 216, 220; squash, 160
adobo, 300
Africa, 3, 19, 32, 36, 41, 48, 51, 54–57, 88, 123, 125, 160–61, 165, 182, 186, 188, 190, 196, 198, 224, 226, 243, 273, 289–90, 297–99, 302–3, 312, 316
agave, 332–33, 343
agouti, 206
agriculture, 3–4, 7, 14, 18–19, 24–25, 34, 36, 39, 49, 56, 62, 75, 179, 198, 211, 217, 342; industrial, 32; slash-and-burn, 26, 47; subsistence, 63. *See also* Neolithic Revolution
aguas, frescas (fruit waters), 317, 334
ahuatle, 207
al-Razi, Muhammad ibn Zakariy , 323
albóndigas (meatballs), 58, 81, 212, 236. *See also* meatballs
alkanet, 51
almonds, 5, 48, 50, 52, 58, 109, 113–14, 147, 171–72, 182, 194–95, 281, 318; bitter, 194; extract, 201; milk, 50–51, 117–18, 194, 200, 222; oil, 290; syrup (orgeat), 195, 312
amaranth, 47, 52, 126, 156

amaretto, 188, 331
ancho chilies, 58, 166, 234, 243, 312; powder, 239
anchovies, 60, 104, 143, 162, 170–71, 228–29, 270, 279, 310, 339, 341
aniseed, 58, 109, 295, 299, 312, 318, 331, 341; extract, 145
annatto. *See* achiote
antojitos, 120
ants, 208; "climbing a tree," 211
Apicius, 20, 41, 43, 48; Apician patina, 131–32
apples, 43, 172, 184–85
apricots, 182–83, 199, 201; kernels, 109, 113, 194–95, 331
aqua vita, 330–31
archaeology, 5, 32–33, 35, 38, 82, 87, 125, 160, 209, 325, 330
armadillos, 34, 47, 53
aroma, xix–xx, 25, 41, 47, 60, 96, 98, 154, 159, 161–62, 169, 173, 184–85, 188, 191, 195, 213–14, 225, 250, 270, 277, 284, 288, 290, 292, 295–96, 298, 301–3, 320–23, 331–35, 340, 344
aromatics (vegetables), xx, 172, 233, 240
artichokes, 64, 152, 162, 173, 210, 289, 310; liqueur, 310
Artusi, Pellegrino, 60, 63, 224, 233
asceticism, 43–45
asparagus, 39, 43, 51, 150, 163, 173, 229
atole, 96, 128, 181, 304, 320, 326

• 351 •

authenticity, 1, 11–13, 23, 28, 137, 247
avocado, 46, 53, 139, 240, 245, 250, 271, 273, 276–77, 342; leaf, 295; oil, 291; soup, 175. *See also* guacamole
Aztecs, 5, 19, 23, 47, 51–54, 57–58, 90, 96, 120, 126, 164–65, 208, 224, 286, 292, 300, 310, 318, 325, 327, 339. *See also* Nahuatl

baccalà, 265–66; alla Vicentina (cod Vicenza style), 279; a la Vizcaina, 266
bacon, 20, 60, 148, 175, 216, 221–22, 233, 238, 246, 285, 332, 342
bacteria, 37, 75, 103, 140, 218–20, 239, 256, 271, 285–86, 305, 316, 304, 326–27, 335, 339, 341
bagna cauda (hot sauce for vegetables), 82, 162, 270, 284, 310, 339
Baja, California, 262, 330
bake, 38, 59, 62, 68, 71, 88–89, 103–4, 107–9, 114, 123, 132–34, 136–41, 143, 147–48, 156, 160–61, 166–68, 171–72, 179, 182, 184–85, 195, 199, 208, 218, 223–24, 226, 229–30, 243, 254–57, 267–68, 270, 272, 274, 279, 287–90, 295, 299–300, 308, 340–41, 343–44
bakers and bakeries, 6, 137, 147, 343
baking powder, 129, 146–47, 202, 259; soda, 129, 143, 148, 259
bamboo shoots, 111, 164, 171, 245, 248
bananas, 22, 192, 318; fried, 202; leaves, 116, 218, 230
baozi, 109, 135
barbecue, 10, 70, 145, 210, 217, 223, 230–31, 246, 273, 304; grill, 82, 87; sauce, xx, 300
barley, 18, 24, 34–35, 39, 103, 126–28, 195, 225, 325–26, 333
basil, 13, 104, 133, 139, 168, 194, 258, 292, 295, 309, 311, 317
batticarne, 78–79, 227, 229
bay berries, 299; leaf, 60, 176, 228, 233, 240, 250, 254, 298–99
bean sprouts, 159, 171, 245
beans, 36, 46, 52, 56, 68, 101, 105, 115–16, 118, 135, 139, 143, 152, 157–60, 170, 196, 245, 294, 326; black, 47, 167, 192, 211, 221, 253; fava, 51; fermented, 231, 308; for baking pie crust, 199; paste, 109, 112, 117, 135, 147; refried, 23, 117, 173–76; soup (ribollita), 105, 176, 340. *See also* soy
beef, 51, 54, 58, 185, 188, 206, 210–12, 224, 227, 234–35, 264, 286, 302, 309, 344; bones, 130; braised, 107; bresaola, 209, 212; broth, 146; carpaccio, 227; casings, 235, 241; dried, 185; ground, 134, 172, 236, 311; marrow, 146; Peking, 229; shredded, 139; steak, 40, 82, 232; steamed, 234–5
beer, 36–37, 49, 103, 173, 305, 325–27, 333, 344
beets, 152; sugar, 305
Beijing, 57, 63
bellini (peach fizz), 183, 336
berries, 18, 34, 78, 179–81, 193, 198
bird's nest, 4, 61
birria (stew), 234
biscotti, 89, 109
bistecca alla Fiorentina (beefsteak Florentine), 82, 211, 232
biznaga (candied cactus), 172
blanch, 68, 147, 150, 155–56, 163–64, 200–201, 222, 225, 248, 302
blancmange, 52, 118, 222–23
blood sausages, 41, 240
boars, 34, 40, 43, 206, 216
boil, 24, 34, 37, 51, 55, 59, 61, 68–69, 79–80, 83, 85–88, 103, 107–11, 116, 122–34, 136, 141–43, 145, 147–48, 150, 152, 161–62, 172–73, 175–76, 188, 192, 196, 199, 201–2, 208, 211–12, 216–17, 220, 222–23, 225, 227, 233–34, 241–42 247–48, 250, 252–55, 257–59, 272, 274–75, 278, 280, 284, 290, 303, 305–6, 313, 317, 320, 334–35, 339–41, 344
bolillos (rolls), 114, 138–39, 255
bollito misto (boiled meats), 202, 212, 227, 241
Bologna, 59, 130, 216; sausage, 219
Boswell, James, 32
bottarga, 264, 267
bowls, 7, 46, 53, 61, 66, 72, 76, 89, 95–98, 102, 115–16, 127, 129, 131, 133, 136–40, 148, 169, 172–73, 199, 202, 229–30, 134, 235, 243, 245–46, 258–59, 278, 309, 311–13; mixing, 84, 256, 281; soup, 253; wooden, 74, 342
braise, 59, 67, 69, 81, 86, 107, 154–55, 157, 161, 171, 175, 210, 212, 217, 220, 222, 225–27, 233–36, 248, 251, 266–67, 270, 272–73, 275, 301–2, 308, 330, 340
branzino, 268, 278
breakfast, 10, 113–16, 120, 127–28, 135, 140, 143–44, 224, 242, 320
bream, 267, 269, 272
bresaola. *See* beef
Brillat-Savarin, Jean-Anthelme, 8
brisavoli, 59

broccoli, 150, 173, 229; Chinese, 151
bruschetta, 105, 139
Buddha's delight, 171
Buddha's hand (citrus), 187
Buddhism, 9, 44–45, 112, 153, 162, 171, 186, 208, 211. *See also* vegetarianism
buñuelos de jeringa, 115, 145. *See also* churros
butter, 48, 59–60, 108, 118, 122–23, 127–28, 130, 133–34, 139–41, 143–44, 146–48, 162, 194, 199, 212–13, 216, 223, 229, 233, 240, 250, 255–56, 259, 286–87, 310, 340, 343; annatto in, 300; knives, 90; yak, 213
buttermilk, 286

cabrito. *See* goats
cacao, 19, 21–22, 46–47, 53, 58, 90–91, 96, 115, 128, 144, 203, 216, 224, 254, 256, 297, 300, 312, 318–20, 323, 342; beans, 58; candy, 62, 196; chocolate sauce, 9, 259; cookies, 109, 146; spread (Nutella), 141, 196
cactus, 164–65, 170, 172, 224, 270; prickly pear, 334
cake, xxi, 39, 71, 89, 104, 108, 124, 126, 139–40, 155, 182, 184, 194, 198, 288, 298; corn, 47, 240–41; moon, 112–13, 145, 225; sponge, 108, 255–56; steamed, 117, 122; tres leches, 258–59
Calabria, 104
calamari fritti (fried squid), 275, 279. *See also* squid
calcium oxide. *See* lime
calories (in diet), 32–33, 36, 101, 115, 158, 194, 284, 289, 325
Campari, 331, 336
Canton and Cantonese cooking. *See* Guangdong
capers, 170, 187, 220, 228, 266, 281
capirotada (bread pudding), 114, 254–55, 304
capon, 51, 222, 241, 255
caponata, 167, 170
capybara, 53
cardamom, 298, 302; black, 234, 302
cardoon, 162, 213, 310
carne asada (roasted meat), 212, 231–32, 245
carp, 263, 270–71, 278, 281
carpaccio (raw beef), 227
carrots, 85, 117, 130, 133, 145, 152, 154, 162, 171, 175–76, 212, 228, 233, 236, 240, 246, 248, 250, 254, 266, 278, 285, 310, 312, 340, 344
carrying capacity (of population and resources), 26
cashews, 197, 245
casu marzu, 208

Catholicism, 43, 49, 53, 58, 105–6, 126, 143, 162, 209, 223, 263, 276, 328
Cato, the Elder, 39, 71, 151, 328
caviar, 207, 264
cavolo nero (black cabbage), 150, 159, 176, 340
celery, 85, 130, 132–33, 152, 154, 162–63, 169–70, 212, 228, 233, 240, 246, 250, 254, 266, 278, 285, 312, 340, 344; juice (nitrate source), 238
ceramic, 19, 36–38, 40, 46, 71, 73–74, 78, 82–83, 87, 89, 95, 97–98, 137, 140, 171, 206, 222, 225, 236, 248, 322, 326, 331, 334, 339–40, 343
ceviche (raw fish), 271, 274, 276–77
champurrado, 96, 128, 320
Chang'an, 48
Charlemagne, 48–49
chayote, 160, 170
cheese, 52, 58–59, 103, 105, 107, 114–15, 118, 120–23, 126, 130, 132–33, 138, 143, 155, 161, 164, 167–68, 170, 172–73, 176, 182, 185–86, 194, 202, 208–9, 212–16, 220, 223–24, 229, 238, 250, 252, 255–58, 280, 286–87, 292, 304, 316, 343–45; tools, 72, 74, 123
cherimoya, 193
cherries, 182–84, 201, 306, 329, 335
chervil, 293
chestnut, 126; honey, 304; wood, 306
chia, 52
Chianti (wine), 159, 329
chicken, 51–52, 55, 58, 72, 117–18, 120, 131, 182, 193, 195–96, 211, 221–24, 227, 252, 287, 301, 308, 312, 344; broth and stock, 128, 130, 136, 145–46, 175, 233, 236, 253–55, 310; burrito, 245; cacciatore, 243–44; encacahuatado, 243; fried, 3 ; salt baked, 243; shredded, 143; soup, 161, 248–50; steamed, 234; stir-fried, 244–45
chickpeas, 51, 157–58
chicle, 193
chilaquiles, 120, 143
chili, xix–xx, 14, 21, 23, 35, 46–47, 53, 56–58, 64, 71, 78, 85, 93, 104, 135, 154, 156, 165–67, 171–72, 174–75, 196, 203, 211, 217–18, 220, 223–25, 229, 234, 236, 242, 247, 250, 268, 273, 276–77, 280, 284, 298–301, 310–12, 326, 342; en nogada, 172; green, 170–72, 224, 231, 240, 280; in chocolate, 318–19; paste, 248–49; pickled, 139; powder, 120, 152, 191, 202, 239, 241, 245, 280, 310
chili sauce, 116, 120, 128, 143, 179, 243, 252–53, 313
chinampas, 52

INDEX

chocolate. *See* cacao
chopsticks, 44, 66, 88, 98, 115, 143, 164, 202, 268
chorizo (Mexican sausage), 117, 123, 220, 224, 239, 307
Christianity. *See* Catholicism *and* Lent
Christmas, 9, 104, 156, 263, 266, 270, 320
churros, 96, 115, 144, 224, 340
cilantro (leaf), 167, 170, 175, 231, 234, 236, 242–43, 250, 276, 291, 295–96, 310–11
cinnamon, 5, 23, 50, 55, 58–59, 128, 145–48, 170, 172, 193, 199–202, 225, 230, 234, 251, 254–55, 281, 287, 297–300, 302, 310, 312, 318, 320, 331, 334, 336
Cinzano, 280, 331
citrus, xix, 50, 182, 187–88, 195, 271; juice, 217; press, 92; rind, 72
cittara, 74–75
clams, 273–74, 278
clay (unfired), 251, 308. *See also* ceramic
cleavers, 72, 91, 111, 131, 222, 236–37, 248
climate change, 18, 27, 36, 50, 64
cloves, 5, 19, 51, 58–60, 172, 201, 230, 239, 252, 254–55, 281, 297, 299, 302, 312, 336, 340
coati, 53
cochineal, 208
cochinita pibil (roasted pig), 218, 230, 340
Cocinero Mexicano, El, 63
cocktails, 198, 331–33, 335–36; Chinese, 337
coconut, 197–98, 203; milk, 69, 78, 198, 204; oil, 291; shrimp, 280; soda, 318
coffee, 22, 25, 62, 96, 138, 198, 224, 320, 322–24; espresso, 96, 256, 324; grinder, 78
collard greens. *See* greens
Columbian Exchange, 54–55, 57
Columbus, Christopher, 55, 124, 192, 297
comals, 19, 52, 85–86, 119–20, 142, 166, 234, 242, 245, 274, 312, 340
conduction, 68
Confucius, 44–45
congee (rice porridge), 116, 127–28, 251, 340
convection, 68
convenience food, xxi, 10, 13, 196
cookies, 109, 113, 124, 194, 199, 298; Chinese peanut, 113, 147; Mexican wedding, 89, 147; sbrisolona, 146
copper, 73, 84, 122, 339
coriander (seed), 41, 58–59, 291, 298–99, 312
corn, 4, 19, 23–24, 34, 36–37, 46–47, 52, 55–56, 58, 60, 63–64, 78, 96, 102, 106, 114–16, 118–22, 125–26, 134, 142, 162, 193, 212, 220, 325, 342; baby, 121, 171, 245; dried, 78, 243; flakes, 280; husks, 69, 86, 135; meal, 127, 137, 146, 274, 320; oil, 290; popcorn, 126; starch, 131, 222, 229, 236, 244–45, 248–49, 275, 278, 281, 320, 344; syrup, 319. *See also* nixtamalization
cornetti, 140–41
Corrado, Vincenzo, 60, 167
Cortés, Hernán, 52, 193, 339
cotechino (skin sausage), 212, 216, 227, 241
cotijo, 215
cotoletta alla Milanese (veal cutlets), 228–29
court cuisine, 15, 17, 44–45
crabapples, 201, 203
crabs, 253, 263, 271–72
cream, 9, 89, 108, 148, 175, 201, 204, 214, 216, 223–24, 256, 259, 286–87, 310; Mexican crema, 257; pastry, 140, 199; sour, 121, 250, 257, 273
Cremona, 201
crostata di frutta, 198–99
crostini, 139
Cultural Revolution, 31, 124
cumin, xx, 41, 218, 230, 234, 236, 239, 242, 247, 250, 252, 299–300, 302, 310
currants, 180–81
custard, 253
cutlery, 97–98
cuttlefish, 272, 274–75, 278
Cynar (artichoke digestive), 331

Daoism, 44–45, 183
dates, 19, 48, 117
deer. See *venison*
demographics, 18, 25–27, 33–34, 36, 46, 48, 50–51, 53–54, 56, 60–63, 115, 122, 124, 211
Dias, Bartolomeu, 55
Díaz, Porfirio, 31
dim sum, 69, 111, 131, 135, 198, 272–73, 281, 284, 321, 340
distinction, food as a marker of, 8–9, 15, 17, 105, 115, 297
dogs, 35, 47, 61, 208–9
dormice, 23, 40–41, 206
Dragon Boat Festival, 117
duck, 34, 53, 70, 81, 182, 186, 201, 224–26, 246, 302; braised, 248; eggs, 128, 147, 223, 225–26, 250–51; fat, 287; liver, 219, 239; Peking, 70, 112; in pumpkin seed sauce, 247; ragú, 246–47

dumplings, 48, 85, 91–92, 95, 107, 109–11, 117, 123, 155–56, 266, 281, 313, 340; Chinese, 131, 135; tortellini, 130. *See also* gnocchi
durian, 25, 190

earthenware, 37, 83–84, 95–96, 134, 140, 158, 253, 279, 342–43
Easter, 44, 108, 210, 213, 223, 263, 340, 343
eels, 263, 270, 272; smoked, 266
egg roll, 112
eggplant, 50, 166–67, 170–72
eggs, 20, 40–41, 43–44, 49, 52, 58, 60, 68–69, 107–9, 114, 116, 118, 128, 130–34, 143, 145–47, 155–56, 161, 173, 222–24, 226, 231, 236, 249–50, 253–56, 266, 280, 340; 1,000-year-old, 250–51; egg custard, 253; fried, 120, 224; frittata, 252; huevos motuleños, 253; insect, 207; poached, 229; quail, 221; salted, 112, 148; scambled, 139, 224; tea, 251–52; and tomatoes, 168, 252–53; tools for, 74, 81, 84, 89; tortilla, 123; turtle, 277; white, 139, 172, 195, 204, 268, 281; yolk, 128, 172, 199, 228. *See also* duck eggs
elotes (corn on the cob), 120
emmer, 103
empanada, 115, 179, 185; de cajeta (pastry with milk fudge), 259
enchilada, 119, 143
environment and influence on cuisine, 3, 7, 18, 26–28, 34, 47–48, 205, 210–11, 226, 276
epazote, 47, 135, 155, 162, 173, 294, 310
escabeche, 268, 271
etiquette, 2, 7, 44, 48, 57, 66, 99
Eucharist, 58, 328
evolution, culinary, 2, 11, 13, 17, 26, 28, 94, 102, 126, 169; human, 32, 178
exploration, 27, 57, 265, 276, 297
exportability (of cuisines), 21–22
extinction (of species), 26, 33

famine, 26, 36, 49, 60, 64, 124, 208
farro, 103
fast food, 11, 14, 23, 344
fasting 43–44, 49–50, 53, 159, 200, 263. *See also* Lent
fegato alla Veneziana (Liver Venetian Style), 240. *See also* liver
fennel (seed), 51, 188, 217, 231, 278, 293, 299, 302, 340; bulb, 310; fronds, 293; pollen, 59, 293
fermentation. *See* bacteria
Fernet (digestive), 331

Ferrara, 58–59
Fertile Crescent, 4, 18, 35–36, 38, 73, 102, 209
festivals, 8, 37, 117, 127, 143, 147, 208, 251, 326, 336
feudalism, 48, 54
Feuerbach, Ludwig, 27
fig peckers, 40, 131–32, 221
figs, 48, 186, 201, 240; leaves, 39; sap, 213
fish, xxi, 10, 18, 26–27, 34, 38–39, 51–52; 111–12, 115, 128, 156, 161, 168, 170, 187–88, 196, 212, 223, 161–281, 295–96, 304, 308; balls, 272, 281; equipment for, 88, 99; freshwater, 57, 117, 263, 281; fried, 279; and Lent, 159; raw, 271, 276; salted, 189, 265; saltwater, 57; sauce, 8, 41, 132, 183, 308, 311, 341; shellfish, 117, 271–76, 278, 303; soup, 105, 272, 298; Westlake sour fish, 278, 281
fishing, 33, 49, 263–64, 276
five spice powder, 211, 234, 239, 302, 340
flan, 89, 198, 223–24, 253–54
flavor balance, xx, 9, 14, 61, 120, 155, 169, 179, 186, 198, 239, 249, 277, 304, 312; combinations, xix, 4–6, 10, 13–14, 23, 60, 179, 200, 301–3, 336; profiles, 1, 4, 21–22, 120, 154, 166, 169, 283–84
Florence, 58–59, 156; alla Fiorentina, 59, 82, 156, 232
focaccia, 104
food, academic study of, xxii, 6, 14, 16; culture, 1–2, 6–7, 43, 57, 101, 122, 162; ideology, 2, 7–9, 11, 41, 43; voice, 2, 9
food mill, 78, 80, 179, 202, 311
foodways, 2, 4, 7, 47, 49, 52, 57
forks, 97–99, 104, 123, 128, 132, 145, 148, 199, 259, 334
fortune cookies, 113
Frangelico, 196, 331
French cuisine, xxi, 3, 8, 13–14, 21, 50, 60, 63, 67, 74, 81, 85, 114, 133, 138, 140, 199, 215, 253, 286–87, 298, 340, 343; oenology, 328; service, 97
frittata. *See* eggs
fritters, 39, 52, 69, 92, 108, 113, 115, 143–44, 155–56, 158–60, 179, 184–85, 192, 266, 340
frogs, 263, 275
fry, 59, 68–69, 87, 127, 136, 143–45, 147–48, 162, 169–75, 192, 199, 202, 236, 241, 243–44, 249–50, 252–53, 269, 279–80, 284–86, 288–89, 291, 310, 312

galangal, 297, 301
Galliano, 331

gaoliang (sorghum alcohol), 218, 331
garlic, xx, 4, 21, 25, 52, 58–61, 85, 130–31, 139, 143, 154, 161, 166–67, 170–72, 186, 211–12, 217–18, 220, 224–25, 228–31, 233–34, 236–37, 239–40, 242, 244–45, 247, 249, 250, 255, 258, 266, 270, 274, 278–80; 284–85, 287, 300, 309–12, 339; press, 92
garum. *See* fish, sauce
gastronomy, 2, 5–6, 8, 12, 23, 59–60, 128, 209, 284, 300, 315, 318; molecular, 21, 94
goose, 81, 184, 224, 226; fat, 287; liver, 226
gooseberry, 180–81, 189
gorgonzola, 214
gelato, 179, 195, 198, 254; al limone (sorbet), 204
Genoa, 51, 55, 59–60, 82, 194, 212, 266, 284, 309
ginger, 4, 21, 51–52, 74, 111, 131, 136, 145, 167, 173, 200, 218, 225, 229, 234, 236, 243–44, 248, 252, 267, 275, 278, 280–81, 291, 296–98, 300–2, 313, 320; pickled, 251
ginkgo nuts, 128, 171, 196
ginseng, 223, 301, 331; tea, 301, 334–35
gluttony, 43
gnocchi, 109, 123, 148, 299
goats, 18, 34–35, 206, 209–10, 234, 322, 340; cheese, 172, 215; kid, 43, 51, 58, 66, 209–11, 255, 340; milk, 213, 259, 286
goji (wolfberry), 128, 179–80
gordita, 212, 240, 341
Grana Padano, 213, 250
grapes, 24, 43, 48, 103, 181–82, 186, 203, 299; juice, 36; must, 109, 181, 306, 342; sapa (or saba), 181, 305; seed oil, 181; verjuice, 50, 303, 345; wine, 338–21
grappa, 331–33
grasshoppers, 208
graters, 74, 123
greens (leafy), 34, 108, 151, 154–57, 162, 169, 171–72, 175–76, 236, 248, 252, 285
gremolata, 212, 233
griddles, 19, 33, 47, 52, 71, 103, 119, 142, 166, 234, 242, 245, 312, 340. *See also* comals
grill, 24, 59–60, 70, 81–82, 87, 105, 121–22, 127, 139, 141, 155, 168, 193, 196, 210, 217, 221–22, 224, 231–33, 245, 268, 270, 273–75
grissini Torinesi, 104, 136–37
guacamole (Avocado Sauce), 21, 53, 78, 232, 295, 310
Guangdong, 113, 143, 168, 191, 208, 231, 239, 251, 253, 267, 274, 290, 304

guava, 185, 193, 317–18; paste, 202
guinea fowl, 56, 224, 226
guinea pigs, 206

hair vegetable (fa cai), 164
halibut, 269–71, 276
Han Dynasty, 30, 38, 44–46, 87, 181, 341, 344
Hangzhou, 278, 321
hazelnuts, 136, 195–96; liqueur, 331; oil, 196, 290; spread, 141
Hernández, Francisco, 164
hibiscus, 317–18, 334
hoja santa, 295, 310
hominy. *See* nixtamalization
honey, xx, 37, 39–41, 47, 183–84, 201, 208, 231, 281, 305, 318, 326
Hong Kong, 63, 181, 209, 262, 272, 311, 317
horchata (rice drink), 195, 317, 333–34
horseradish, xx, 154, 241
horsemeat, 206, 209, 334
hot and sour soup, 248–49, 296
huachinango a la Veracruzana (Veracruzan red snapper), 281–82
Huangdi, 46, 125–26
huevos motuleños (Mexican omelet), 253
huitlacoche, 23, 121, 162
humors, 10, 42, 59–60, 128, 182, 188, 283, 291–92, 309
Hunan, 13, 64, 123, 126, 188, 234, 251, 280, 304
hunting, 18, 26, 32–36, 49, 72, 206–7, 220–21, 224, 245, 264; hunter style, 243
hybrid cooking methods, 69, 71
hybrid cuisines, xix, 13, 22–23, 50, 57–58; and medicine, 292
hybridization (plants), 118–19, 184

ibn Battuta, Abu Abdullah Muhammad ibn Abdullah al-Lawati al-Tanji, 278
Ice Age, 32, 36
ice cream, 9, 20, 179, 181, 195, 201–2, 259; Eskimo, 5; maker, 204. *See also* gelato
iguana, 47, 275
India, 3, 5, 19, 21, 44, 48, 55, 57–58, 63, 115, 117–18, 144, 166, 191–92, 224, 281, 286, 289, 295, 297–98, 300–3, 320
Indians (Native American), 4, 18, 33, 54, 56, 63, 76, 83, 114, 122, 158, 160, 192, 263–64, 316, 326, 342
insalata di mare (seafood salad), 277–78
insects (as food), 23, 34, 207–8

Instacure, 237, 239, 241, 341. *See also* nitrates
Iran. *See* Persia
iron (cookware and implements), 45, 66, 68, 70–71, 73, 80–82, 85, 87, 107, 245, 340–41; Age 71–73
irrigation, 19, 25, 36, 45, 52, 270
Islam, 5, 50, 55, 126, 182, 210, 278, 287, 323

jackfruit, 192
jalapeño chilies, 166, 170, 236, 276, 281, 310, 313
Jalisco, 234, 332
jellyfish, 267
Jesus Christ, 106, 190, 209, 254, 328
jicama, 47, 152
Jinhua ham, 220
Judaism, 9, 43, 55, 162, 210, 226
jujubes, 117, 187, 189–90
juk. *See* congee
juniper berry, 220, 241, 299, 307, 331
Juvenal (Decimus Iunius Iuvenalis), 88, 43

ketchup, 21, 168, 273
kiwi (fruit), 181, 189, 199
Kongzi (Confucius), 44–45
knives, 72–75, 97–99, 104, 136, 139, 170–71, 227, 231, 240, 252, 256–57, 259, 268, 280, 310; banishing, 44, 88, 98, 211
kosher, 2, 43, 208, 213, 227
kumis (fermented milk), 209, 213, 256–57, 316, 341

La Varenne, François Pierre, 287
lactobacilli. *See* bacteria
lamb, 51, 60, 209–11, 234
Landa, Diego de, 47
Laozi, 44
lap cheong (Chinese sausage), 154, 219–20, 239
lard, 58, 109, 113, 117, 138–39, 142–43, 145–46, 148, 158, 174, 216, 220, 222, 233, 241, 243, 284–86, 310, 312
lardo, 216, 246, 280, 285, 341
lasagna al forno, 51, 89, 107–8, 132–33, 213
Lateran Council, 106
Latini, Antonio, 60, 167
lavender, 294
lemon, 60, 187–88, 228–29, 274–75, 280, 317, 331, 336, 340, 345; juice, 10, 92, 128, 161, 173, 204, 227–28, 233, 240, 246, 257, 268, 273, 278–79; peel (zest), 199, 201, 204, 212, 233–34, 255, 259. *See also* citrus
lemonade, 292

lemongrass, 302
lengua (tongue), 240. *See also* tongue
Lent, 44, 50, 53, 117, 159, 194, 200, 213, 227, 254–55, 261–63, 265–66, 279
lentils, 158, 226, 241
lettuce, 48, 139, 152, 155, 170, 232, 245, 281; salad, 169; stir-fried, 169
lily buds, 171, 296; bulbs, 128, 155
lime (calcium oxide), 37, 119, 134, 142, 251, 274
lime (fruit), 92, 121, 187–88, 234, 242; juice, 120, 152, 170, 193, 203, 230–31, 250, 253, 271, 276, 280–81, 295, 310–11, 333–35; leaves, 302
limoncello (lemon liqueur), 188, 331, 336
lion's head meatballs, 236
liver: calves, 240; chicken, 139; duck, 219, 225, 239; higaditos (liver omelet), 252; goose, 226; pork, 220, 230; Venetian style, 240
lobster, 271, 273
localism, locavore, 8–9, 15, 17, 23, 43, 94, 118, 122, 179, 185, 276–77, 291, 330, 336, 341, 344
Lombardy, 49, 175, 233, 275
loquat, 189
lotus leaves, 243; root, 128, 153; seed, 113, 117, 135, 147–48, 196, 296
Lu Yu, 48, 320
luffa (loofa), 161
lychee, 188–89, 322, 337

Macao, 122, 192
macaroni, 10, 52, 59, 91, 108
mackerel, 264, 268–69, 276, 311, 341
maguey, 208, 326, 332, 343; worm, 208
maize. *See* corn
Malthus, Thomas, 26
mamoncillo, 188
manchego, 215
mandoline, 171, 314
mango, 191, 203, 280, 310, 317, 334; pickle, 21
mangosteen, 191, 193
Manila, 56, 266, 276, 298; clams, 278
manioc, 46–47, 52, 124
manners. *See* etiquette
mantou, 109, 135
Mantua, 58, 109, 146
Mao Zedong, 64; favorite red-cooked pork, 234
margaritas, 332–33, 335
marrow. *See* beef
Marsala (wine), 69, 212, 223, 330
Martial (Marcus Valerius Martialis), 8

Martino of Como, 51, 117

marzipan, 60, 194

masa harina, 119

mascarpone, 256

mastic, 60, 195

Maximilian, Emperor of Mexico, 31, 138

Maya, 5, 26, 30, 47–48, 96, 156, 180, 218, 318, 326

maw (fish swim bladder), 266

McCormick reaper, 62

measles, 56

meatballs, 27, 41, 60, 92, 117, 212, 214, 285; albóndigas, 58, 212, 236; lion's head, 236; polpette, 235–36

medicine and food, 42, 46, 60, 78, 126, 154, 156, 160–63, 166, 177–79, 180, 186, 194, 198, 206, 208, 266, 292–95, 301, 303, 309, 317–19, 334–45

medieval cuisine, 4–5, 50–52, 57, 59, 79, 114, 117, 144, 172, 179, 182, 194, 226, 266, 271, 297–300, 302–4, 325, 245

Mei, Yuan, 61

menudo (tripe soup), 242–43, 248

mescal, 208, 333

Messisbugo, Christoforo, 59, 280–81, 287

Mestisaje, 58, 342

metate, 52, 78, 120, 135, 142, 312, 318, 342

mezzaluna, 74, 292, 342

Middle East, 4, 16, 18–19, 37, 50, 58, 73, 115, 158, 181–82, 186, 190, 194, 198, 213, 286, 289, 298–300, 323

migliaccio, 124, 216

Milan, 59, 118, 146, 175, 212, 214, 228–29, 233

millet, 18, 24, 34, 36, 46, 48, 51, 57, 109, 116, 124–25, 127, 216, 306, 326

milpa, 63, 118, 342

minestrone, 175–76

Ming Dynasty, 31, 56–57, 61, 320, 342

mint, 41, 293; peppermint, 320

Modena, 109, 130, 306–7, 336

molcajete, 78, 135, 310, 342

molé poblano, 11, 58, 224, 312, 324

molé verde, 295; with tomatillos, 310

molinillo, 90–91, 318, 342

Moluccas, 5, 19, 51, 297, 299

monasticism, 43–45, 112, 171

monosodium glutamate, xix, 308, 345

mooncakes, 112–14, 145, 225

mortadella, 51, 68, 130, 219, 241

mortar, 34, 75–76, 78, 93, 135, 147, 167, 170, 172, 202, 228–29, 243, 247, 255, 279, 281, 309–10, 312, 342–43

mostarda di frutta, 201–2, 228, 241, 299

Moutai (alcohol), 331–32

mouthfeel, xix–xx, 154, 161, 169, 196, 284, 342

mozzarella, 21, 104, 133–34, 137, 168, 171–72, 210, 213–15, 229, 257–58

mu shu, 112, 296

Mughals, 5, 58, 300

mulato chilies, 58, 166

mulberry, 178, 306

mushrooms, xix, 112, 118, 121, 151, 161–62, 171, 175, 224, 243–44, 248, 253, 266–67, 280, 307, 309, 345

mussels, 43, 273–74, 278

Mussolini, Benito, 62

mutton. *See* sheep

myrtle berries, 299

NAFTA (North American Free Trade Agreement), 118

Nahuatl (Aztec language), 47, 53, 58, 78, 120, 123, 155, 165, 167, 294, 310, 324

Naples, 22, 60, 104, 138, 167, 186

Native Americans. *See* Indians

Neanderthals, 32

negroni, 335–36

Neolithic Revolution, 30, 35, 210

New Guinea, 19, 303

New Year's Day, 66, 110–11, 117, 147, 164, 171, 241, 268

nitrates, 219, 237–39, 241

nixtamalization, 37, 52, 55, 78, 102, 119, 122, 127–28, 135, 142, 217, 274, 326, 341–43. *See also* lime (calcium oxide)

nocino (walnut liqueur), 195, 331, 336

noodles. *See* pasta

nopalitos. *See* cactus

nutella, 141, 196

nutmeg, 5, 19, 50, 59, 74, 130, 146, 200, 246, 254–55, 297–300, 312

nutrition, 2, 9, 26, 101, 149, 177, 207–8, 283, 297, 309; anti-nutritional procedures, 159; balance, 150; malnutrition and deficiencies, 37, 55; theory, 41, 59, 285. *See also* humors

nuts, 5, 19, 27, 33–34, 50, 58, 93, 104, 109, 177, 186, 193–98, 216, 224, 259, 309; oils, 290; paste, 113

Oaxaca, 193, 208, 215, 333
octopus, 272–75, 278
okra, 56, 64, 161
olive, 40, 149, 170, 186–87, 220, 226, 238, 266, 281; mill, 76; oil, 39, 58, 103–5, 127, 133, 136–37, 144, 146, 151, 155, 159, 162, 169–71, 173, 175–76, 186, 202, 222, 227–28, 231, 233, 226, 240, 243–45, 252, 258, 266, 268, 270, 273, 275, 278–79, 281, 287–90, 307, 309–11, 339; press, 39, 76–77; wood, 89
Olmecs, 30, 38, 46–47, 318, 342
onions, 24, 78, 117–18, 123, 130, 133, 145–46, 150–51, 154–55, 158, 162, 166, 170, 172, 174–76, 199–200, 212, 217, 223–24, 228–31, 233–36, 240, 242–46, 250, 252–54, 266, 271, 276–81, 285, 310–12, 320, 240, 344
orange, 187–88, 218, 281; blossom honey, 305; flower water, 188; juice, 59, 230, 268; liqueur, 333, 335; peel, 104, 188, 201, 211, 225, 317, 320, 331; soda, 318
oregano, xx, 39, 133, 139, 171–72, 174–76, 230, 237, 239, 241–42, 245, 281, 292–93, 300, 311; Mexican, 295
organic, 9, 207, 307, 336
ossobuco (braised veal shanks), 146, 212, 233–34
oven, 68–71, 83–84, 87–88, 103–4, 109, 132, 134, 138–41, 147–48, 171, 199, 202, 217, 231, 234, 236, 247, 250, 254–55, 259, 279; brick (wood-burning), 7, 38, 137–40, 225, 231; convection, 68; electric, 67, 71; gas, 71; horno, 71, 88, 341; microwave, 14, 171; solar, 67; steam, 69; underground (pib), 218, 230, 340
oysters, 25, 43, 273–74; sauce, 308

paella, 117; pan, 85
paletas (popsicles), 181, 203–4
panata, 128
pancakes, 103, 112, 123, 126, 156, 173, 216, 225
pancetta, 60, 108, 148, 175, 216, 219, 233, 285, 312, 342
pandoro, 104, 139–40, 299, 343
panettone, 104, 139, 343
panforte, 104, 299
panna cotta (cooked cream), 201, 259
panzanella, 105
papaya, 193, 231, 310, 334
Parmigiano-Reggiano, xix, 24, 72, 74, 118, 127, 130, 146, 148, 213–14, 227, 235, 279, 307, 309
parsley, 130, 133, 161, 163, 172, 212, 228, 233–35, 240–41, 247, 250, 254, 266, 273, 278–79, 292–93, 298, 309–10, 339
pasilla chilies, 58, 166, 312
pasta, xxi, 51, 85, 103, 106–8, 118, 128–30, 132–34, 156, 159, 166, 168, 212, 214, 217, 226, 235, 246–47, 268, 273, 287, 299, 340, 343; black, 275; in broth, 250; dried dough, 63; Chinese, 129–30; equipment for making, 72, 74, 90–92; sauce, xxi, 108, 220, 309, 311
pastelitos de boda (wedding cookies). *See* cookies
pastry, 39, 43, 59–60, 108, 112–14, 140, 181, 183, 185, 194, 196, 202, 251, 259, 285, 287, 289, 290, 300, 324; equipment, 90–92, 133, 136, 145, 340, 343; knives, 72; pasta frolla, 199; pate a choux, 108, 115, 143–44, 340; puff, 108, 113
pastry cream. *See* cream
pate a choux. *See* pastry
patronage, 14, 21, 38
Pavia, 250
peaches, 12, 182–83, 194, 199; bellini, 306; pickled, 200–201; soda, 318
peacock, 221
peanut, 56, 64, 117, 128, 171, 196, 273, 312; butter, 196, 318; cookies, 147; oil, 143, 169, 202, 229, 245, 289–90; sauce, 243
pears, 43, 172, 184–85; minestra, 199–200; wood, 307
peas, 116–17, 145, 148, 160, 171, 253
pecans, 195
peccary, 46–47, 53, 216
peel, baker's, 89, 137–38, 343
peking beef, 299–30
pepitas. *See* pumpkin, seeds
pepper, 41–42, 51, 55, 58–59, 131, 133, 155, 170, 172, 174–76, 193, 200, 218, 220, 227–31, 233–36, 245–46, 278, 284, 292, 295, 297–99, 310–12, 318; corns, 68, 132, 219, 237, 241, 243, 247, 255; long, 51, 298; melegueta, 51, 298; mill, 78; white, 236, 249, 281, 313. *See also* Sichuan, pepper
peppers. *See* chili
Persia, 4–5, 48, 58, 117–18, 135, 144, 156, 183, 186, 195, 300, 323
persimmons, 185–86
pesce lesso (poached fish), 278
pesto Genovese, 194, 292, 309
pheasant, 43, 53, 161, 221, 245
pico de gallo (tomato salsa), 167, 284, 310

Pienza, 214
pigeon, 221, 245
piloncillo. *See* sugar
pine nuts, 60, 132, 170, 172, 187, 194, 216, 236, 240, 266, 271, 309
pineapples, 22, 192, 310
Pisanelli, Baldassare, 59
pistachio, 48, 195, 219
pizza, 22–23, 59, 71, 168, 186–87, 209, 213, 217, 258, 270, 288, 292; equipment, 71, 88–89, 343; Margherita, 13, 22, 104, 137–38
pizzoccheri della Valtellina, 126, 130
plague, 26 , 31, 51, 56
plantain, 192, 202
Platina (Bartolomeo Sacchi), 52
plums, 182–83, 188, 200–201
poach, 49, 68, 132, 185, 200, 222–23, 228–29, 240, 272, 278–79, 281, 340, 343
polenta, 24, 55, 102, 122, 127, 233, 266, 279, 343; buckwheat, 126; millet, 51, 124
Polo, Marco, 106, 278
pollo encacahuatado (chicken in peanut sauce), 243
polpette. *See* meatballs
pomegranate, 172, 187, 335
Popol Vuh, 47
population. *See* demographics
porcelain, 21, 46, 57, 63, 83, 320, 343
porchetta (stuffed roasted pig), 218, 230–31
pork, 10, 51–52, 54, 57–59, 68–70, 95, 109, 112, 116–17, 130–31, 151, 172, 183, 186, 206, 211–12, 216–20, 230–31, 234–37, 239, 293, 300, 302, 310–11; barbecued, 145; buns, 135–36; fat, 284–85, 287, 341; loin, 184, 229; mu shu, 112, 296; neck bones, 248; prohibition, 226; red-cooked, 234; roast (char siu), 231; shredded, 139; skin, 241. *See also* cochinita *and* porchetta
porpoise, 264
Portugal, 55, 57, 122, 144, 192, 265, 297, 299. *See also* Macao
posole. *See* nixtamalization
posolli (or posole, drink), 47, 326
prejudice (culinary), 4, 25, 207
prosciutto, 74, 104–5, 107, 130, 137, 139, 182, 186, 206, 212, 218, 220, 226, 268, 311
prosecco, 173, 183, 330, 336, 343
Puebla, 11, 58, 123, 172, 342
pulque, 37, 54, 326–27, 332, 343
pumpkin, 160, 259; ravioli, 183; seeds, 58, 160, 182, 196, 226, 247, 273, 310, 312; seed oil, 291

Qing Dynasty, 31, 60, 63, 122
quails, 53, 132, 221, 246; eggs, 251
queso fresco (fresh cheese), 215, 253, 257, 343
quinces, 185, 188, 202, 228

rabbit, 34, 53, 186, 220
ragù, 133; Bolognese, 108, 168, 212, 311; di anatra (duck sauce), 246–47
raisins, 58, 60, 114, 126, 132, 171–72, 180–81, 184, 187, 224, 236, 240, 259, 266, 271, 280–81, 312; wine, 41, 132
raspberry, 179, 318
ravioli, 51, 52, 90–91, 107, 111, 160, 183, 214
recipes, 3, 5–7, 10–14, 16–17, 20, 22, 24, 26–27, 50–51, 59, 61, 63, 66–68, 93, 97, 99, 102, 114, 132, 198, 217, 224, 254, 264, 276; ancient, 38–41; copyright, xxi; family, 8, 88; format, xx; medicinal, 117, 292, 318; pastry, 108; precision, 67
red snapper, 281–82
Renaissance, 16, 31, 52, 58, 108, 179, 182, 186, 336
Revolution, Agricultural (Neolithic), 30, 32, 34–36, 210; Agricultural of year 1000, 50; Cultural, 31, 64, 124; Green, 63; Industrial, 27, 31, 61, 67
ribollita, 105, 150, 159, 176, 340
rice, 18, 36–37, 45–46, 48, 56–58, 64, 66, 69, 78, 95, 98, 102, 106, 115–18, 160, 175, 182, 192, 195, 224, 236, 273, 275, 285, 295, 304, 317, 333–34; black, 306; fermented, 271; flour, 146, 234–35, 258; fried, 145; hulls, 251; paddies, 206, 225, 263, 275; pudding, 203; Spanish, 23, 146; starch, 91, 109, 111, 118, 222; straw, 225; wine, 37, 111, 131, 136, 229, 231, 234, 236, 239, 243, 278, 325–26, 344; wine vinegar, 200, 305–6, 313. *See also* congee *and* risotto
ricotta, 60, 108, 133, 148, 156, 213, 215, 252, 257–58
risotto, 102, 118, 160–61, 233, 273–75, 287, 299, 343; alla Milanese, 118, 146
rituals, 7, 9, 33, 37, 44, 53, 57, 106, 162, 324, 326–27
roast, 33, 37, 40, 49, 52, 59, 66, 68–71, 80–82, 88, 116, 123, 126, 143, 152, 155, 161, 166–67, 172, 194–96, 206, 210–12, 216–18, 221–26, 228, 230–31, 245–46, 255, 270, 273, 275, 284–85, 292, 299, 301–2, 323–24, 340
Rome, 104, 162, 210, 218, 266, 270, 316; ancient, 8, 23, 39, 44, 182–83, 190, 303, 305; Renaissance, 52, 58–59
romeritos, 155, 266
rose dew alcohol, 331
rosemary, 137, 156, 217, 220, 224, 231, 246, 258, 292–93

rosewater, 52, 60, 109, 199, 222, 256, 294
roux, 133, 287, 343
rue, 293

sacrifice, 6, 57, 209; animal, 209, 211; human, 53
saffron, 51, 117, 140, 146, 199, 252, 255, 281, 291, 298–300
sage, 108, 130, 148, 175, 212, 220, 241, 246, 293
Sahagun, Bernard, 53
salad, 9, 152, 155–56, 161–64, 167–69, 181, 188, 193, 196, 211, 213, 267, 271, 292; bread, 105; cactus, 164–65, 170; dressing, 228, 288, 290, 306–7; fish, 271, 273–74, 277; forks, 99
salamander, 275
salame (salami), 59, 104, 154, 202, 206, 209, 218–20, 236–39, 252, 299, 343; duck, 226; Felino, 236–37
salmon, 18, 264, 268–69, 271
salsa (Mexican), 21, 23, 53, 60, 120, 154, 167, 179, 230, 232, 245; mango, 280; tomatillo, 155, 167; verde 120, 240. *See also* guacamole *and* pico de gallo
salt, 307–8
salt-baked chicken, 243
sanguinaccio, 216
sapa (or saba). *See* grapes
sapodilla, 193
sapote, 193
sardines, 24, 263, 270
Sardinia, 131, 208, 214, 261–62
sausage, 34, 68, 82, 92, 117, 122–23, 139, 154, 166, 212, 216–20, 226–27, 307, 341; blood, 41, 240; skin, 241
savory (herb), 293
sbrisolona, 109, 146–47
scallions, 95, 112, 116, 136, 145, 154, 174, 225, 234, 236, 243, 248, 253, 267, 275, 281, 296, 313; pancakes, 173
scallops, 273–74, 276; dried, 266, 311; heng yang spicy, 280
Scappi, Bartolomeo, 20, 59, 61, 72, 224, 255, 287
schiacciata, 104, 182
sea bass, 268–69, 278
sea urchins, 43, 275
seaweed, 164, 309
Seneca, Lucius Annaeus, 43
serrano chilies, 166, 247, 310
sesame, 58, 169, 185, 198, 229, 267, 310, 312; balls, 147; oil, 46, 111, 131, 136–37, 171, 229, 236, 245, 253, 275, 281, 289; paste, 117, 135

Shang Dynasty, 30, 38
Shanghai, 110, 135, 236; braised duck, 248
Shanxi, 110
Shaoxing (rice wine), 116, 174, 244, 248, 253, 275, 285, 326, 344
shark's fin, 4, 266
sheep, 18, 34–35, 209–10, 213–14, 286; tail fat, 284. *See also* lamb
shellfish, 117, 271–76, 278, 303
Shengzhi, 45
shrimp, 60, 111–12, 131, 193, 249, 253, 271–73, 276, 278, 311; chips, 124; coconut, 198, 280; dried, 156, 266; paste, 272
Sichuan, 56, 110, 123, 166, 211, 225, 248, 299; pepper, 243, 301–2, 313, 340
Sicily, 24, 38, 50, 165, 170, 187–88, 261, 263, 304
Silk Road, 27, 48, 190, 334
silkworms, 178, 208
skin (animal), 34, 72, 136, 209–10, 212, 216, 223, 225, 227, 230–31, 234, 240–41, 247–48, 270, 278, 280–81, 331
slavery, 6, 39–40, 49, 55, 123, 303; trade, 56, 188, 243
Slow Food, 8, 23, 344
smallpox, 56
smoke, as flavoring, xix–xx, 70, 82, 88, 154, 165–66, 183, 190, 194, 214, 217, 219–20, 225, 239, 246, 258, 266–70, 273, 284–85, 322
snails, 47, 271, 274; fork, 99
snapper (red), 268, 271, 276, 278, 281
snipe, 241, 245–46
soda, 317–18, 333; Italian, 335–36
Song Dynasty, 31, 111, 313, 320, 344; Southern, 278
sorghum, 125, 218, 306, 325–26, 331, 337
sour cream. *See* cream
sous-vide, 68, 344
soy, 308; bean curd sticks, 171; fermented paste (jiang), 308; oil, 290; sauce, tofu, 45
Spain, 5, 15, 19, 47–48, 52–58, 63, 74, 79, 81, 84–85, 94–95, 97–98, 108, 113–14, 117, 119, 122–24, 126, 138, 140, 144, 156, 160, 164–67, 172, 181, 184–88, 194–95, 202, 208–10, 212, 215, 220–21, 223–24, 239–40, 253, 255, 264, 266, 271, 275, 286, 289, 292, 300, 307, 318–19, 326, 330, 333, 339, 341–42
Spanish rice, 23, 102, 117, 146
speck, 70, 220
spelt, 103, 224
spices, 5, 15, 19, 21, 39, 50–51, 55, 58–60, 71, 78, 104, 114, 126, 133, 148, 184–85, 187, 192, 199–201, 206, 217–18, 224–25, 234, 280, 283, 287, 291–92,

296–303, 309, 312, 318–19, 325, 330–31, 335–36, 340
spinach, 48, 50, 59, 108, 148, 156–57, 252
spirulina, 52, 164
spits, 52, 70, 80–82, 218, 221, 226, 246, 299
squash, 4, 46, 52, 60, 160–61, 182; blossoms, 173; seeds, 35, 326
squid, 267, 271, 273–75, 278–80
star anise, 234, 243, 248, 251, 298, 302, 340
steam, 41, 47, 52, 68–70, 85–88, 109–12, 115–119, 122, 125, 131, 134–36, 143, 151, 161–62, 167, 172, 186, 211, 222–23, 225, 234–35, 239, 245, 253, 268–70, 272, 274, 281, 285, 290, 296, 308, 311, 320, 335, 340–41
steel (cookware and implements), 73, 75, 84, 86–87, 203, 328, 340; rollers, 62, 78
Stefani, Bartolomeo, 60
stew, 12, 37, 41, 53, 67–69, 78, 83, 97, 121–23, 132, 142, 151, 154, 156, 159–62, 167, 175, 182, 192, 196, 206, 210–12, 216–17, 220, 222–23, 234, 243, 266, 274–75, 279, 295, 299, 341, 344
stir-fry, 69, 87, 89, 95, 121, 171, 229, 231, 235, 244, 290
stoneware, 83, 152
stracciatella alla Romana, 254
strawberries, 12, 163, 181, 199, 203–4, 255, 318, 334
street food, 14, 23, 82, 104, 109–10, 113, 120, 139, 143, 173, 181, 191, 195–96, 203, 281, 317
Strega (liqueur), 331
sturgeon, 60, 263–64
suet, 286
sugar, 5, 19, 22, 37, 50, 52, 59, 62, 74, 108–9, 117–18, 124, 135–38, 140, 143–48, 170–72, 181, 190, 193–94, 199–200, 203–4, 222, 225, 237, 239, 241, 248, 253, 255–57, 259, 266, 271–72, 278, 280–81, 287, 289, 302–5, 307, 309, 313, 318, 326, 332, 334–36; brown, 144, 334; caramelized, 234, 305; piloncillo, 128, 300, 304, 320, 343; powdered, 140, 147; spun, 185; syrup, 141, 144, 201–3, 336
suppli al telefono, 118
sustainability, 2, 25–26
sweet potatoes, 26, 34, 46–47, 52, 56, 60, 116, 122, 123–24; camotes fritos, 148
swordfish, 263–64
Szechuan. *See* Sichuan

taboo, 7, 23, 205, 226
tacos, 11, 23, 119, 121, 123, 143, 152, 155, 158, 162, 165–66, 174, 212, 217, 223, 231–32, 240, 268, 270

Taillevent (Guillaume Tirel), 50
Taiwan, 331, 334
taleggio, 214
tamalero, 86
tamales, 47, 69, 78, 86, 102, 114, 117, 119, 134–35, 157–58, 162, 192, 223, 266, 285, 295, 342
tamarind, 266, 302–3, 318, 334
Tang Dynasty, 47–48, 181, 186–87, 192, 194, 213, 295, 320, 330
tang hu lu (candied bottle gourd), 203
tannin, xix, 320, 328, 330, 344
tansy, 293
tarragon, 293, 295
Taxco, 208
tea, xix, 21, 48, 62–63, 115, 179, 188, 251, 301, 305, 320–22, 334–35, 340, 344; bubble, 334; cup, 96; iced, 317; lapsang souchong, 194; leaves for smoking, 225; medicinal, 164; strainer, 80; Tibetan, 213, 287
tea eggs, 223, 251–52
tench, 269; stuffed, 280–81
tequila, 332–33, 335, 343
terroir, 2, 24, 328, 344
thrushes, 131–32, 221
thyme, xix, 233, 246, 250, 254, 258, 293, 298
Tibet, 123, 213, 286–87, 307, 316
tiger nut (chufa), 333
tiramisu, 109, 255–56
tofu, 45, 159, 171, 174–75, 225, 248
tomatillo, 53, 155, 167–68, 240, 247, 252; molé verde, 310
tomatoes, xx–xxi, 13, 23–24, 27, 53, 56, 60, 64, 85, 139, 143, 146, 149, 167–70, 175–76, 220, 224, 233–34, 244, 250, 253, 258, 266, 276, 281, 285, 310–11, 342, 344; paste, 247, 312; puree, 117, 175, 243, 312; sauce, 21, 67, 78, 104–5, 108, 133, 137, 151, 166, 171, 187, 212, 228, 236, 252, 270–75, 292, 295, 299, 302, 304, 311–12, 330; stuffed, 172; with eggs, 252–53. *See also* ragù *and* salsa
tongue, as meat, 227–28, 240–41; organ of taste, xix, 154, 284, 291, 301
tortellini, 80, 95, 107–8; in brood, 130–31, 222
tortilla, 24, 47, 52–53, 58, 63, 78, 85, 98, 102, 114, 119–20, 141–43, 208, 224, 232, 234, 240, 242, 247, 253, 273, 285, 310, 326, 340–42; chips, 23, 120, 143, 271, 311, 344; press, 93; soup, 250; Spanish, 123; stale, 58, 224, 312; wheat flour, 115, 245
Tostones de Plátano. *See* plantain
totopos. *See* tortilla chips

trade, 3–6, 15, 19, 22, 27–29, 32, 39, 44, 46, 48–49, 50–51, 55–58, 61, 63, 73, 102, 179, 181–82, 190, 198, 266, 276, 297–98, 301, 309
tres leches cake, 258–59
tripe, 242–43
trout, 268–69
truffles, 122, 161, 310, 330
Tull, Jethro, 62
tuna (fish), 39, 186, 263–64, 271, 276–77; belly, 70; sauce, 212, 228
tunas (cactus fruit), 164, 334
Turin, 104
turkey (bird), 34, 47, 53, 56, 58–59, 64, 179, 223–24, 226, 230, 312
Turkey (country), 18–19, 35, 122, 135, 165
turmeric, 298, 300
turnips, 51, 150–52, 224, 228, 291
turtles and tortoises, 47, 275
Tuscany, 104, 159, 161, 182, 211, 217–18, 285, 288, 304, 329–30, 341. *See also* Florence

uccelli arrosto (roast birds), 66, 245–46
umami, xix, 309, 345
Umbria, 161, 217

vanilla, 19, 128, 300, 318, 320, 334; extract, 145, 147, 253, 256, 259, 300; pod (bean), 199, 259
veal, 51–52, 59, 130, 146, 211–12, 221–22, 227–29, 233, 235, 255, 264
vegetarianism, 2, 7, 9, 45, 112, 162, 164, 167, 171, 196, 211, 213
velveting, 68, 222
Venice, 52, 58–59, 227, 266, 274, 323, 336, 341
venison, 34, 36, 46–47, 206
Veracruz, 38, 46, 123, 196, 243, 262, 273, 281, 295, 342
verjuice. *See* grapes
vermouth, 173, 280, 331, 336
Vespucci, Amerigo, 55
vinegar, xx, 59, 111, 155, 169–70, 181, 183, 206, 216, 218, 225, 228, 234, 239–40, 243, 245, 248–49, 266, 268, 270–71, 273–74, 280–81, 284, 288, 298, 303, 305–7, 339–40, 345; balsamic, 181; black, 136, 278, 306, 313; coagulant for cheese, 257; pickles, 152; pineapple, 193; raspberry, 179; rice wine, 115–16, 173, 200–202, 218, 313; Tasters (painting), 45
vitello tonnato, 212, 228

walnuts, 5, 172, 195, 280; liqueur, 331, 336; oil, 290
water, 316–17
water buffalo, 210–11, 213
watermelon, 182, 203, 318, 334, 337
wedding cookies. *See* cookies
West Lake sour fish, 278
whale, 264–65
wine (grape), xix, 2, 8, 18, 24, 36–37, 39, 41, 48, 54, 57–60, 69, 106, 131–33, 136, 146, 169, 173, 181, 183, 185, 201, 212, 220, 222, 233, 237–40, 244, 247, 266, 272–74, 279–81, 311–12, 315, 328–31, 336, 342–44. *See also* rice, wine
woks, 19, 45–46, 67, 69, 80, 84–90, 116, 145, 156, 169, 202, 225, 229–30, 236, 243, 245, 252–53, 268, 278, 290–91
wood (fuel), 66, 68, 71, 85, 88, 104, 115, 137, 139, 225, 230–32, 239
wood ear fungus, 128, 162–63, 171, 248
wormwood, 293, 331

xiaolongbao (steamed pork buns), 135–36
XO sauce, 311

Yangmei (yumberry), 181
yeast, 6, 103, 135–38, 140, 143, 256, 325–26, 328, 343
yin and yang, 46, 292, 345
youtiao (fried crullers), 113, 143–44
Yuan, Qu, 117
Yucatan, 47, 69, 156, 218, 253, 262, 268, 340

zabaglione, 69, 223
zampone, 212, 219
zeppole, 69, 108, 143–44
Zhen He, 57
Zhongjing, Zhang, 111
zongzi, 117
zucchini, 160, 170–72, 223, 236
zuppa alla Pavese, 223, 250
zuppa Inglese, 255–56

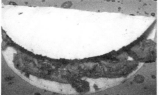

About the Author

KEN ALBALA is professor of history at the University of the Pacific in Stockton, California, where he has taught courses on food history, the Renaissance and Reformation, and the history of medicine for the past eighteen years. He also teaches in the gastronomy program at Boston University and his food history class is being filmed for The Great Courses company. He is the author of many books on food including *Eating Right in the Renaissance* (2002), *Food in Early Modern Europe* (2003), *Cooking in Europe, 1250–1650* (2005), *The Banquet: Dining in the Great Courts of Late Renaissance Europe* (2007), *Beans: A History* (2007; winner of the 2008 International Association of Culinary Professionals Jane Grigson Award), and *Pancake* (2008). He has also coedited *The Business of Food* (2007) and *Human Cuisine* (2008) and edited two other collections of essays: *Food and Faith in Christian Culture* (2011) and *A Cultural History of Food: The Renaissance* (2012). The *Routledge International Handbook of Food Studies*, which he edited, was published in 2012. Albala has also edited three food series for Greenwood Press with thirty volumes in print, and his four-volume *Food Cultures of the World Encyclopedia* was published in 2011. Albala is also coeditor of the journal *Food, Culture and Society* and general editor of the new AltaMira Studies in Food and Gastronomy. He has also coauthored a cookbook on historic cooking techniques entitled *The Lost Art of Real Cooking* (2010), the sequel of which is *The Lost Arts of Hearth and Home* (2012). He is researching a history of theological controversies over fasting in the Reformation Era.